Contents

Preface to first metric edition

Due to the adoption of SI units (Système International d'Unités) as the primary system of weights and measures it has become necessary to bring the book up to date in this respect. The text has been rewritten in SI units and no mention has been made of Imperial or other units. In this way it is hoped that the student will be able to follow the principles of the subject without having to master various unit systems and face the complications caused by conversion from one set of units to another. These other unit systems will exist for some time to come and conversion of units will be a necessary common exercise, but it is felt that these aspects are best dealt with by the teacher or by a different form of book.

Besides meeting the needs of the Ordinary National Certificate course the book has been widely used by students in Diploma and Technician courses. Since some deletions have had to be made because of the change in units the opportunity has been taken to add some material to bring the book more completely into line with the Ordinary National Certificate and Diploma syllabuses. Two main additions are a section on Impact of Jets in the chapter on Fluids in Motion, and some work on the Method of Sections as applied to Frameworks. It has been found necessary also to increase the number of worked and unworked examples in order to illustrate and give practice in the new system of units.

J.H.
M.J.H.

Preface to second edition

In this second edition the opportunity has been taken to bring symbols into line with current practice. New topics and examples have been added and parts of the text up-dated. A main change is in the section on Dynamics, which has been reshaped, considerably amplified throughout, and now includes a simple treatment of aircraft and rocket work. Discussions on the engineering aspects of the subject have been retained to give interest and background to the principles involved.

The text meets the mechanical engineering and 'applied mechanics' requirements of the Business Technician Education Council's Certificate and Diploma courses in Engineering, and courses of similar level. It is hoped also that the book will continue to be useful as a supporting text to students in the early stages of higher diploma and degree courses and in comparable courses overseas.

Since my co-author Mr. M. J. Hillier has been working abroad for many years the revisions and alterations in this edition have been done by me.

I am indebted to those users of the book in many parts of the world who offered constructive criticisms, and to their colleagues for help and advice.

J.H.

CHAPTER 1

Statics

Statics is the study of forces on bodies *at rest* or in *steady motion*. The student at this stage should already be familiar with the elementary principles and theorems relating to forces in equilibrium and the following notes are intended as revision but with an emphasis on the application of these principles to engineering problems.

1.1 Mass, force and weight

The *mass* of a body is the quantity of matter it contains.

A *force* is simply a push or a pull and may be measured by its effect on a body. A force may change or tend to change the shape or size of a body; if applied to a body at rest the force will move or tend to move it; if applied to a body already moving the force will change the motion.

A particular force is that due to the effect of gravity on a body, i.e. the *weight* of a body.

These three quantities – mass, force and weight – are dealt with fully in Chapter 5, but it is necessary here to specify the units and the essential relationship between mass and weight.

The base SI unit of mass is the *kilogram* (**kg**); other units of mass are:

$$1 \text{ } megagram \text{ } (\textbf{Mg}) \text{ or } tonne \text{ } (\textbf{t}) = 10^3 \text{ kg}$$

$$1 \text{ } gram \text{ } (\textbf{g}) = 10^{-3} \text{ kg}$$

$$1 \text{ } milligram \text{ } (\textbf{mg}) = 10^{-6} \text{ kg}$$

The derived SI unit of force is the *newton* (**N**) defined as *that force which, when applied to a body having a mass of one kilogram, gives it an acceleration of one metre per second squared.* From Newton's second law of motion (*see* page 96) we have

force = mass × acceleration

F = $m \, a$

where F is the applied force, m the mass of a body, and a the acceleration produced in the body. Thus in SI units $F = 1$ N, $m = 1$ kg and $a = 1$ m/s^2, i.e.

$$1 \text{ (N)} = 1 \text{ (kg)} \times 1 \text{ (m/s}^2)$$

Other units of force used are:

$$1 \text{ } kilonewton \text{ (\textbf{kN})} = 10^3 \text{ N}$$

$$1 \text{ } meganewton \text{ (\textbf{MN})} = 10^6 \text{ N}$$

$$1 \text{ } giganewton \text{ (\textbf{GN})} = 10^9 \text{ N}$$

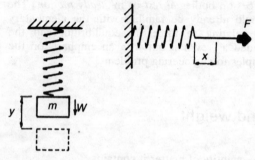

Fig. 1.1

The acceleration of any body towards earth in free fall is $g = 9.8$ m/s^2, hence the weight W of a body of mass m is:

force = mass × acceleration

W $= mg$

If the mass m is in kilograms, then

$W = m \times 9.8$ N

If the mass m is in megagrams (tonnes), then

$W = m \times 1{,}000 \times 9.8$ N

$ = m \times 9.8$ kN

(Fig. 5.5, Chapter 5, shows the relationship between the weight W of a body and its mass m.)

Although defined in dynamic terms, a force may also be measured statically by the weight of the mass it will just support or by comparing its effect with the weight of a standard mass. Thus if a mass m is suspended from a spring (Fig. 1.1) the extension is said to be due to the force of gravity on the mass, i.e. to its weight $W = mg$. If y is the extension produced by this force W and a force F on the same spring produces an extension x then the value of F is measured in terms of W by simple proportion: thus

$$\frac{F}{W} = \frac{x}{y}$$

Proper specification of a force requires knowledge of three quantities:

1. its magnitude
2. its point of application
3. its line of action.

Since a force has magnitude, direction and sense it is a *vector quantity* and may be represented by a straight line of definite length.

Dead loads

In statics the vertically downward force due to the weight of 'dead loads' must always be taken into account, and this force of gravity acts through the centre of gravity of the load. In a structure such as a bridge the dead load is the weight of the bridge framework itself plus the cladding and road surface. (Note that a train travelling over the bridge is a 'live' load.) A load may be given in force units, i.e. N, kN, MN, or GN. A 'load' may also be specified in mass units, i.e. kg or Mg (tonne), and in this case the corresponding *weight* of the mass must be found before carrying out calculations involving forces.

1.2 Forces in equilibrium: triangle of forces

Statics is the study of forces in *equilibrium* ('in balance'). A single force cannot exist alone and is unbalanced. For equilibrium it must be balanced by an equal and opposite force acting along the same straight line. Thus in Fig. 1.2 the load of 1 kN *on* the tie is balanced at the

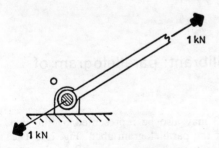

1 kN

1 kN

Fig. 1.2

joint O by an equal and opposite force of 1 kN exerted *by* the joint *on* the tie. Thus forces may be said to exist in pairs. Nevertheless a single force may also be balanced by any number of other forces.

For three forces in the same plane to be in equilibrium:

(*a*) they must have their lines of action all passing through one point, i.e. they must be *concurrent*;

(*b*) they may be represented in magnitude and direction by the three sides of a triangle *taken in order*, i.e. by a *triangle of forces*.

The condition that all three forces must pass through one point is particularly useful in solving mechanics problems. For example, the light jib crane shown in Fig. 1.3 (*a*) is in equilibrium under the action of three forces. The jib carries a load W at A; the free end is supported by a cable in which the tension is T; the end C is pinned to the wall by a joint which allows free rotation of the jib at C. The reaction F of the joint on the jib is completely unknown; the magnitude of T is unknown but its direction must be that of the cable. Since the three forces are in balance their lines of action must pass through one point, i.e. where the lines of action of W and T intersect (point Z, Fig. 1.3 (*a*)). The line of action of F is therefore found by joining C to Z.

Since the magnitude of W is known and the directions of the three forces have been determined the triangle of forces can now be drawn, Fig. 1.3 (*b*).

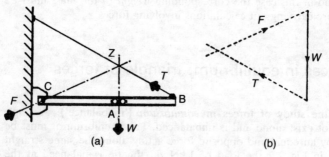

(a)   (b)

Fig. 1.3

1.3 Resultant and equilibrant: parallelogram of forces

The forces W and T of Fig. 1.3 may also be represented by the two sides **ab** and **ad**, respectively, of the parallelogram **abcd**, Fig. 1.4. The diagonal **ac**, taken in the sense **a** to **c**, is the *resultant R* of the two forces *W* and *T* acting together. This resultant force is equivalent to, and may replace completely, these two forces. The resultant **ac** may be balanced by an equal and opposite force **ca** called the *equilibrant*. This is in fact the force *F* at the pin-joint, Fig. 1.3 (*a*). This construction, by which two forces are replaced by a single equivalent force, is known as the *parallelogram of forces*. It can only be used if the two forces are specified in both magnitude *and* direction.

1.4 Resolution of forces

Since the forces represented by **ab** and **ad** in Fig. 1.4 may be replaced completely by a single force **ac**, it is often useful to carry out the reverse process, i.e. to replace a single force by two other forces in any two convenient directions. These two forces are then known as the *components* of the single force. Physically this is equivalent to finding the effects of the single force in the two chosen directions.

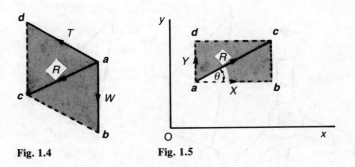

Fig. 1.4   **Fig. 1.5**

The most convenient choice of directions in which to *resolve* a force is in two directions at right angles. Fig. 1.5 shows a force $R = $ **ac** resolved into forces $X = $ **ab** and $Y = $ **ad** along the two perpendicular directions O*x* and O*y*, respectively. Since the three forces shown do not represent independent forces the components of *R* are shown in broken lines. Let *R* make an angle θ with the O*x* direction, then

$$\textbf{ab} = \textbf{ac} \cos \theta$$

i.e. $X = R \cos \theta$

and $\textbf{ad} = \textbf{ac} \sin \theta$

or $Y = R \sin \theta$

1.5 Polygon of forces

If more than three forces act at the same point and are in equilibrium, they may be represented in magnitude and direction by the sides of a polygon *taken in order*. 'Taken in order' refers to the order of drawing the sides of the polygon and not to the order in which the forces are taken from the space diagram.

Suppose the four forces 1, 2, 3 and 4 acting at the joint shown in Fig. 1.6 (*a*) to be in balance; they may then be represented by the four sides of the polygon **abcd**, Fig. 1.6 (*b*). This is a closed polygon since the forces are in equilibrium. If the forces are not in balance the polygon will not close and the required closing line gives the equilibrant or the equal and opposite resultant force, depending on the sense in which it is taken. Fig. 1.6 (*c*) shows the force polygon assuming unbalance. The forces 1, 2, 3, 4 are represented by lines **ab**, **bc**, **cd** and **de**, respectively. To close the polygon and maintain a balance of forces requires the equilibrant **ea** taken in the sense **e** to **a**. The resultant of the original set of four unbalanced forces is given by the line **ae** and acts in the sense **a** to **e**.

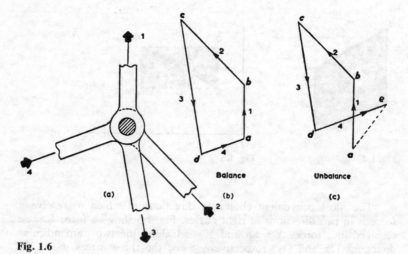

Balance

Unbalance

(a)   (b)   (c)

Fig. 1.6

1.6 Moment of a force

The *moment* of a force F about a point O (Fig. 1.7) is the product of the force and the perpendicular distance x of its line of action from O, thus:

moment of F about O $= Fx$

If the force F is in newtons and the distance x in metres then the units for moment are written *newton-metres* (**N-m**). For larger values it is usual to retain the metre for distance and use the higher multiples of the newton, e.g. MN-m.

Fig. 1.7

1.7 Couple

A pair of equal and opposite parallel forces, which do not act in the same straight line, form a *couple*, Fig. 1.8. The total moment M of the two forces F about any point O in the plane of the couple is

$$M = F(d + x) - Fx$$
$$= Fd$$

This moment is independent of the distance x and is therefore *the same about any point in the same plane*. The turning effect of the couple is also the same wherever it may be placed in the plane.

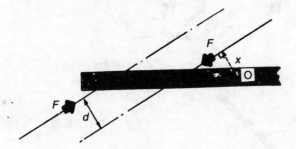

Fig. 1.8

The magnitude Fd of a couple is known as its *moment* or *torque*, although the term torque is usually restricted to a moment tending to twist a shaft.

1.8 Principle of moments

Consider any system of coplanar forces which do not act at a point. Take moments about any arbitrary point and let clockwise moments be positive and anticlockwise moments be negative. Then, if the body is in equilibrium there can be no unbalanced moment about this point and for balance of moments we must have:

clockwise moments = anticlockwise moments

or *the algebraic sum of the moments of all the forces about the same point is zero.*

1.9 Resolution of a force into a force and a couple

Consider the arm OA fixed at O and subjected to a force F at A which acts perpendicularly to OA (Fig. 1.9 (*a*)). We wish to know the effect of force F at O. Suppose therefore two equal and opposite forces F to be introduced at O acting parallel to the existing force F at A (Fig. 1.9 (*b*)). The system of forces is not upset and the resultant force

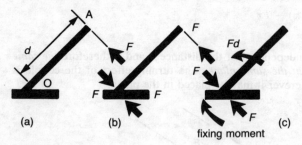

fixing moment

Fig. 1.9

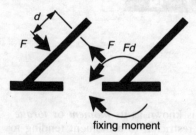

fixing moment

Fig. 1.10

is unaltered since the two new forces are self-cancelling. However, it can now be seen that the effect of F at A is equivalent to a single force F at O, together with a couple of moment Fd tending to turn the arm anticlockwise (Fig. 1.9 (c)). The effect therefore of a single force F on a body at a point offset from the line of action of the force is to produce at that point both a force *and* a couple. Thus in Fig. 1.9 (a) the support must exert a reaction F and a *fixing* or resisting moment Fd to counteract the effect of the applied force.

A pure couple on the other hand will not introduce this single out-of-balance force at any point. In Fig. 1.10 the two forces F form a pure couple; they are self-balancing and no force is produced at the bearing (unlike the previous case). The couple due to the two forces is of course out of balance and can only be balanced by an equal and opposite couple in the same plane. This result is independent of the position of the couple. In this case the balancing couple is provided by the fixing moment at the support.

1.10 The general conditions of equilibrium

We now require to consider the balance of *any* system of forces, in a plane, which do *not* all act through the same point. Since any force may be replaced by a similar force at any other point, together with a couple, then each of the forces may be considered as acting at any one point, provided that allowance is made for all the couples produced. For complete equilibrium of the system* therefore, there must be no unbalanced force *or* couple. *For force balance a polygon of forces may be drawn and must close for equilibrium. For couple balance the algebraic sum of the moments of all the forces about any point must be zero.*

Alternatively it is often convenient to resolve all the forces in the same two mutually-perpendicular directions. Consider the force system shown in Fig. 1.11. The forces may be resolved in the two directions parallel to Ox and Oy. Let X and Y be the algebraic sum of the components of all the forces in the Ox- and Oy-directions, respectively; and let M be the algebraic sum of the moments of all the forces about any chosen point O. Then the resultant force R is given by Fig. 1.11 (b):

$$R^2 = X^2 + Y^2$$

---

*A body 'at rest' or 'in equilibrium' can be taken to mean *not being accelerated*. Thus a body in *steady motion*, i.e. moving at constant velocity, is not accelerating and there is therefore no resultant force acting (*see* paragraph 5.7) so that the forces on the body can be considered as 'in dynamic equilibrium' and treated as a problem in statics.

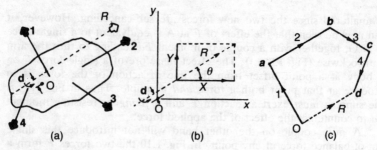

Fig. 1.11

and the line of action of the resultant is at an angle θ to Ox given by:

$$\tan \theta = \frac{Y}{X}$$

The resultant couple M is the same about any point in its plane hence it may be obtained by calculating the algebraic sum of the moments of all the forces about any point O.

The conditions of equilibrium are therefore:

$R = 0$ (i.e. $X = 0$, and $Y = 0$)

and $M = 0$

Otherwise, if R and M are not zero, to find the *position* of the resultant R we calculate the distance d of its line of action from any chosen point O, Fig. 1.11 (*a*). The resultant R may be replaced by a force acting at O, together with a couple of moment $R \times d$ about O. This couple is equal in magnitude and sense to the resultant couple M, i.e.

$$R \times d = M$$

This determines the distance d from O of the line of action of R.

If $M = 0$, then $d = 0$ and the resultant R passes through the chosen point O.

If M is not zero, then as R becomes very small the distance d becomes very large so that $R \times d$ is always equal to M.

Note: If the force polygon is drawn, Fig. 1.11 (*c*), the closing line **Od** gives the resultant R in magnitude, sense and direction but *not* in position. To obtain the position we must take moments about some point O.

1.11 Free-body diagram

In the solution of problems in statics it is often useful to isolate completely a single body from its surroundings, removing its supports and holding devices, and by means of a *free-body diagram* show the magnitude and directions of all the forces and couples acting on the body at a particular instant. Only those forces acting *on* the body are shown including external loads and couples, reactions at supports, and forces due to gravity. A direction for an unknown force might have to be assumed in solving a problem but the sign obtained will indicate if the correct direction has been chosen. Several bodies taken together may be isolated and a free-body diagram constructed for the system. For example, Fig. 1.12 (*a*) shows the jib crane already dealt with in paragraph 1.2. The jib carries a load W and the resulting tension T in the supporting cable is known in direction only. The third force maintaining the jib in equilibrium is the reaction F at the pin-joint, and this force is unknown both in magnitude and direction. Fig. 1.12 (*b*) is the free-body diagram for the jib with the joint removed. It shows the forces F, W and T, acting *on the jib*, the direction of F being assumed until a solution is obtained.

In constructing a free-body diagram it is essential to know the *kind* of force exerted between bodies in contact, and through various connections and supports.

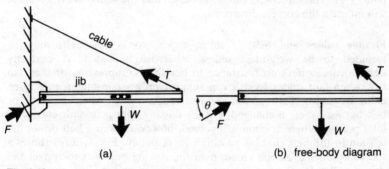

(a) (b) free-body diagram

Fig. 1.12

1.12 Contact forces; supports and connections

Smooth surfaces A *perfectly smooth surface* is one which offers no resistance to sliding parallel to the surface. *The force (or reaction) R exerted by such a surface must be at right angles (normal) to the surface.* If it were not normal the reaction would tend to resist or assist sliding. In practice it is not possible that there are no frictional forces resisting sliding, but in many cases it may be a fair approximation to reality. The assumption of smoothness, meaning the complete absence of friction, simplifies the solution to many practical problems.

Roller or ball support   Fig. 1.13 (*a*) shows a smooth rigid roller or ball on a smooth flat curved surface. The reaction of the surface must be along the normal to the surface of the roller at the point of contact. This is the case whether or not the roller moves under the load.

Knife-edge support   The direction of the reaction of a smooth surface to a knife-edge contact is normal to the surface, Fig. 1.13 (*b*). A *simple support* for a beam is one in which the beam rests on a knife-edge.

Smooth pin-joint   Fig. 1.13 (*c*) shows two links attached by a pin or hinged joint. If the joint surfaces are perfectly smooth *there is no resistance to rotation* and the links are free to rotate relative to each other. Each link can then transmit a force *only along its length*.

Rigid wall fixing   Fig. 1.13 (*d*) shows an *encastré* or *built-in* beam, where the wall holds the beam rigidly fixed in direction. Whatever the forces and moments acting on the beam, the reaction at the wall can be represented by its vertical and horizontal components *V* and *H* respectively. There will also be a *fixing-moment M* at the wall (see para. 1.9). The directions can be assumed and the signs finally obtained will indicate the correct directions.

Flexible cables and belts   Cables, ropes, cords and belts may be assumed to be weightless unless otherwise indicated. A perfectly flexible cable offers no resistance to bending, compression or shear so that when taut under load it can support only a constant *tensile* force along its length. This tension remains constant even when the cable or belt has its direction changed, e.g. by passing over a smooth gravity or idler pulley. Where friction is involved, however, as in a belt drive, the tension in the belt changes passing over a pulley. Fig. 1.13 (*e*) shows a simple belt drive. A belt passes over the *driving* pulley A then over the *driven* pulley B. The friction between belt and pulley alters the tension in the belt, being *P* on the tight side where it is pulled onto the pulley, and *Q* on the slack side leaving the pulley. For the driven pulley or *follower* B, the tight side tension in the belt leaving is *P*, and in the slack side going on it is *Q*.* Note that we are concerned here only with the 'static' problem of forces maintaining a belt drive system in equilibrium.

Example   *Fig. 1.14 (a) shows the tensions in the tight and slack sides of a rope passing round a pulley of mass 40 kg. Calculate the resultant force on the bearings, and its direction.*

---

*The relationship between *P* and *Q* can be shown to depend on the angle of contact on the driving pulley and on the coefficient of friction between belt and pulley.

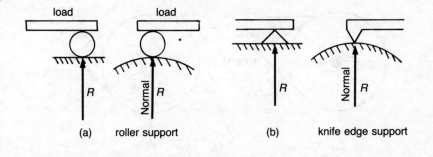

(a) roller support

(b) knife edge support

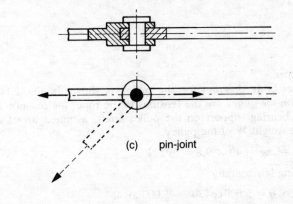

(c) pin-joint

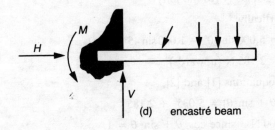

(d) encastré beam

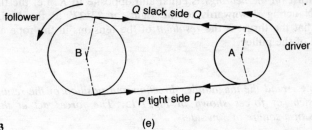

Fig. 1.13

(e)

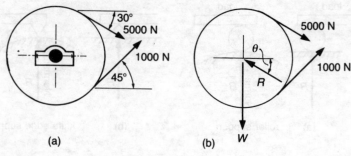

(a)   (b)   W

Fig. 1.14

SOLUTION
The free-body diagram for the pulley is shown in Fig. 1.14 (*b*). The forces *on the pulley* are the tensions in the rope, the reaction *R* exerted by the bearing support on the pulley shaft, assumed to act as shown, and the weight *W* of the pulley.

$$W = mg = 40 \times 9.8 = 392\ N$$

Resolving horizontally:

$$R \cos \theta = 5{,}000 \cos 30° + 1{,}000 \cos 45°$$

$$= 5{,}035\ N \text{ (to the left)} \qquad \qquad \ldots [1]$$

Resolving vertically:

$$R \sin \theta = 5{,}000 \sin 30° - 1{,}000 \sin 45° + 392$$

$$= 2{,}185\ N \text{ (upwards)} \qquad \qquad \ldots [2]$$

hence, from equations [1] and [2],

$$R^2(\cos^2 \theta + \sin^2\theta) = 5{,}035^2 + 2{,}185^2$$

i.e. $R = 5{,}500\ N$ since $\cos^2 \theta + \sin^2 \theta = 1$,

and $\theta = 23° 27'$

The thrust *on the bearings* is equal and opposite to *R*, i.e. the thrust is **5,500 N** acting downwards to the right, at **23° 27'** to the horizontal. Note that the thrust is the *resultant* of the tensions in the rope and the weight of the pulley.

Example   *Find the magnitude, direction and position of the resultant of the system of forces shown in Fig. 1.15. The forces act at the four corners of a square of 3 m side.*

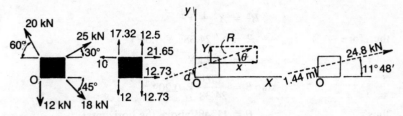

Fig. 1.15

SOLUTION

Force (kN)	Vertical component (kN)	Moment of vertical component about O (kN-m)
20	$+20 \sin 60° = 17.32$	0
12	-12	0
18	$-18 \sin 45° = -12.73$	$+12.73 \times 3 = 38.2$
25	$+25 \sin 30° = 12.5$	$-12.5 \times 3 = -37.5$
Totals	$Y = +5.09$	$+0.7$

Force (kN)	Horizontal component (kN)	Moment of horizontal component about O (kN-m)
20	$-20 \cos 60° = -10$	$-10 \times 3 = -30$
12	0	0
18	$+18 \cos 45° = +12.73$	0
25	$+25 \cos 30° = +21.65$	$+21.65 \times 3 = 64.95$
Totals	$X = +24.38$	$+34.95$

The horizontal and vertical components of each force together with the moments of these components about point O are shown in the above table. Note that *both* the vertical and horizontal components of an inclined force may have moments about any point, as in the case of the 25 kN force here. Upward vertical forces and horizontal forces to the right are positive, clockwise moments are positive.

$$R^2 = X^2 + Y^2$$

therefore
$$R = \sqrt{(24.38^2 + 5.09^2)}$$
$$= \mathbf{24.8\ kN}$$

$$\tan \theta = \frac{5.09}{24.38} = 0.209$$

thus
$$\theta = \mathbf{11° 48'} \text{ above the horizontal}$$

Total moment about O $= + 0.7 + 34.95$

$$= + 35.65\ \text{kN-m (clockwise)}$$

This moment is equal to that of the resultant force R about O. If d is the perpendicular distance of the line of action of R from O then

$$R \times d = 35.65$$
$$d = \frac{35.65}{24.8} = \mathbf{1.44\ m}$$

Therefore R must act along a line 1.44 m from O as shown in Fig. 1.15, such that it produces a *clockwise* moment about O and is inclined at 11° 48' to the horizontal. This solution may be checked by drawing the force polygon to obtain the magnitude and direction of R. To find the total moment about O, measure from a scale drawing the perpendicular distance of the line of action of each force from O.

Example   *The jib crane shown in Fig. 1.16 (a) carries a mass M tonne at C. If the maximum permissible loads in the tie and jib are 25 and 36 kN respectively find the safe value of the load M.*

SOLUTION
M tonne $= M \times 1,000$ kg. The weight of this mass is

$$W = M \times 1,000 \times 9.8\ \text{N}$$
$$= 9.8\ M\ \text{kN}$$

Fig. 1.16 (*b*) shows the triangle of forces for joint C where **oc**

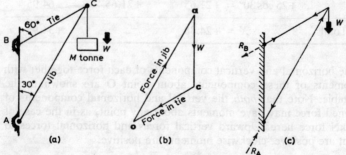

Fig. 1.16

represents the force in the tie, **oa** the force in the jib, and **ac** the load
W. If **oc** is drawn to represent 25 kN (the maximum permitted force in
the tie) then, by construction or calculation:

ac = 25 kN

hence

$W = 9.8 M = 25$ kN

$M = 2.55$ tonne

If **oa** is drawn to represent 36 kN, the maximum permitted force in
the jib, then, by construction or calculation (take AB = 1 m):

ac = 20.8 kN

hence

$W = 9.8 M = 20.8$ kN

$M = 2.13$ tonne

The safe value of the load is therefore **2.13 tonne** since the
permitted forces in the tie and jib are not exceeded.

The sense of each force is found by putting arrows on the force
diagram the 'same way round', starting with the known sense of the
load *W* which is from **a** to **c**. Thus the force in the tie at C is from **c** to
o, i.e. from C to B, and the force in the jib at C is from **o** to **a**, i.e.
from A to C.

Note that Fig. 1.16 (*b*) also represents the triangle of forces for the
external forces on the crane, i.e. the reactions at A and B and the load
W. Since there are only two forces acting at joint A these must be
equal and opposite. Thus the reaction at A, R_A, is equal and opposite
to the force in the jib (Fig. 1.16 (*c*)). Similarly the reaction at B, R_B, is
equal and opposite to the force in the tie. This follows because the load
is applied *at the joint* C (*see* page 4). Note that Fig. 1.16 (*c*) is the
free-body diagram showing the external forces on the crane.

Example *A beam carries a dead load of 200 kg and is subject to a
vertical force of 2 kN and to an inclined force of 1 kN acting at the
points shown in Fig. 1.17. The beam is 'encastré', i.e. built-in to a wall,
at each end, and due to the fixing there are moments of 2 kN-m and
1.6 kN-m acting in the directions shown. Find the reactions R, L and H.*

SOLUTION
Fig. 1.17 is the free-body diagram for the beam. Consider the horizontal forces acting on the beam, then

$$H = 1 \times \cos 60° = \textbf{0.5 kN}$$

The weight of the 200-kg mass = $200 \times 9.8 = 1,960$ N

Taking moments about the right-hand end and equating clockwise
and anticlockwise moments (working in newtons):

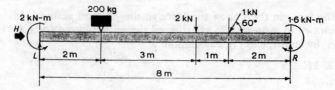

Fig. 1.17

$$L \times 8 + 1{,}600 = (1{,}000 \times \sin 60^\circ \times 2) + (2{,}000 \times 3)$$
$$+ (1{,}960 \times 6) + 2{,}000$$
$$L = 2{,}486.5 \text{ N}$$
$$= \mathbf{2.49 \text{ kN}}$$

Similarly, taking moments about the left-hand end:

$$R \times 8 + 2{,}000 = (1{,}960 \times 2) + (2{,}000 \times 5)$$
$$+ (1{,}000 \times \sin 60^\circ \times 6) + 1{,}600$$
$$R = 2{,}339.5 \text{ N} = \mathbf{2.4 \text{ kN}}$$

As a check:

net upward vertical force = net downward vertical force

$$L + R = 1{,}960 + 2{,}000 + 1{,}000 \sin 60^\circ$$
$$= 4{,}826 \text{ N}$$
$$2{,}486.5 + 2{,}339.5 = 4{,}826$$

Note that the moment of 2 kN-m at the left-hand end is a constant couple applied at every section of the beam in an anticlockwise direction. Similarly the moment of 1.6 kN-m at the right-hand end is applied in a clockwise direction at every section. This type of fixing moment or couple may be imagined as being provided by two equal and opposite forces parallel to the beam.

Problems

1. Fig. 1.18 shows a link AB which is maintained in equilibrium by three forces at A, G and B. The force at A acts along line XX. The force at G is 100 N and acts at 50° to the link as shown. Find the magnitude and sense of the force at A and the magnitude, direction and sense of the force at B. AG = GB = 500 mm.

    (40.8 N upwards; 87.5 N, 26° to horizontal, upwards in direction B to A)

2. The jib of a crane, Fig. 1.19 is 30 m long and weighs 70 kN. It is pin-jointed at one end and supported at the other end by a cable which maintains the jib in the position shown. The centre of gravity of the jib is 12 m from the pinned end. Find the pull in the cable and the reaction at the pin joint.

    (22.9 kN; 68 kN)

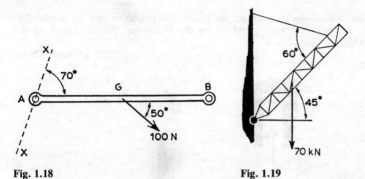

Fig. 1.18   **Fig. 1.19**

3. Calculate the resultant force on the gusset plate shown, Fig. 1.20 and the angle made by its line of action with the vertical.

(227 kN, 5° 24′)

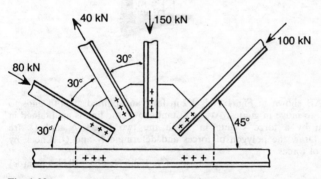

Fig. 1.20

4. Fig. 1.21 shows the forces acting on the handle and dipper of a power shovel. T = thrust in handle = 250 kN, W = weight of handle and dipper = 20 kN, F = cutting force at rock face = 242.5 kN. Find the rope pull P and the angle θ.

(147.5 kN, 72° 51′)

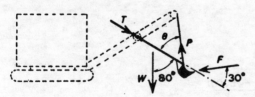

Fig. 1.21

5. Fig. 1.22 shows the elements of a small press used to produce blocks of compressed material. A piston 100 mm diameter is forced downwards by air pressure to operate a toggle mechanism controlling a plunger 50 mm square

in section. If the pressure on the material is not to exceed 1.2 N/mm² find the maximum pressure permitted on the piston and the side thrust on the plunger.

(135 kN/mm²; 529 N)

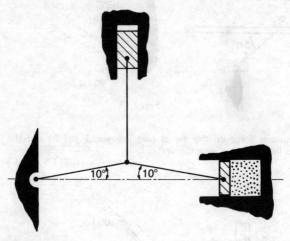

Fig. 1.22

6. The link AB shown in Fig. 1.23 is 1.2 m long and pinned at both ends to blocks free to move in guides. At the instant shown the link is maintained in equilibrium by a force system in which the two forces *N* and *Q* are unknown. Draw the polygon of forces and determine *N* and *Q*. Check by resolution of forces.

(*N*, 153.5 N; *Q*, − 50 N, upwards)

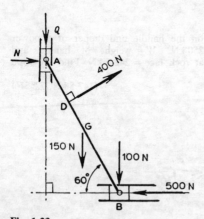

Fig. 1.23

7. The pulley and shaft shown in Fig. 1.24 have a mass of 102 kg and the tensions in the sides of the belt passing round the pulley are 2000 N and

500 N. Find the magnitude and direction of the resultant force on the bearing.

(2.54 kN, 17° 8' to horizontal)

8. For the force system shown in Fig. 1.25 calculate: (*a*) the resultant force; (*b*) the angle which the line of action of the resultant makes with the O*x*-axis; (*c*) the total moment about O; (*d*) the point at which the resultant cuts the O*x*-axis. The figure is marked off in 1 m squares.

(24.4 kN; 28° 36'; 85.2 kN-m, 7.3 m to right of O)

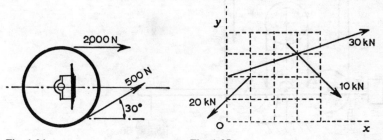

Fig. 1.24   **Fig. 1.25**

9. The forces which keep an aircraft in steady level flight are as shown in Fig. 1.26, i.e.: (i) the weight *W* acting at the centre of gravity, G; (ii) the lift *L* acting vertically upwards through the centre of pressure, C, 200 mm behind the centre of gravity; (iii) the thrust *T* acting horizontally forwards; (iv) the drag or resisting force *D* acting horizontally backwards, its line of action 800 mm below that of the thrust; (v) a small vertical balancing force at the tail-plane, its line of action being 9 m behind the centre of gravity. In a particular case when $W = 12g$ kN, and $D = 15$ kN, find the tail-plane load in magnitude and direction and the magnitude of the lift. Assuming the position of the centre of gravity to be fixed find the tail plane load (a) when the centre of pressure and the centre of gravity coincide, (b) when the line of action of the lift is 200 mm in front of the centre of gravity?

(4.04 kN downwards, 121.64 kN; 1.33 kN downwards; 1.25 kN upwards)

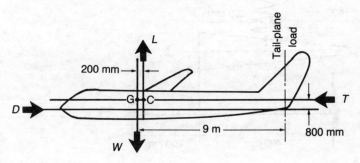

Fig. 1.26

10. Part of a transmission dynamometer is shown in Fig. 1.27. Pulley A is the *driving* pulley. Pulleys B and C are jockey pulleys mounted on a beam

pivoted at D about which point the complete beam is balanced when at rest. A, B and C are each 400 mm diameter and DE = 900 mm. Draw the free-body diagram for the beam and find the slack side tension at the driving pulley when the tight side tension is 1200 N and the beam is horizontal. What is the reaction at D in magnitude and direction?

(649 N; 3.21 kN vertically downwards)

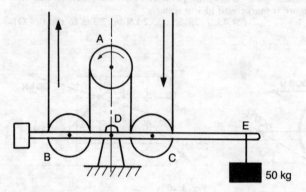

Fig. 1.27

11. In the linkage shown in Fig. 1.28 the bar CD weighs 12 N and is of uniform section, pin-jointed at C. It is connected by a light vertical link BE to a thin triangular flat plate AB which also weighs 12 N. In the position shown the plate is symmetrical about a horizontal line through the pin-joint at A and bar CD is horizontal. The system is supported by a helical spring and carries a concentrated mass of weight 20 N at D. Draw the free-body diagrams for the plate and bar and find the force in link BE, the tension in the spring, and the reactions at the pin-joints.

(39 N; 70.5 N; A, 19.5 N, C, 7 N, both vertically downwards)

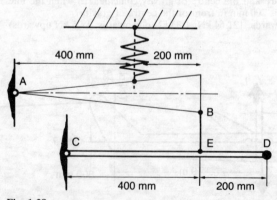

Fig. 1.28

12. A wall crane consists of a horizontal joist AB pinned at end A and supported at end B by a tie inclined at 30° to the joist and attached to the wall at a point vertically above A. The permissible loads in the tie and joist

are 20 and 30 kN respectively. Find the maximum safe load in tonnes that can be carried at the midpoint of the joist. For this safe load, what is the reaction at A in magnitude and direction?

(2.04 tonne; 20 kN upwards at 30° to the horizontal)

13. Analysis of the loading on a beam shows that it is subject to couples of 5, 10 and 20 kN-m acting in the planes shown in Fig. 1.29. The beam carries a mass of 1 Mg which may be assumed to be supported at a single point 2 m from the right-hand end, and is also subjected to a force of 2 kN inclined at 45° to the beam as shown. Find the vertical force L and the magnitude and directions of the vertical and horizontal forces required at the right-hand end to maintain equilibrium.

($L = 2.71$ kN; vertical force $= 8.505$ kN, upwards;

horizontal force $= 1.414$ kN, to the left)

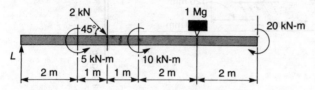

Fig. 1.29

14. The beam shown in Fig. 1.30 is supported by a smooth pin-joint at one end and by a smooth roller at the other end. There are two point loads as shown and the loading over a length of 6 m varies uniformly from 0.2 to 0.4 tonne/m. Neglecting the weight of the beam, find the reactions at each support.

(pin, 9.4 kN vertically upwards, 1 kN horizontally to the left; roller, 10.37 kN vertically upwards)

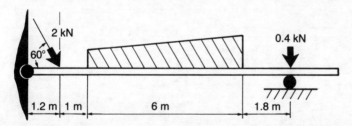

Fig. 1.30

15. Fig. 1.31 shows a governor used to control a hanging link at the end of the arm AC. The movement of AC is controlled by the movement of the central sleeve which in turn is controlled by the radial movement of the rotating balls. As the governor spindle rotates the balls move in or out thereby lowering or raising the sleeve. The ball arms of the bell-crank levers are at right angles. The balls are directly connected by *two* parallel springs. For the position shown, the system is in equilibrium and the forces acting are: (i) an outward radial force of 530 N on each ball; (ii) the weight of the arm AC, 30 N, acting at its centre of gravity G; (iii) the weight of the central sleeve 24 N; (iv) the weight of the link at A, 45 N. Treating the problem as one of

statics and using the free-body diagrams for the bell-crank lever and the arm AC, find the tension in each spring.

(282 N)

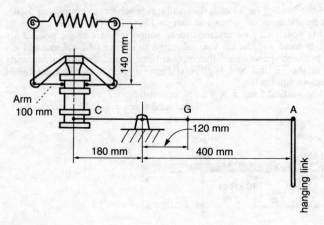

Fig. 1.31

CHAPTER 2

Frameworks

A framework is an assembly of bars connected by hinged or pinned joints and intended to carry loads at the joints only. Each hinge joint is assumed to rotate freely without friction, hence all the bars in the frame exert direct forces only and are therefore in tension or compression. A tensile load is taken as positive and a member carrying tension is called a *tie*. A compressive load is negative and a member in compression is called a *strut*. The bars are usually assumed to be light compared with the applied loads. In practice the joints of a framework may be riveted or welded but the direct forces are often calculated assuming pin joints. This assumption gives values of tension or compression which are on the safe side.

Fig. 2.1 shows a simple frame for a wall crane. In order that the framework shall be *stiff* and capable of carrying a load, each portion such as *abc* forms a triangle, the whole frame being built up of triangles. Note that the wall *ad* forms the third side of the triangle *acd*. The forces in the members of a pin-jointed stiff frame can be obtained

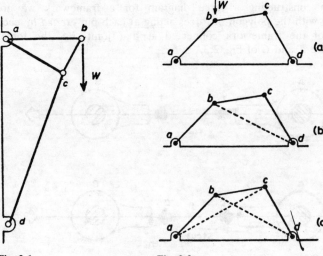

Fig. 2.1   **Fig. 2.2**

by the methods of statics, i.e. using triangle and polygon of forces, resolution of forces, and principle of moments. The system of forces in such a frame is said to be *statically* determinate.

The four bars shown in Fig. 2.2 (*a*) do not form a stiff frame since they would collapse under load. This latter arrangement may be converted into a stiff frame by adding a fifth bar *bd* as shown, Fig. 2.2 (*b*); then both *abd* and *bdc* form complete triangles. However, if both *bd* and *ac* are joined by bars the result, Fig. 2.2 (*c*), remains a frame but is said to be *overstiff*. The forces in the members cannot then be obtained by the methods of statics alone and the structure is said to be *statically indeterminate* or *redundant*. Another example of an indeterminate structure is a beam built-in at one end and propped at the other end. To find the forces in such a case information must be available about the *deflection* of the propped end. Redundant structures are beyond the scope of this book.

2.1 Forces in frameworks

The forces in a stiff or perfect frame can be found by using a force diagram, since the forces in each bar are simply tensile or compressive. Fig. 2.3 shows a tie and a strut under load. The tensile forces acting *on* the tie at each joint are each balanced by equal and opposite internal forces exerted by the tie on the pin joint. Thus the tie appears to be pulling inwards on the two pins at A and B. Similarly the strut appears to be pushing outwards on the pin ends with a force *C* to resist the compressive loads.

When constructing a force diagram for a framework we are concerned with the polygon of forces acting at each pin *exerted by* each member of the framework connected at that joint, i.e. the forces required are *T* and *C* of Fig. 2.3.

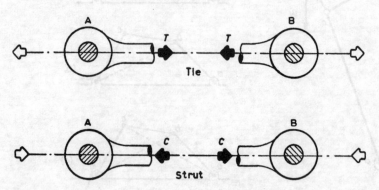

Fig. 2.3

The force diagram is started at any joint at which at least one force is known in magnitude and direction, and where there are not more than *two* unknown forces. The complete diagram is then built up of the force polygons for each joint in succession. The forces are described using *Bow's notation* in which each space between two forces is lettered separately. A force is thus denoted by the two space letters on either side of the force. A joint is conveniently described by using the letters of the spaces meeting at the joint. Thus in Fig. 2.4 the load of 10 kN is described as a force **ab** taken in the sense of the load. The joint at which this force acts is called the joint ABD. It is also necessary to consider the forces at a joint in a definite order (clockwise) and maintain the same order for *every* joint of the framework. Thus the force in the bar DB at the joint ABD is **bd** and the force in the same bar at the joint CDB is **db**.

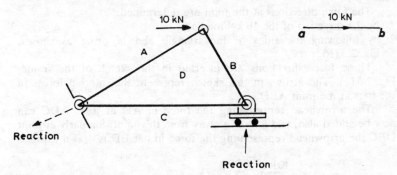

Fig. 2.4

2.2 Wind loads on trusses

Under wind pressure a truss would move sideways unless pinned at one or both ends. For long span trusses one end is usually pinned and the other end rests on rollers. This allows contraction and expansion of the frame with temperature change. When calculating forces in the members of a frame it is only the components of the wind load perpendicular to the face of the truss which are usually considered, together with any vertical or other 'dead loads'.

Example *The framework shown Fig. 2.5 (a) is loaded by a 10 kN horizontal force at the apex. The frame is pinned at the left-hand support and rests on rollers at the right-hand support. The rollers may be assumed frictionless. Find the magnitude and nature of the force in each bar.*

SOLUTION

The load of 10 kN is known in magnitude and direction at the joint ABD. Hence the forces in bars DA and BD may be obtained from the force diagram for this joint. The force diagram is then continued for the forces at the other joints as shown below. Note that at the roller support the reaction must be vertical whereas at the pinned support the reaction is completely unknown.

Joint ABD, Fig. 2.5 (b)

Draw **ab** to represent the 10 kN load.

From **b** draw a line **bd** of unknown length parallel to the bar BD to represent the force in BD.

From **a** draw a line **ad** to represent the unknown force in DA. The lines **bd** and **ad** intersect at **d** to complete the triangle of forces for the joint.

The force directions at the joint are determined:
1. by the direction of the 10 kN load **ab**;
2. by following the sides of the triangle **abd** *in order* as shown, Fig. 2.5(*b*).

These force directions are inserted in the sketch of the frame, Fig. 2.5(*e*). The arrows in this sketch represent the internal forces in the bars at the joint ABD.

The arrowhead representing the force in AD at joint ADC can now be added also, and is drawn away from this joint. Similarly at joint DBC the arrowhead representing the force in bar BD is drawn towards

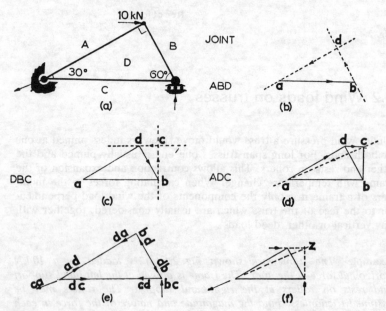

Fig. 2.5

the joint. In each bar the pair of arrowheads should now be pointing in opposite directions.

Joint DBC, Fig. 2.5 (c)
For clarity triangle **abd** is redrawn.

At this joint the known force is **db** drawn in a direction opposite to that at the top joint.

Reaction **bc** is vertical and is so drawn from point **b**.

The force in bar CD is represented by a horizontal line **cd** through **d**.

The intersection of lines **bc** and **cd** at point **c** completes the triangle of forces.

The directions of the forces at the joint follow the sequence **db**, **bc** and **cd** as shown, Fig. 2.5 (c).

These directions are now inserted in the frame sketch, Fig. 2.5 (e).

The only unknown force is now the reaction at joint ADC.

Joint ADC, Fig. 2.5 (d)
The known forces at this joint are **ad** and **dc** in the directions indicated by the arrowheads. The line joining **c** and **a** represents the reaction CA at the hinge support.

Fig. 2.5 (d) now represents the complete force diagram for the frame.

It is unnecessary to draw separate diagrams. The whole force diagram should be built up in one picture while following the sequence indicated above. It is advisable, however, to fill in on a sketch of the frame the directions of the forces at each joint as they are found, and to prevent confusion arrows should be omitted from the force diagram. Those shown in Fig. 2.5 (d) are given for guidance only.

Results
    Reaction BC   **4.33 kN** vertical
    Reaction CA   **10.9 kN** at **23°** to horizontal.

Member	Tension (kN)	Compression (kN)
AD	**8.66**	—
BD	—	**5**
CD	**2.5**	—

Note: As a check on the results the reactions may also be found as follows. Since the frame is right-angled let the sides be: DB = 1 m; DC = 2 m. Then, taking moments about the pinned support, and remembering that the reaction BC at the roller support is vertical:

$$BC \times 2 = 10 \times 1 \cos 30°$$

$$BC = 4.33 \text{ kN}$$

Further, since the whole frame is subject to three external forces only, i.e. the applied load of 10 kN and the two reactions, then the lines of action of these forces must meet in the same point. This point is Z, Fig. 2.5 (*f*). A line joining the pinned support to Z gives the line of action of the reaction at this support. Thus the triangle of forces can now be drawn and the reaction *CA* found.

If the right-hand support had not been on rollers the direction of the reaction there would have been indefinite unless the 10 kN load were vertical. This may be verified by finding the reactions for a 10 kN vertical load at the joint ABD. When the load is vertical the support reactions will be vertical also, and may be found by taking moments about each support in turn. However, when the load is not vertical it will be found that this cannot be done. Such a problem is said to be statically indeterminate and cannot be solved by the methods of statics alone.

Example   *The sloping sides of the symmetrical roof truss shown, Fig. 2.6 (a), are at 30° to the horizontal and the bars BG, CH are of equal length. The reaction at joint KEF is vertical. For the loading shown find the magnitude and nature of the force in each member. All loads are in kilonewtons.*

SOLUTION
It is convenient to begin by calculating the reaction *EF* at joint KEF and then with this information draw the force triangle for joint KEF. Assuming each sloping bar to be of unit length then each horizontal bar is of length $2\cos 30°$. Taking moments about joint ABGF:

$$EF \times 4\cos 30° = (20 \times 1) + (10 \times 2) + (20 \times 2\cos 30°)$$

thus   Reaction *EF* = **21.55 kN**

Joint EFK, Fig. 2.6 (b)
Draw **ef** = 21.55 kN vertically upward. Complete the triangle of forces by drawing **fk** horizontally and **ek** parallel to EK to meet **fk** in **k**. The directions of the forces at the joint follow the sequence **ef**, **fk**, **ke** and the directions of the internal forces in the bars are now shown in the diagram of the truss, Fig. 2.6 (*a*).

Joint EKJ, Fig. 2.6 (c)
Draw **ej** from **e** parallel to JE and **kj** from **k** parallel to KJ. The intersection point **j** is found to coincide with **k**. Hence the force in KJ is zero. (This result could have been seen by inspection. Since no load acts at joint EKJ and the forces in JE and EK are in line, these two forces must be equal and opposite.)

Joint CDEJH, Fig. 2.6 (d)
The known triangle of forces **efk** is redrawn for clarity. **de** is drawn to represent the known downward load DE of 20 kN. The known load **cd** of 10 kN is drawn parallel to CD to end at **d**. The line **jh** is drawn vertically through **j** parallel to JH. The line **hc** is drawn through **c**

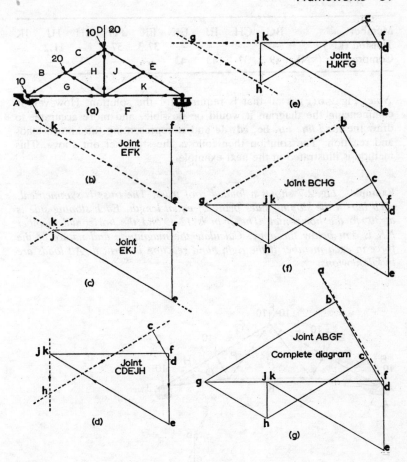

Fig. 2.6

parallel to HC to cut **jh** in **h**, thus completing the polygon. The directions of the forces at the joint follow the closed sequence **cd**, **de**, **ej**, **jh** and **hc**. These directions are now inserted in the diagram of the truss, Fig. 2.6 (*a*).

Similarly for the remaining joints as shown in Fig. 2.6 (*e*) to (*g*). Joining **f** to **a** in Fig. 2.6 (*g*) gives the reaction FA and completes the force diagram. Thus

Reaction FA = **38.9 kN** at **29°** to vertical

Results

The magnitude of the force in each member is found by scaling off from the force diagram and the nature of each force is indicated by the direction of the arrowheads in the truss diagram.

Member	BG	CH	EJ	EK	FK	FG	GH	HJ	JK
Tension (kN)	—	—	—	—	37.3	57	—	11.2	—
Compression (kN)	49	37	43	43	—	—	23	—	—

Note: Fig. 2.6 (*g*) is all that is required for the solution. However, in commencing the diagram it would be possible, and more accurate, to draw the *load-line* **ab**, **bc**, **cd**, **de** and **ef** representing the known loads and reaction. The solution then follows the steps set out above. This method is illustrated in the next example.

Example   *Fig. 2.7 shows a loaded roof truss. The truss is symmetrical, each bar in the sloping sides being of equal length. Each sloping side is at 30° to the horizontal. The span is 12 m and the horizontal member NK is 3 m below the apex. Calculate the magnitude and nature of the force in each member if the right-hand reaction is vertical. All loads are in kilonewtons.*

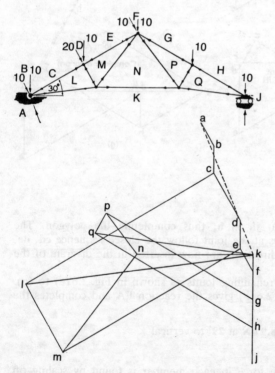

Fig. 2.7

SOLUTION

Since reaction JK is vertical its magnitude may be found by taking moments about the left-hand support. Assuming sloping bars of unit length:

$$JK \times 4\cos 30° = (20 \times 1) + (10 \times 2) + (10 \times 1 \times \cos 30°)$$
$$+ (10 \times 2 \times \cos 30°) + (10 \times 3 \times \cos 30°)$$
$$+ (10 \times 4 \times \cos 30°)$$

thus $JK = \mathbf{36.5\ kN}$.

Load line

Draw the load line **abcdefghj** to represent the known loading in magnitude and direction. Draw **jk** upward from **j** to represent the vertical reaction of 36.5 kN. Then line **ka** represents the reaction KA. Thus

$$KA = \mathbf{52.5\ kN}\ \text{at}\ \mathbf{22\tfrac{1}{2}°}\ \text{to vertical}$$

Joint HJKQ

The polygon for this joint is completed by drawing **kq** parallel to KQ and **hq** parallel to HQ to meet in **q**. The force directions follow the sequence **hj**, **jk**, **kq** and **qh**.

The student should now follow this procedure for the other joints in the truss starting with joint GHQP.

Results

Member	CL	EM	GP	HQ	KQ	KN	KL	LM	MN	NP	PQ
Tension (kN)	—	—	—	—	59.4	43.2	35.7	—	45.2	18.8	—
Compression (kN)	81	75.6	63	68	—	—	—	28.4	—	—	8.6

Problems

1. Fig. 2.8 shows a simple roof truss. A wind load normal to the longer sloping side is assumed to be equivalent to a 10-kN load at each pin joint. The reaction at the right-hand joint may be taken vertical. Determine the reactions and the nature and magnitude of the force in each member.

   (Members: AE, −10; BE, 0; DE, −5; reactions: CD, 8.66; DA, 13.25 at 41° to, and above, the horizontal. *Note*: a negative sign denotes compression, otherwise the member is in tension)

2. In the symmetrical truss shown, Fig. 2.9, a load of 40 kN at the vertex acts normal to one sloping side. The reaction at the right-hand support may be assumed vertical. Find the magnitude of the support reactions, and the magnitude and nature of the force in each member.

   (Reactions: BC, 23.1; CA, 23.1 kN; members: AD, −23.1; AE, −23.1; DE, 0; DC, 40; EF, 0; BF, 46.2; FG, 0; BG, −46.2; GC, 40 kN)

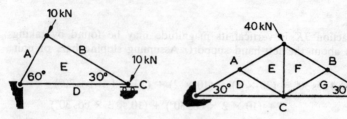

Fig. 2.8   **Fig. 2.9**

3. The truss shown, Fig. 2.10, is made up of three equilateral triangles loaded at each of the two lower panel pins. It is supported by a pin joint at the wall on the right-hand side and by the tension in the cable on the left. Determine: (a) the tension in the cable; (b) the reaction at the wall; (c) the nature and magnitude of the force in each bar.

((a) 19.6 kN; (b) 19.6 kN; (c) EB = BG, = −22.64; EF = FG = 0; DF, 5.67; EA, 11.27; GC, 11.27 kN)

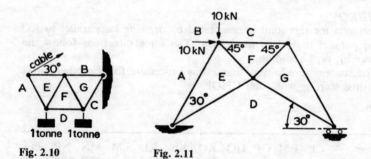

Fig. 2.10   **Fig. 2.11**

4. The framework shown, Fig. 2.11, is loaded by 10 kN horizontal and vertical loads at the top left-hand joint. The reaction at the right-hand support is vertical. Find: (a) the magnitude and direction of each support reaction; (b) the nature and magnitude of the force in each bar of the framework.

((a) 7.9 kN vertically upward; 10.2 kN at 12° to, and above, the horizontal; (b) AE, −13.7; ED, 19.5; EF, 2.8; FC, −19.0; FG, 17; GC, −13.7; GD, 7.9)

5. In the roof truss of Fig. 2.12 the pin joints mid-way along the long sides are joined by a horizontal bar and are connected also to a pin joint at the

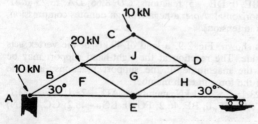

Fig. 2.12

mid-point of the horizontal link spanning the points of support. The right-hand reaction is vertical. Find: (*a*) the magnitude and direction of the support reactions; (*b*) the nature and magnitude of the force in each member.

(Reactions: 11.6 kN vertical; 30.5 kN at 49° to, and above, the horizontal; members: BF, −28.8; EF, 39.5; FG, −11.4; GJ, −20.4; CJ, −5.8; JD, −11.8; DH, −23.2; GH, 11.4; HE, 20 kN)

6. In the roof truss shown, Fig. 2.13, the longer sloping sides are at 30° to the horizontal; the span is 10 m and the horizontal member is 2.5 m below the apex. The two shortest sloping bars are mid-way between each long side and are at right angles to it. The wind load is assumed concentrated at the pin joints of one long side. The left-hand support reaction is vertical. Find the reactions at the supports and the force in each member, stating whether it is in tension or compression.

(Reactions: *DE*, 61.3; *EA*, 23.1 kN; members: EF, 51.5; AF, −59; FG, 0; AG, −59; GH, 7.9; EH, 46.5; HJ, 59.0; JB, −83; JK, −40; KC, −83; KE, 104 kN)

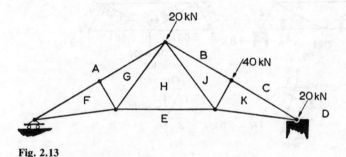

Fig. 2.13

2.3 Analytical methods: method of sections: method of resolution

Where the forces in only a few members are required it is sometimes convenient to calculate the values rather than to construct a complete force polygon. The equations of static equilibrium can be applied to the whole framework for external forces, to each joint for external and internal forces, and to sections of the framework for external and internal forces.

If a number of members meet at a joint and the magnitude and directions of all but two of the forces are known then the unknown forces can be found by resolving forces horizontally and vertically. External forces such as loads or reactions and the internal forces in the members are considered. This is known as the *method of resolution* or *method of joints* and can be tedious unless only a few members meet at the joint.

A framework may be imagined to be cut through by a plane to isolate a section of the framework, and the equilibrium of the section isolated may be considered (in a free-body diagram) under the action of the external loads and reactions and the *internal* forces in the *cut* bars. By resolving forces vertically and horizontally or by taking moments about a convenient point the forces in the cut bars may be found. This is known as the *method of sections* and is useful to check on values obtained from a complex force polygon. The method is particularly convenient for finding the forces in the diagonal members of certain frameworks. Not more than *three* bars must be cut by the plane.

The following examples illustrate the methods of calculation.

Example   *The framework shown in Fig. 2.14 (a) carries a mass of 2 tonne at the lower middle joint. Find the forces in bars 1, 2, 3 and 4 and state whether the bars are in tension or compression.*

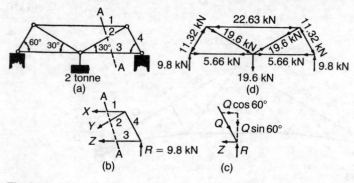

Fig. 2.14

SOLUTION
Let the horizontal members each be of length 2 m, then bar 2 = $\sqrt{3}$ m, bar 4 = 1 m, and the height of the frame is $\frac{1}{2}\sqrt{3}$ m. The weight of the 2-tonne mass is $2 \times 9.8 = 19.6$ kN. From symmetry the reaction R at the right-hand support is 9.8 kN.

Cut the frame by an imaginary plane AA, shown in Fig. 2.14 (*a*), and let X, Y, Z be the internal forces in the cut bars 1, 2, 3 respectively. If we assume that the cut bars are all in tension then the forces X, Y and Z will act as shown in the free-body diagram, Fig. 2.14 (*b*), i.e. forces X and Y are pulling away from the top right-hand joint and force Z is pulling away from the lower right-hand support. To find force Y resolve external and internal forces on the frame section, thus in the vertical direction:

$$Y \sin 30° = R = 9.8$$

$$Y = +\ \mathbf{19.6\ kN}$$

Y is positive therefore the assumed direction is correct and bar 2 is **in tension**.

To find force X we may take moments about the joint carrying the 2-tonne mass for the forces on the frame section Fig. 2.14 (*b*), thus eliminating the forces Y and Z which pass through this joint:

$$X \times \frac{\sqrt{3}}{2} + R \times 2 = 0$$

$$X = -\frac{9.8 \times 4}{\sqrt{3}} = -22.63 \text{ kN}$$

X is negative (therefore the direction of the force must be reversed) and bar 1 is in **compression**.

To find force Z, resolve forces horizontally for the frame section, thus:

$$X + Y \cos 30° + Z = 0$$

$$-22.63 + 19.6 \times 0.866 + Z = 0$$

$$Z = +5.66 \text{ kN}$$

Z is positive, therefore bar 3 is in **tension**.

To find the force in bar 4, we may consider the forces acting at the lower right-hand support joint. There are three forces R, Z, and the force in bar 4, say Q. Fig. 2.14 (c) shows the free-body diagram for this joint. Resolving vertically,

$$Q \times \sin 60° = R = 9.8$$

$$Q = 11.32 \text{ kN}$$

and since R acts upwards, Q must act downwards, i.e. towards the joint so that bar 4 is in **compression**. Alternatively, resolving forces horizontally,

$$Q \times \cos 60° = Z = 5.66$$

$$Q = 11.32 \text{ kN}$$

The forces in the remaining bars can be found from symmetry and the complete solution is shown in Fig. 2.14 (d). In practice the correct directions of the forces in the bars can often be obtained by inspection.

Example *Find the forces in bars 1, 2 and 3 of the framework shown in Fig. 2.15 (a).*

SOLUTION

By taking moments, the reactions at the supports will be found to be $R_1 = 13$ kN and $R_2 = 14$ kN. Cut the frame through bars 1 and 2 and consider the equilibrium of the section of the frame to the left of the cutting plane. Fig. 2.15 (b) shows the free-body diagram for the frame section isolated. The net upward force is $13 - 12 = 1$ kN, and the only bar that can provide a downward force is the diagonal, bar 1, i.e. the vertical component of force X in bar 1 must be 1 kN. Thus:

$$X \cos 45° = 1$$

$$X = 1.414 \text{ kN}$$

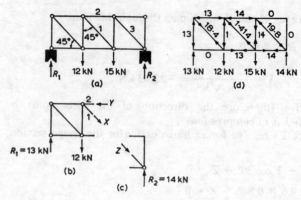

Fig. 2.15

and X must act downwards as shown, i.e. away from the joint, so that bar 1 is in **tension**.

Taking moments about the joint at which the 15 kN load acts, assuming the lower members to be of unit length the force X is eliminated, and we have to consider the external loads 13 kN, 12 kN and the internal force Y,

$$\text{net clockwise moment} = 13 \times 2 - 12 \times 1 = 14 \text{ kN-m}$$

This moment must be balanced by an anticlockwise moment due to a force Y in bar 2, thus:

$$Y \times 1 = 14 \text{ kN-m}$$

$$Y = \mathbf{14 \text{ kN}}$$

Force Y therefore acts towards the joint as shown and bar 2 is in **compression**.

To find the force in bar 3 cut the frame by a plane as shown in the free-body diagram Fig. 2.15 (c). Resolve forces vertically for the section to the right of the plane, and let the force in bar 3 be Z, then:

$$Z \cos 45° = R_2 = 14$$

$$Z = \mathbf{19.8 \text{ kN}}$$

Again Z must act downwards to balance the upward force R, so that bar 3 is in **compression**.

It can be seen that the correct directions of the forces and hence the nature of the forces can be found by studying the directions of the known forces and moments. Alternatively as in the previous example all forces in the bars can be assumed to be tensile and a negative sign in the answer will indicate a compressive force. The student should complete the analysis of the framework. The complete solution is shown in Fig. 2.15 (d).

Example *The frame of Fig. 2.16(a) is made up of members of equal length, rests freely on rollers at B and is supported by a pin-joint at A. Find the forces in the members marked 1 and 2.*

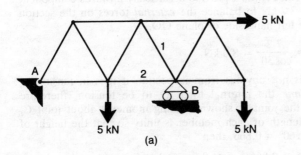

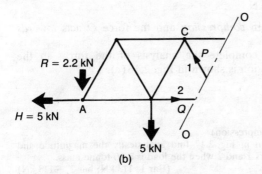

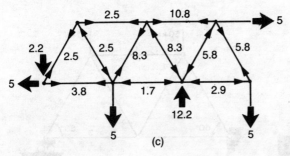

Fig. 2.16

SOLUTION
The support reactions at A will be found to be $R = 2.2$ kN vertically *downwards* and $H = 5$ kN, horizontally to the left, and at B, 12.2 kN vertically upwards.

To find the forces in bars 1 and 2 cut the frame by plane OO, as shown in the free-body diagram Fig. 2.16(*b*). If *P* is the internal force in bar 1, *by inspection*, it must act towards the top joint as shown and be in **compression** since it must provide a vertically upwards component force of 2.2 + 5 = 7.2 kN to balance the *external* forces on the section of the frame to the left-hand side of plane OO.

Therefore $P = \dfrac{7.2}{\cos 30°} = 8.3$ kN

For bar 2 it is not obvious from inspection if it is in tension or compression. *Assume* the internal force *Q* to be tension, therefore acting *away* from the joint as shown. Taking moments about joint C, and assuming the length of each member is unity so that the height of the frame is $1 \times \sin 60° = 0.866$, then

$$Q \times 0.866 = 5 \times 0.866 - 5 \times 0.5 - 2.2 \times 1.5$$

i.e. $Q = -1.7$ kN

negative, hence bar 2 is in *compression* and the force *Q* acts *towards* the joint.

The student should complete the analysis of the forces in the frame. The complete solution is shown in Fig. 2.16(*c*).

Problems

(Negative answers denote compression)
1. For the framework shown in Fig. 2.17 find analytically the magnitude and nature of the forces in bars 1 and 2 when the load is a 10-tonne mass.

(Bar 1, 113 kN; bar 2, −113 kN)

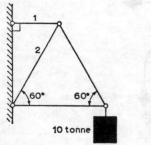

Fig. 2.17

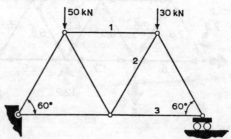

Fig. 2.18

2. For the framework shown in Fig. 2.18 find analytically the magnitude and nature of the forces in bars 1, 2 and 3. All bars are of equal length.

(Bar 1, − 23.12 kN; bar 2, 5.78 kN; bar 3, 20.2 kN)

3. The framework shown in Fig. 2.19 carries masses of 2 and 3 tonne as shown. Find analytically the magnitude and nature of the forces in all members of the frame and state the reaction at the supports.

(BG, 38.2; GA, −27; CF, 24.5; FG, 3.43; DE, 31.4; EF, −3.43; EA, −22.1 kN; reaction AB = 27 kN; reaction DA = 22.1 kN)

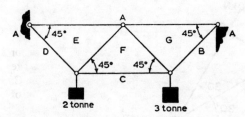

Fig. 2.19

4. For the framework shown in Fig 2.20 find analytically the magnitude and nature of the forces in all the members of the frame and state the reactions at the supports.

(AB, −66.7; CD = vertical AJ = horizontal AJ = 0; BC, 94.3; BE = HG = FH = 33.3; HJ = EF = −47.1; EG, 66.7; AF, −33.3; AC = −66.7 kN; reaction at X, 66.7 kN vertical; reaction at Y, 33.3 kN vertical)

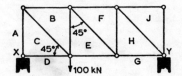

Fig. 2.20

5. In the Warren girder shown in Fig. 2.21 all bars are of equal length and the loads are vertical. Find the magnitude and nature of the forces in the members A, B and C. Both reactions are vertical.

(A, 46.2; B, −23.1; C, −69.3 kN)

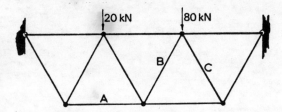

Fig. 2.21

6. Find the reactions at the supports A and B in the framework of Fig. 2.22. Using method of sections and method of joints find the nature and magnitude of the forces in the bars. (For convenience, take $h = 1$)

(A, 11.55 kN vertically upwards, 15 kN horizontally to the left; B, 14.43 kN vertically upwards; CB, −10; CD, 0; DB, −11.55; DA, −5.77; BA, 15 kN)

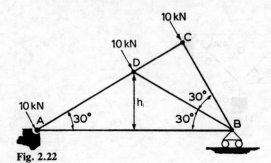

Fig. 2.22

CHAPTER 3

Friction

When one body slides on another and the surfaces are pressed together a *friction force* tangential to the surfaces has to be overcome before relative motion can take place between the bodies.

Friction conditions may fall into one of the following categories:

dry and clean;
greasy or boundary;
fluid or viscous;
pure rolling.

Dry friction is mainly dealt with here but a discussion of the mechanism and types of friction is given in paragraph 3.7.

Calculations involving friction between two dry and clean surfaces in contact are based upon the following *experimental* facts:

1. If a force is applied tending to move one body over another the opposing friction force brought into play is tangential to the surfaces in contact and is just sufficient to balance the applied force.
2. There is a limit beyond which the friction force cannot increase. When this limit is reached sliding is about to start and the corresponding friction force is termed the *limiting* value.
3. The limiting friction force is proportional to the *normal* load pressing the two surfaces together and is independent of the area of contact.
4. The ratio of the limiting friction force F and normal reaction N is a constant which depends only on the nature of the pair of surfaces in contact.

     Thus

$$\frac{F}{N} = \mu, \text{ a constant}$$

or

$$F = \mu N$$

μ is called the *coefficient of static or limiting friction*.

In Fig. 3.1 when sliding is about to start the pull P is equal to the limiting friction force F, i.e.

$$P = F = \mu N$$

Before sliding starts P is equal to the friction force which is less than the limiting value μN.

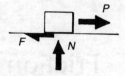

Fig. 3.1

5. After sliding starts the direction of the friction force is *opposite* to that of the *resultant* motion. The friction force is again given by $F = \mu N$, where μ is now called the *coefficient of sliding* or *kinetic friction*. The kinetic value is usually slightly less than the static or limiting value. It is also approximately independent of the speed of sliding. Otherwise sliding friction obeys the same laws as limiting friction.

When the pull P remains equal to μN during sliding the body moves at a steady speed. When P is greater than μN the body is accelerated and this particular case is dealt with in Chapter 5.

These laws of friction are approximately true and are sufficient for most engineering purposes. Their limitations are discussed in paragraph 3.7.

3.1 Friction on a rough inclined plane (Fig. 3.2)

Consider a body maintained at rest on an inclined plane under the action of its own weight and the friction force only. The force tending to move the body down the slope is the component of the weight, $W \sin \theta$, and N is the normal reaction of the plane on the body supporting the normal component of the weight, $W \cos \theta$. The friction force acts up the slope to maintain the body at rest. Resolving forces normal and parallel to the plane

$$N = W \cos \theta$$

$$F = W \sin \theta$$

These two equations hold true whether motion is impending or not. Only if motion is about to take place does the friction force F have its maximum or limiting value μN. For *impending* motion, therefore:

$$F = \mu N = \mu W \cos \theta$$

and $F = W \sin \theta$

hence

$$\mu W \cos \theta = W \sin \theta$$

i.e. $\qquad \mu = \tan \theta$

This particular angle, $\theta = \tan^{-1}\mu$, is called the *angle of repose*. If θ is less than the angle of repose the body remains at rest and $F < \mu N$. If θ is greater than this critical angle the body slides down the plane and $F = \mu N$, where μ is now the coefficient of *sliding* friction.

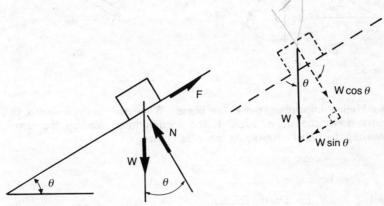

Fig. 3.2

Force applied parallel to the plane (Fig. 3.3)

A number of possibilities exist as to the state of the body when subject to a pull parallel to the plane. The body may be at rest without motion impending, about to move or actually moving with constant or changing speed. Change of speed is dealt with in later chapters.

Consider the case when a force P is applied to *pull* the body up the plane. Assume $\theta > \tan^{-1}\mu$.

(i) Motion impending up the plane If the body is about to move up the slope the forces acting are shown in the free-body diagram, Fig. 3.3 (*a*), and the force diagram is shown in Fig. 3.3 (*b*). The friction force has its limiting value, acting down the slope. Resolving forces as before:

$$N = W \cos \theta$$

$$P = F + W \sin \theta$$

and $\quad F = \mu N$

If P is increased above the value satisfying these equations the body will actually travel up the slope. If the speed is *constant* the equations for static equilibrium apply but μ should be the sliding value.

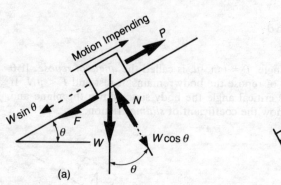

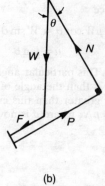

Fig. 3.3

(a)　(b)

(ii) Motion impending down the plane When the body is about to move down the slope, Fig. 3.4, the friction force acts up the slope assisting the pull P. Resolving forces as before:

$$N = W \cos \theta$$

$$P = W \sin \theta - F$$

and $F = \mu N$, for limiting friction.

Note that P may have any value between the two extreme values for motion about to take place up or down the slope. For any intermediate value of P the body is at rest, motion is *not* impending, and the friction force F is less than the limiting value μN.

We have dealt only with the particular case of a force applied parallel to the plane, acting upwards. The force may be applied in any direction to the body and it is then necessary to take into account the components of P parallel and perpendicular to the plane in arriving at the equation for balance of forces. In case (ii) if $\theta < \tan^{-1}\mu$ then P acts downwards.

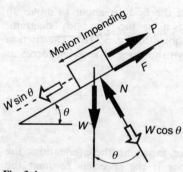

Fig. 3.4

Example *A body of weight 50 N is at rest on an inclined plane, Fig. 3.5. A force P is applied horizontally as shown. If $\mu = 0.4$ find the range of values of P over which the body will remain at rest.*

SOLUTION

The extreme values of P will be required when the body is on the point of moving up or down the plane. The forces acting on the body when motion is impending *up the slope* are shown in the free-body diagram, Fig. 3.6. Resolving forces normal to the plane:

$$N = 50 \cos 30° + P \sin 30°$$
$$= 43.3 + 0.5P$$

Resolving forces parallel to the plane, and taking $F = \mu N$ for limiting friction:

$$P \cos 30° = F + 50 \sin 30°$$

i.e. $0.866\,P = 0.4\,N + 25 = 0.4\,(43.3 + 0.5\,P) + 25$

hence $P = 63.5 \text{ N}$

When motion is impending *down the plane* the friction force is reversed, acting up the plane. Resolving forces:

$$N = 43.3 + 0.5\,P$$

and $P \cos 30° = 50 \sin 30° - F$

i.e. $0.866P = 25 - 0.4\,(43.3 + 0.5\,P)$

hence $P = 7.2 \text{ N}$

Thus, for any value of P between **7.2 N** and **63.5 N** the body remains at rest.

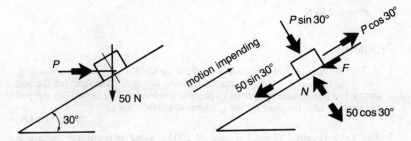

Fig. 3.5   **Fig. 3.6** Free-body diagram

Example *A 1 Mg load is pulled steadily up a track inclined at 30° to the horizontal by a force P inclined at 20° to, and above the track. Calculate the value of P if $\mu = 0.15$.*

SOLUTION

Pull P has components $P\cos 20°$ and $P\sin 20°$ parallel and perpendicular to the track, respectively, Fig. 3.7. $W = 9800 \, N$, and $F = \mu N = 0.15 \, N$.

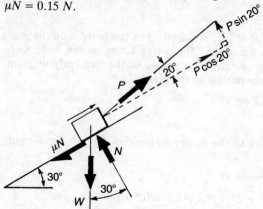

Fig. 3.7

Resolving forces parallel to the track:

$$P\cos 20° = 0.15 \, N + 9,800 \sin 30°$$

$$P = 0.16 \, N + 5,215$$

Resolving perpendicular to the track:

$$N = 9,800 \cos 30° - P \sin 20°$$

$$= 9,800 \times 0.866 - P \times 0.342$$

$$= 8,490 - 0.342 \, P$$

Solving for P, gives

$$P = \mathbf{6,230 \, N}$$

Problems

1. A load of mass 1350 kg lies on a gradient inclined at 60° to the horizontal. For static friction $\mu = 0.5$, for kinetic friction, $\mu = 0.4$. Calculate: (*a*) the pull parallel to the gradient required to prevent the load sliding down; (*b*) the pull required to haul the load up the gradient at uniform speed.

   (8.14, 14.1 kN)

2. The force required to haul a load of 500 kg along a horizontal surface is 1.2 kN. Find (*a*) the force parallel to a track of slope 20° required to haul the load up the incline; (*b*) the force required to lower it down the incline at steady speed. Assume the coefficient of friction to be the same in all cases.

   (2.8 kN, 548 N)

3. A 1500-kg boat is winched steadily up a slip inclined at 25° to the horizontal. If $\mu = 0.5$ for the surface contact of the boat and slip find the force in the winch cable, which is parallel to the slip.

   (12.9 kN)

4. A body of mass m on a rough plane inclined at 20° to the horizontal is moved steadily up the plane by a force of 200 N applied upwards and parallel to the plane. When the force is reduced to 75 N the body slides steadily downwards. Find the values of m and μ.

(41 kg; 0.17)

5. A body of mass 100 kg is at rest on a plane inclined at 20° to the horizontal. A force P is applied to pull the body up the plane, directed at an angle of 20° to the plane i.e. at 40° to the horizontal. If $\mu = 0.2$, calculate the value of P: (a) when the body is just about to move up the plane; (b) when the body is on the point of sliding down the plane.

(515 N; 173 N)

6. A trolley of mass 800 kg is about to move up a 20° slope due to the pull of a 2-tonne counterweight guided by a pulley, Fig. 3.8. The tensions in the rope are P and Q as shown. Neglecting friction at the trolley wheels find P. If $\mu = 0.2$ for the surfaces in contact at the counterweight find Q.

(2.68 kN; 9.6 kN)

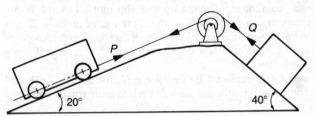

Fig. 3.8

3.2 The angle of friction and total reaction

In all cases considered so far the body is in equilibrium under the action of the four forces P, W, N and F. The method of resolution of forces will solve all problems of this type but two equations are required for a solution. However, a simpler method of calculation exists for certain problems. This makes use of the resultant reaction R of the friction force F and the normal reaction N, together with the angle ϕ between R and N. This angle ϕ is known as the *angle of friction*.

Consider a body *about to move* to the right (Fig. 3.9). The force R is the resultant of N and F. Since the latter are at right angles the angle ϕ between R and N is given by

$$\tan \phi = \frac{ab}{oa}$$

$$= \frac{F}{N}$$

$$= \frac{\mu N}{N} \text{ for limiting friction}$$

$$= \mu$$

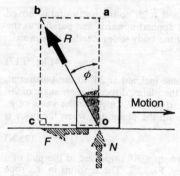

Fig. 3.9

The direction in which R must be drawn is determined by the fact that its tangential component F must oppose the motion of the body, that is, R is always drawn backwards to the direction of motion. R acts only at angle ϕ to N for limiting friction; if R lies inside the angle of friction the force of friction is less than the limiting value and *slipping cannot take place*.

When N and F are replaced by one force R, the forces P, W and R form *three* forces in equilibrium and the triangle of forces can then be drawn.

It should be noted particularly that in the *absence of friction* the only force between two surfaces is *normal* to the surfaces.

3.3 Application of angle of friction to motion on the inclined plane

Fig. 3.10 shows a body being moved up a plane by means of a horizontal force P. R is the resultant force exerted by the plane on the body. It acts at an angle ϕ to the normal and, when slipping is just about to start, $\tan \phi = \mu$. Since motion is up the plane, R must have a component down the plane to provide the resisting friction force and is therefore directed as shown. The pull P, weight W and resultant R form a triangle of forces and, since P and W are at right angles:

$$P = W \tan (\theta + \phi)$$

Fig. 3.11 shows the case when the body is just about to move down the plane against a resisting horizontal force P. R now has a component up the plane and is directed backwards to the direction of motion at angle ϕ to the normal. From the triangle of forces

$$P = W \tan (\theta - \phi)$$

In this case the angle of the plane θ is greater than the angle of friction ϕ.

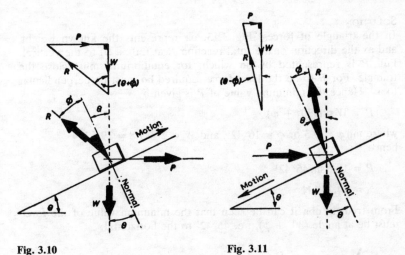

Fig. 3.10

Fig. 3.11

When the body is about to move down the plane and force P assists the motion (Fig 3.12), R is directed backwards to the normal as before, and hence

$$P = W \tan(\phi - \theta)$$

The angle of inclination of the plane θ is in this case less than the angle of friction ϕ.

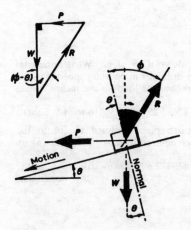

Fig. 3.12

Example *A casting of mass 2 tonne is to be pulled up a slope inclined at 30° by a force at an angle to the slope. If the coefficient of friction is 0.3 find the least force required and its direction to the horizontal.*

SOLUTION

In the triangle of forces, Fig. 3.13, **oa** represents the known weight, and **ax** the direction of the total reaction R at $(30° + \phi)$ to the vertical. Pull P is represented by **bo** which, for equilibrium, must close the triangle. For P to be the least force required **bo** must be perpendicular to **ax**. Hence the minimum value of P is given by

$$P = W \sin(30° + \phi)$$

where $\tan \phi = 0.3$ or $\phi = 16°\,42'$, and $W = 2 \times 9.8 = 19.6$ kN hence

$$P = 19.6 \sin 46°\,42'$$

$$= \mathbf{14.25\ kN}$$

From triangle **oba** it can be seen that the minimum value of the force must be at angle $(30° + \phi)$, i.e. **46° 42′** to the horizontal.

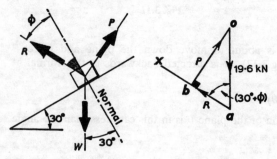

Fig. 3.13

Problems

1. A load of 300 kg will just start to slide down a 25° slope. What horizontal force will be required to haul the load up the slope at constant speed?

    What is the *least* force required to haul the load up the incline? State the direction of this least force.

    (3.5 kN; 2.25 kN at 50° to the horizontal)

2. A load of mass 100 kg rests on a rough plane inclined at 20° to the horizontal. If $\theta = 0.5$ find the *least* force required to move the load: (*a*) up the plane; (*b*) down the plane.

    (712 N upwards at 46.6° to the horizontal; 112.6 N downwards at 6.6° to the horizontal)

3.4 Wedges

The wedge is an application of the inclined plane, an example of friction being used to advantage. It is used as a splitting device, to apply a large force to lift or adjust a heavy load with small displacement, or to change the direction of an applied force. In effect, it is a simple machine. Double wedges can be arranged in various ways to achieve greater force and control. A wedge raising a load, Fig. 3.14(a), may involve three pairs of friction surfaces and the tilting of the load against its vertical guides. Such problems are complicated but may be simplified for our purpose by assuming only the inclined surface of the wedge to be rough, and the wedge and load to be guided by smooth rollers. Also, problems involving wedge friction are most easily dealt

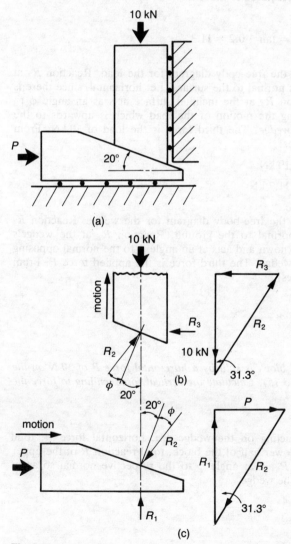

Fig. 3.14

with using the angle of friction together with the triangle of forces. In general, the less friction there is and the flatter the wedge the smaller the value of the force needed to lift a given load. Reducing the angle of the wedge, however, means that the wedge has to travel a greater distance for a given lift.

Example   *Fig. 3.14 (a) shows a 20°-wedge of negligible weight used to raise a load of 10 kN guided by frictionless rollers. The wedge sits on frictionless rollers and μ for the inclined surfaces is 0.2. Find the least value of the force P required to move the load upwards.*

Solution
Angle of friction $\phi = \tan^{-1} 0.2 = 11.3°$

Forces on the load
Fig. 3.14 (b) shows the free-body diagram for the load. Reaction R_3 at the vertical guide is normal to the surface, i.e. horizontal, since there is no friction. Reaction R_2 at the inclined surface acts as an angle ϕ to the normal opposing the motion of the load which is upwards to the left *relative to the wedge*. The third force is the load of 10 kN. From the triangle of forces:

$$R_2 \cos 31.3° = 10 \text{ kN}$$

hence $R_2 = 11.7 \text{ kN}$

Forces on the wedge
Fig. 3.14 (c) shows the free-body diagram for the wedge. Reaction R_1 at the ground is normal to the ground. Reaction R_2 at the wedge's inclined surface is known and acts at an angle ϕ to the normal opposing the motion of the wedge. The third force is the applied force P. From the triangle of forces,

$$P = R_2 \sin 31.3°$$

$$= 11.7 \sin 31.3°$$

$$= \mathbf{6.1 \text{ kN}}$$

Example   *A wood block is split by a horizontal force P of 30 N on the wedge shown (Fig. 3.15). Calculate the vertical force tending to force the wood apart. μ = 0.4.*

Solution
The three forces acting on the wedge are: horizontal force P, total reaction Q of the lower half of the block, total reaction R of the upper half. Both Q and R act at angle ϕ to the respective normal so as to oppose motion of the wedge.

$$\tan \phi = \mu = 0.4$$

thus $\phi = 21° 49'$

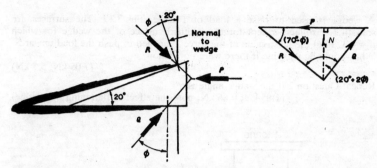

Fig. 3.15

From the triangle of forces, using the sine rule,

$$\frac{Q}{\sin(70° - \phi)} = \frac{P}{\sin(20° + 2\phi)}$$

therefore

$$Q = \frac{30\sin(70° - 21°49')}{\sin(20° + 43°38')}$$

$$= 25\ N$$

The force N tending to separate the two halves of the block is the vertical component of Q (or R), i.e.

$$N = Q\cos\phi$$

$$= 25 \times \cos21°49'$$

$$= 23.2\ N$$

Problems

1. The catch shown in Fig. 3.16 acts through the wedge forcing apart two steel balls against the springs S. If the coefficient of friction between wedge and balls is 0.3, calculate the force on each spring when the vertical load on the wedge is 100 N.

   What would be the spring force if friction were negligible?

   (47, 86.6 N)

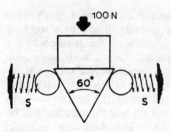

Fig. 3.16

2. A wedge is used to raise a load of 1 tonne, Fig. 3.17. The surfaces are smooth (on rollers) except for the inclined surface of the wedge for which $\mu = 0.1$. Find the least value of the force P needed to push the load upwards. What would be its value if there were no friction?

(7.05 kN; 5.7 kN)

3. Repeat Question 2 if the wedge angle is 5°.

(1.85 kN; 0.86 kN; note the effect of flattening the wedge)

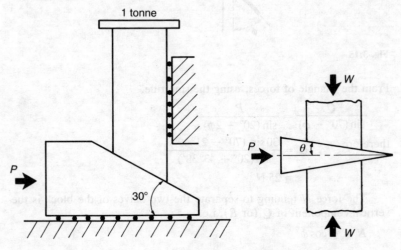

Fig. 3.17   **Fig. 3.18**

4. A wedge is driven into a block of wood against a resisting load W, Fig. 3.18. If the friction angle is ϕ and the half-angle of the wedge is θ, show that the force needed to force the wood apart is $P = 2W\tan(\theta + \phi)$ and the force needed to pull the wedge out is $P = 2W\tan(\phi - \theta)$

3.5 Toppling or sliding

The problems on friction so far have been basically of a 'body on a plane' type where we have ignored the possibility of the body toppling over rather than sliding; this possibility depends on the position of the centre of gravity of the body. For example, when a body such as a crate, Fig. 3.19 topples about an edge in contact with a surface there must be an unbalanced moment about the edge. Prior to toppling, the reaction of the surface to the body will act across the contact surfaces but at the moment of toppling the body will tilt and the reaction will be concentrated at the tilting point or line. The following example illustrates the approach when it is necessary to ascertain if a body will first slide or topple under load.

Example *A uniformly loaded box of mass 45 kg is at rest on horizontal ground when pushed by a gradually increasing load P, applied at 1.2 m above the ground, Fig. 3.19 (a). If µ = 0.4, determine whether the box will first slide or tip over and the least force required.*

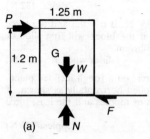

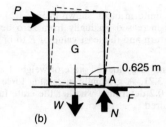

Fig. 3.19

SOLUTION

The forces acting on the box are the weight W, friction force F, normal reaction N, and the force P.

For sliding to take place, Fig. 3.19 (a), then

$$P \gtreqless F$$

The normal reaction N is equal to W and acts upwards through the centre of gravity G; F has its limiting value $\mu N = \mu W$, therefore the *least* value of P required is

$$P = \mu W = 0.4 \times 45 \times 9.8 = 176.4 \text{ N}$$

When the box is about to tip at corner A the reaction of the ground is at the edge A. Taking moments about A, Fig. 3.19 (b):

$$P \times 1.2 = W \times 0.625$$

$$= 45 \times 9.8 \times 0.625$$

therefore

$$P = 230 \text{ N}$$

The box therefore will slide first when the force reaches **176.4 N**.

Notes: 1. The closer the centre of gravity is to the tilting corner the smaller the value of the force required to cause toppling.

2. When the box is on an inclined plane the height of the centre of gravity above the plane is also critical.

3. If the force P in Fig. 3.19 is inclined to the horizontal its vertical component affects the reaction N, and the horizontal component causes the sliding or toppling.

Problems

1. A uniform block of mass 50 kg has a base of square section 0.6 m side. When at rest on horizontal ground it is pulled by a gradually increasing force P,

applied horizontally, 1.3 m above the ground, the point of application being below the height of the block. If $\mu = 0.25$ determine whether the block first slides or tips over and the least force required.

(topples; 113 N)

2. Solve Question 1, if the force P is applied at 40° to and above the horizontal.

(slides; 132 N)

3. The uniform block shown in Fig. 3.20 weighs 850 N. If $\mu = 0.3$ and the force P is increased gradually from zero determine if the block will first slide or overturn and the least value of P to break equilibrium.

(slides when $P = 245$ N)

4. A uniformly loaded crate of weight 1 kN is at rest on a rough inclined plane, Fig. 3.21. A gradually increasing force P is applied horizontally as shown. If $\mu = 0.2$ determine whether the crate first slides or topples and the least force required.

(topples; 726 N)

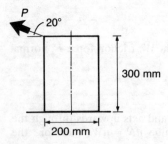

Fig. 3.20

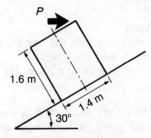

Fig. 3.21

3.6 The ladder problem

When ladders or poles are propped against a wall, there are friction forces and normal reactions at both the ground and the wall. Assuming the weight of the ladder is known there are therefore four unknown forces. By resolving forces vertically and horizontally and taking moments about a suitable point only *three* equations can be obtained. Further information is necessary and this is usually provided by assuming a value for the coefficient of friction for the surfaces, thus enabling the friction forces to be obtained in terms of the reactions when slip is about to take place. The number of unknown forces is thereby reduced. Otherwise, if it is assumed that the wall is smooth the friction force at the wall is eliminated. It is assumed that the ladder is in a vertical plane perpendicular to the wall so that the direction of possible motion is known. The following example illustrates the method of solution.

Example *A uniform ladder weighing 200 N is propped against a smooth wall as shown in Fig. 3.22. The ladder is in equilibrium and*

*μ = 0.4 for the ground. If a man of weight 800 N starts to climb the
ladder, how far up the ladder will he reach before the ladder starts to
slip?*

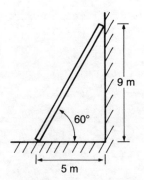

Fig. 3.22

SOLUTION
Fig. 3.22 shows the free-body diagram for the ladder when it is about
to slide and the man has reached a horizontal distance x from the foot
of the ladder. The weight of the ladder acts at its midpoint, and the
friction force F at the ground has its limiting value μN_1, where N_1 is
the normal reaction at the wall. Resolving forces vertically and horizontally:

$$N_1 = 200 + 800 = 1000 \text{ N}$$

and $N_2 = F = \mu N_1 = 0.4 \times 1000 = 400$ N

Taking moments about the foot of the ladder,

$$N_2 \times 9 = 200 \times 2.5 + 800 \times x$$

Substituting in this equation for N_2 gives

$$x = \textbf{3.9 m}$$

Example *A uniform ladder 8 m long, weighs 220 N, rests on rough
ground, and is propped against a vertical rough wall at an angle θ to the
horizontal. If μ = 0.4 for the ground and the wall surfaces, find the
value of θ when slip is about to take place.*

SOLUTION
Fig. 3.23 shows the normal reactions, friction forces and self-weight
acting on the ladder. Resolving forces vertically and horizontally, and
taking moments about the top of the ladder we obtain three equations,
thus (working in newtons and metres):

$$N_1 + 0.4 N_2 = 220$$

$$N_2 = \mu N_1 = 0.4 N_1$$

and $\quad N_1 \times 8 \cos \theta = 220 \times 4 \cos \theta + 0.4 \, N_1 \times 8 \sin \theta$

Solving these equations gives

$$\theta = \mathbf{46.4°}$$

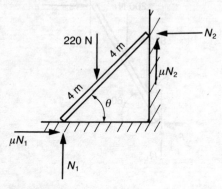

Fig. 3.23

Problems

1. A uniform ladder rests against a smooth wall with its lower end on rough ground. If $\mu = 0.45$, find the inclination of the ladder to the vertical when it is on the point of sliding.

   (42°)

2. A uniform ladder rests on rough ground and is propped against a rough vertical wall. If μ is 0.5 for the ground and 0.25 for the wall, find the inclination of the ladder to the horizontal when it is on the point of sliding.

   (41.2°)

3. A ladder of negligible weight is propped against a rough vertical wall, the coefficients of friction being the same at the wall and ground. The top of the ladder is 10 m above the ground and the bottom is 6 m from the base of the wall. If a man climbs the ladder, find the minimum value of the coefficient of friction so that he reaches the top of the ladder before slip takes place.

   (0.6)

4. A ladder 10 m long is propped against a rough wall at an angle of 30° to the vertical. Neglecting the mass of the ladder, find how far a man of weight W can climb up the ladder before slipping takes place. For the wall $\mu = 0.4$, for the ground $\mu = 0.25$.

   (4.85 m)

3.7 Further notes on friction and lubrication

Dry friction

The simplest explanation of the friction effect when dry clean surfaces rub together is that it is due to surface roughness. When studied closely even the most apparently smooth surfaces consist of 'hills' and 'valleys'. There is a tendency for each surface to shear the tips of the irregularities of the other. Since it is only the projecting 'hills' or high spots which are actually bearing on one another the area of true contact is very much less than the apparent area of contact; this is shown in Fig. 3.24. At average loads the area of true contact is proportional to the load applied and is almost independent of the apparent area of contact. Hence the friction force, which is determined by the area of true contact, is proportional to the load applied and almost independent of the apparent area of contact: the ratio of friction force to load, μ, is therefore constant and for a given pair of materials independent of the load. However, for very great loads the area of true contact may not increase in simple proportion to the load but more rapidly. In practice therefore μ may increase with the load. Also as surfaces become worn the value of μ changes.

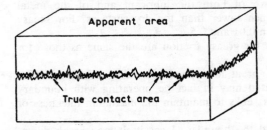

Fig. 3.24

It would appear that dry friction would be reduced by improving the smoothness of surfaces. For example, surfaces of smooth wood slide more easily on each other than surfaces of emery paper. However, this is only true up to a point, for smooth surfaces will have a greater area of true contact than rough surfaces. Owing to the attraction between the surface molecules of the materials there tends to be cohesion or binding together of the surfaces and the greater the area of true contact the greater is this tendency for cohesion. This condition ultimately leads to the surfaces seizing together. For example, two highly-polished dry-metal surfaces will tend to seize together very rapidly under load.

Fluid friction (viscous friction)

When there is an excess of lubricant present two solid surfaces may be separated by a film of fluid so that friction depends wholly upon the lubricant and not on the nature of the surfaces. The force necessary to produce relative motion is that required to shear the lubricant film. The friction force in fluid friction increases with the velocity of sliding. In contrast to dry friction the friction force is proportional to the total or apparent area of contact.

Fluid friction only exists when there is motion, otherwise the lubricant is squeezed out by the load. In practice all bearings running under design conditions should have fluid-film lubrication.

Boundary friction (greasy friction)

It should be realised that for perfectly clean and dry surfaces the coefficient of static friction is often greater than unity and may be very high indeed. However, such conditions are not usually met with in engineering practice. Unless specially cleaned all surfaces possess a very thin film of grease and this may only be an 'adsorbed' film perhaps 0.0003 mm or 0.3 micrometres thick. This attaches itself to the bearing surface and may prevent metal-to-metal contact. Cohesion, therefore, between smooth surfaces occurs between relatively weak grease molecules rather than strong metal ones. The coefficient of friction now depends upon the nature of both the lubricant and of the metal surfaces, but is very much lower than for dry surfaces. For greasy friction μ may be between 0.01 and 0.5.

The laws of friction for greasy friction are the same as those for dry friction.

Under heavy loads, or at low speeds of sliding, bearings which appear profusely lubricated may in fact be operating with boundary lubrication. The engineer seeks to maintain the maximum thickness of the oily boundary layer.

Under excessive load the boundary layer itself may break down. Contact takes place between high spots on the metal surfaces and the high rubbing temperatures which occur may result in local melting and seizure.

Rolling resistance

A cylinder rolling on a flat plane encounters no friction resistance to motion providing there is no sliding and that neither cylinder nor plane deform under load. In practice, of course, both surfaces will deform to some extent. Assuming the cylinder to be hard and the plane soft, the deformation is as shown in Fig. 3.25. For example, a rotating cylinder pressed into a rubber surface travels forward in one revolution a distance which may be 10 per cent less than its circumference. Negligible slip occurs between cylinder and surface, but energy is lost due to the stretching of the surface of the rubber along the line ABCD. Owing to the deformation of the surface the reaction R has a horizontal component **ab** opposing the motion. This horizontal component is an apparent friction force, R_r.

The rolling resistance is very little affected by lubricant films. Lubricant may reduce wear but does not reduce the rolling resistance since there is no sliding friction. The coefficient of 'friction' for rolling

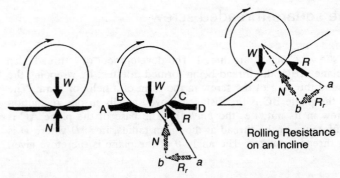

Rolling Resistance
on an Incline

Fig. 3.25

is about 0.001 or less, depending on the hardness of the surfaces in contact. Rolling resistance may be assumed to be directly proportional to the normal load and the ratio of rolling resistance to normal load is called the *coefficient of rolling resistance*, but this coefficient is little used.

In designing machines the attempt is made to replace sliding by rolling friction wherever possible. Hence the use of ball or roller bearings in place of plain bearings, although in roller bearings sliding friction may occur owing to their method of construction. For example, it is usually necessary to enclose the rollers in a cage. The effect is probably small since the cage carries little load. Again with ball bearings it is often necessary to allow the balls to run in a groove in the ball race (Fig. 3.26). Sliding now takes place between the ball and the sides of the groove and this contact between the surfaces may be heavily loaded. Lubrication may therefore be necessary to reduce sliding friction as well as wear and to protect the bearing against corrosion. For further notes on rolling resistance of cars and trains see paragraph 5.10.

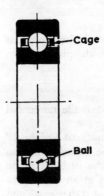

Fig. 3.26

3.8 The square-threaded screw

Fig. 3.27 shows a single-start thread. The development of a thread is an inclined plane ABC, the thread being formed in effect by wrapping the plane around the *core* of the screw in the form of a helix or spiral. The height of the plane BC is the distance moved axially in one revolution of the screw in its nut, i.e. the *pitch, p*. The base of the plane AC is the circumference of the thread at the mean radius, i.e. πD where D is the mean thread diameter. The angle θ of the plane is therefore given by

$$\tan \theta = \frac{p}{\pi D}$$

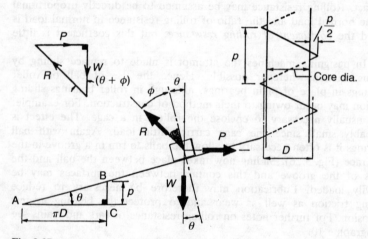

Fig. 3.27

For a double-start thread the distance moved axially by the screw in its nut in one revolution is the *lead l*, which is twice the pitch. Thus:

$$\tan \theta = \frac{l}{\pi D} = \frac{2p}{\pi D}$$

Let W be the axial load on the screw or nut and P the tangential force required at the *mean thread radius* to turn the screw.

Turning the screw is equivalent to moving a mass of weight W along the inclined plane by a horizontal force P. The forces acting on the screw are W, P, and the total reaction R of the inclined plane formed by the thread of the nut. The reaction R acts at an angle ϕ to the normal where ϕ is the angle of friction and $\tan \phi = \mu$, the coefficient of friction between screw and nut.

Consider two cases according as the load is being raised or lowered.

(a) Raising load

When the load is raised by the force P the motion is up the plane and the reaction R acts to the left of the normal as shown (Fig. 3.27). From the triangle of forces

$$P = W \tan(\phi + \theta)$$

The *torque* T required to rotate the screw against the load is

$$T = P \times \tfrac{1}{2}D$$

$$= \tfrac{1}{2}WD \tan(\phi + \theta)$$

The *efficiency* of the screw is equal to:

$$\frac{\text{work done on load } W \text{ in 1 revolution}}{\text{work done by } P \text{ in 1 revolution}} = \frac{W \times \text{lead } (l)}{P \times \pi D}$$

But

$$\frac{P}{W} = \tan(\phi + \theta) \text{ and } \frac{l}{\pi D} = \tan\theta$$

Hence

$$\text{efficiency} = \frac{\tan\theta}{\tan(\phi + \theta)}$$

Alternatively

$$\text{efficiency} = \frac{\text{force } P \text{ required without friction } (\phi = 0)}{\text{force } P \text{ required with friction}}$$

$$= \frac{W \tan\theta}{W \tan(\phi + \theta)}$$

$$= \frac{\tan\theta}{\tan(\phi + \theta)}$$

Note that this efficiency is *independent of the load*. However, the above theory has neglected the weight of the screw itself. If the constant weight of the screw were to be taken into account it would be found that the efficiency would increase with the load as in other load lifting machines. For notes on efficiency generally see paragraph 10.15.

(b) Load being lowered

When the load is being lowered the reaction R must lie to the right of the normal (Fig. 3.28). If the angle of friction ϕ is greater than the angle of the plane θ (the usual condition) then it can be seen from the triangle of forces that the force P must be applied to *help* lower the load. The angle between R and W in the triangle of forces is now $(\phi - \theta)$ so that

$$P = W \tan(\phi - \theta)$$

and the torque required is

$$T = P \times \tfrac{1}{2}D$$

$$= \tfrac{1}{2}WD \tan(\phi - \theta)$$

When ϕ is less than θ, R is still to the right of the normal, opposing the motion, but the triangle of forces must now be as shown in Fig. 3.29 with force P applied to *resist* the downward movement of the load. Under this condition the load would just be about to move downwards. If P were not applied in this direction the load would *overhaul*, that is, move down under its own weight.

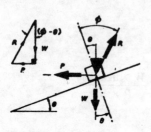

Fig. 3.28   **Fig. 3.29**

From the triangle of forces (Fig. 3.29) when P is resisting:

$$P = W \tan(\theta - \phi)$$

and

$$T = \tfrac{1}{2} WD \tan(\theta - \phi)$$

When the load is lowered or falling, the efficiency of the screw has little physical significance. When it overhauls, the load is the effort. When the load is lowered by a force P it assists the effort. When the load falls against a restraining force P, this force is, in effect, a resistance. The term *reversed efficiency* is sometimes used in this latter case; it is the ratio of the work done against P in one revolution of the screw to the corresponding work done by the load W.

3.9 Overhauling of a screw

When the load moves down and overcomes the thread friction by its own weight it is said to overhaul. When it moved down against a resisting force P we found that

$$P = W \tan(\theta - \phi)$$

When the load overhauls, $P = 0$, therefore

$$\tan(\theta - \phi) = 0$$

or $$\theta - \phi = 0$$

i.e. $$\theta = \phi$$

Hence when the angle of the inclined plane is just equal to the angle of friction the load will overhaul.

The efficiency of such a screw when *raising* a load is given by

$$\eta = \frac{\tan \theta}{\tan (\theta + \phi)}$$

but $\theta = \phi$; hence

$$\eta = \frac{\tan \phi}{\tan 2\phi}$$

Table 3.1 gives the efficiency of the screw which will just overhaul in this manner for various values of the angle of friction ϕ and coefficient of friction μ. It is seen that as the angle ϕ tends to zero,

Table 3.1

ϕ (degrees)	$\mu = \tan \phi$	Efficiency (%)
1	0.0175	50 approx.
10	0.1763	48.4
30	0.5774	33.3
45	1.000	0

as for a frictionless screw, the efficiency tends to a limiting value of 50 per cent. Note, however, that this is *not* the greatest efficiency a screw may have; for if $\phi = 0$ and θ is *not* equal to the angle of friction, the efficiency is

$$\eta = \frac{\tan \theta}{\tan (\theta + \phi)}$$

$$= 1 \text{ or } 100 \text{ per cent}$$

In general, if the angle of the inclined plane is greater than the angle of friction the load will accelerate downwards.

The efficiency of a screw is often only about 25 per cent but may be increased by separating the sliding surfaces between screw and nut by steel balls. Thus sliding friction is replaced by rolling friction, as in a recirculating ball nut.

Example   *A screw-jack carries a load of 4 kN. It has a square-thread single-start screw of 20 mm pitch and 50 mm mean diameter. Calculate the torque to raise the load and the efficiency of the screw. What is the torque to lower the load? Take $\mu = 0.22$.*

SOLUTION

$$\tan \theta = \frac{p}{\pi D}$$

$$= \frac{20}{\pi \times 50}$$

$$= 0.1276$$

hence $\theta = 7° 16'$

$$\tan \phi = \mu = 0.22$$

hence $\phi = 12° 24'$

Torque to raise load $= \frac{1}{2}WD \tan(\phi + \theta)$

$$= \frac{1}{2} \times 4000 \times 0.05 \times \tan 19° 40'$$

$$= 36 \text{ N-m}$$

$$\text{Efficiency} = \frac{\tan \theta}{\tan(\phi + \theta)}$$

$$= \frac{0.1275}{0.3574}$$

$$= 0.357 \text{ or } \textbf{35.7 per cent}$$

Torque to lower load $= \frac{1}{2}WD \tan(\phi - \theta)$

$$= \frac{1}{2} \times 4000 \times 0.05 \times \tan 5° 8'$$

$$= \textbf{9 N-m}$$

Example *A turnbuckle has right- and left-hand square threads of 10 mm pitch, mean diameter 40 mm, μ = 0.16. The turnbuckle is used to tighten a wire rope, Fig. 3.30. If the tension in the rope is constant at 12 kN find the turning moment required.*

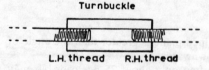

Turnbuckle

L.H. thread   R.H. thread

Fig. 3.30

SOLUTION

For *each* thread, $\tan \phi = \mu = 0.16$; hence $\phi = 9° 5'$

$$\tan \theta = \frac{p}{\pi D}$$

$$= \frac{10}{\pi \times 40}$$

$$= 0.0796$$

hence

$$\theta = 4°33'$$

Torque to overcome friction on each thread

$$= \tfrac{1}{2}WD \tan(\phi + \theta)$$
$$= \tfrac{1}{2} \times (12 \times 10^3) \times 0.04 \tan 13°38'$$
$$= 58.2 \text{ N-m}$$

Total torque for *two* threads $= 2 \times 58.2$

$$= \mathbf{116.4 \text{ N-m}}$$

Example *A double-start square-thread screw drives the cutter of a machine tool against an axial load of 450 N. The external diameter of the screw is 52.5 mm and the pitch is 5 mm. If the coefficient of friction for the thread is 0.15 find the torque required to rotate the screw.*

SOLUTION
Mean thread diameter $= 52.5 - 2.5 = 50$ mm

$$\tan \theta = \frac{2p}{\pi D} = \frac{2 \times 5}{\pi \times 50} = 0.0637$$

hence $\quad \theta = 3°39'$

$$\tan \phi = 0.15,$$

hence $\quad \phi = 8°32'$

$$\tan(\theta + \phi) = \tan 12°11' = 0.216$$

Alternatively,

$$\tan(\theta + \phi) = \frac{\tan \phi + \tan \theta}{1 - \tan \phi \tan \theta}$$
$$= \frac{\mu + \tan \theta}{1 - \mu \tan \theta}$$
$$= \frac{0.15 + 0.0637}{1 - 0.15 \times 0.0637}$$
$$= 0.216$$

Torque, $T = \tfrac{1}{2}WD \tan(\theta + \phi)$

$$= \tfrac{1}{2} \times 450 \times 0.05 \times 0.216$$
$$= \mathbf{2.43 \text{ N-m}}$$

Example *A load of mass 2 tonne is raised by a single-start square-thread screw jack. The load bears on a collar and does not revolve with the screw. The screw pitch is 12 mm and the outside diameter of the thread is 56 mm. The efficiency of the jack at this load is 15 per cent. The torque to overcome friction at the bearing collar is estimated to be*

*120 N-m. Find the torque to overcome thread friction and the coefficient
of friction for the screw thread.*

SOLUTION

$$W = 2 \times 1,000 \times 9.8 = 19,600 \text{ N}$$

$$\text{Efficiency} = \frac{\text{work done on load } (W) \text{ per revolution}}{\text{work done by torque } (T) \text{ per revolution}}$$

$$= \frac{W \times \text{pitch}}{T \times 2\pi}$$

therefore
$$0.15 = \frac{19,600 \times 0.012}{T \times 2\pi}$$

hence
$$T = 250 \text{ N-m}$$

and torque to overcome thread friction = 250 − bearing collar torque

$$= 250 - 120$$

$$= \textbf{130 N-m}$$

therefore

$$\tfrac{1}{2}WD \tan(\theta + \phi) = 130 \quad D = 56 - \tfrac{1}{2} \times 12 = 50 \text{ mm}$$

$$\tan(\theta + \phi) = \frac{130 \times 2}{19,600 \times 0.05} = 0.265$$

$$\theta + \phi = 14° 51'$$

$$\tan\theta = \frac{p}{\pi D} = \frac{0.012}{\pi \times 0.05} = 0.0763$$

hence
$$\theta = 4° 22'$$

Therefore
$$\phi = 14° 51' - 4° 22' = 10° 29'$$

and
$$\mu = \tan\phi = \textbf{0.185}$$

Problems

1. The helix angle of a screw thread is 10°. If the coefficient of friction is 0.3
   and the mean diameter of the square thread is 72.5 mm, calculate (a) the
   pitch of the thread, (b) the efficiency when raising a load of 1 kN, (c) the
   torque required.

   (40 mm, 35 per cent, 18.3 N-m)

2. Find the torque to raise a load of 6,000 N by a screw jack having a
   double-start square thread with two threads per centimetre and a mean
   diameter of 60 mm. $\mu = 0.12$. What is the torque required to lower the
   load?

   (31.4 N-m, 12 N-m)

3. A nut on a single-start square-thread bolt is locked tight by a torque of
   6 N-m. The thread pitch is 5 mm and the mean diameter 6 cm. Calculate (a)

the axial load on the screw in kilograms, (b) the torque required to loosen the nut. $\mu = 0.1$.

(161 kg; 3.47 N-m)

4. Calculate the pitch of a single-start square-thread screw of a jack which will just allow the load to fall uniformly under its own weight. The mean diameter of the thread is 8 cm and $\mu = 0.08$. If the pitch is 15 mm what is the torque required to lower a load of 3 kN?

(20 mm; 2.4 N-m)

5. A double-start square-thread screw has a pitch of 20 mm and a mean diameter of 100 mm; $\mu = 0.03$. Calculate its efficiency when raising a load.

(80.5 per cent)

6. A lathe saddle of mass 30 kg is traversed by a single-start square-thread screw of 10 mm pitch and mean diameter 40 mm. If the vertical force on the cutting tool is 250 N find the torque at the screw required to traverse the saddle. The coefficient of friction between saddle and lathe bed, and for the screw thread, is 0.15.

(0.376 N-m)

7. A stop valve in a horizontal pipeline consists of a plate of 120 mm diameter which moves in vertical guides. If the coefficient of friction between valve and guides is 0.3 and the pressure upstream of the valve is 3.75 MN/m², calculate the vertical force required just to move the valve when fully closed.

If the valve is raised by a screw having a square thread, 10 mm pitch and mean diameter 40 mm, calculate the torque on the screw. The coefficient of friction between screw and nut is 0.2.

(12.7 kN; 72.2 N-m)

8. A wire rope is tightened by means of a turnbuckle having right- and left-hand square threads of 6 mm pitch. The mean diameter of the thread is 22 mm. Find the turning moment to tighten the rope at the instant the pull in the rope is 7.5 kN. $\mu = 0.12$.

(34.6 N-m)

9. The bush A is drawn from the shaft B by the screw operated extractor shown in Fig. 3.31. If the radial pressure between bush and shaft is 3 MN/m² and the coefficient of friction is 0.2 what is the force required to draw the bush? If the screw pitch is 5 mm and the mean diameter of the square thread is 20 mm what is the torque required on the screw? For screw and nut $\mu = 0.15$.

(13.6 kN; 31.6 N-m)

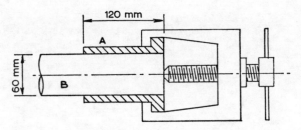

120 mm

A

60 mm

B

Fig. 3.31

10. A load of 6 kN is lifted by a jack having a single-start square-thread screw of 45 mm core diameter and pitch 10 mm. The load revolves with the screw and the coefficient of friction at the screw thread is 0.05. Find the torque required to lift the load. Show that the load will overhaul and find the tangential force required at 240 mm radius to keep the load from descending.

   (17.2 N-m; eff. = 56% > 50%, $\theta > \phi$, therefore *not* self-locking; 8.6 N)

CHAPTER 4
Velocity and Acceleration

4.1 Average speed

The *average speed*, v_{av}, of a body is defined as the distance travelled s divided by the time taken t; thus

$$v_{av} = \frac{s}{t}$$

The SI unit for speed is *metre per second* (**m/s**). Another unit commonly used is *kilometre per hour* (**km/h**) and in machine tool work speeds may be stated in units such as *metre per minute* (**m/min**).

Note that

$$\mathbf{1\ km/h} = \frac{1 \times 1,000}{3,600} = \frac{1}{3.6}\ \mathbf{m/s}$$

4.2 Constant speed

If the distance travelled is the same in successive intervals of time then the speed is said to be *constant*.

4.3 Varying speed

When the speed is not constant but changes continuously we require to state exactly what we mean by the *speed at a point*. Consider therefore a body which travels a distance s m after a time t s and suppose that s is given by the equation

$$s = 2t + t^2$$

Table 4.1 gives the values of s and of the average speed v_{av} corresponding to time intervals $t = 0, 1, 2, 3, 4$ seconds.

When $t = 0$, the average speed given by the formula s/t would appear to be $0/0$, which has no meaning. However, if the average speed be plotted against the *time interval* a smooth curve (in this case a straight line) is obtained which cuts the speed axis at a speed of 2 m/s (Fig. 4.1). Thus at point A when the time interval is zero the *average* speed is 2 m/s.

Table 4.1

Time t (s)	0	1	2	3	4
Distance s m travelled from time $t = 0$	0	3	8	15	24
Average speed (m/s) $v_{av} = s/t$	?	3	4	5	6

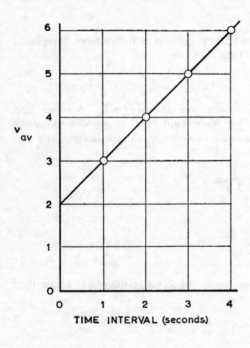

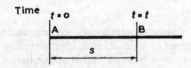

Fig. 4.1   Average speed–time graph

The ratio (distance travelled)/(time taken) therefore has a real meaning even when the time interval is zero. It is then known as the speed at a point A. Thus, speed at a point, v, = limiting value of the ratio

$$\frac{\text{distance travelled}}{\text{time taken}}$$

when the time interval is zero and is the *rate of change* of distance with respect to time, i.e. ds/dt in the notation of the calculus.

The speed at time $t = 0$ is therefore obtained by first differentiating the expression $s = 2t + t^2$ and then setting $t = 0$. Thus:

$$v = \frac{ds}{dt} = \frac{d}{dt}(2t + t^2)$$

$$= 2 + 2t$$

$$= 2 \text{ m/s} \quad \text{when } t = 0$$

4.4 Velocity

The *velocity* of a body is defined as a vector quantity, of magnitude equal to its speed, and direction tangent to the path of motion of the body. Velocity therefore is completely specified only by stating both magnitude *and* direction.

The distinction between speed and velocity and changes in those quantities is illustrated in Fig. 4.2. This shows the speed and velocity of a body at successive points A and B on a track AB. It should be noted that if the speed changes then the magnitude of the velocity changes accordingly, Fig. 4.2 (*b*), but the velocity may be altered by a change in direction without change in speed, Fig. 4.2 (*c*).

Uniform velocity is motion along a straight line at constant speed.

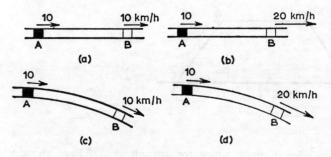

Fig. 4.2 (*a*) Constant speed; uniform velocity
      (*b*) Change in speed and velocity
      (*c*) Constant speed; vector change in velocity
      (*d*) Change in speed and velocity

4.5 Motion in a straight line

Average and uniform acceleration

The *average acceleration* a_{av} of a body moving in a straight line is defined as the change in speed or velocity divided by the time taken. If u is the initial velocity, v, the final velocity, and t, the time taken, then

$$a_{av} = \frac{\text{change in velocity}}{\text{time taken}} = \frac{v - u}{t}$$

The SI unit of acceleration is *metre per second per second* (m/s^2) and this is the only form of unit used.

If the velocity increases by equal amounts in equal intervals of time then the acceleration a is said to be *constant*.

Uniform acceleration is motion in a straight line with constant acceleration. Thus:

$$a = a_{av}$$
$$= \frac{v - u}{t}$$

or $v = u + at$

Fig. 4.3 shows a typical speed-time graph (or velocity-time graph for motion in a straight line) for a body moving with uniform acceleration. Uniform acceleration from rest to a maximum speed is shown by OA; the speed is maintained constant at the maximum value, shown by AB; finally the body decelerates uniformly to rest, BC. Note that provided the acceleration is uniform the v–t graph is made up of straight lines. Note also that the area under the graph OABC represents the total distance travelled.

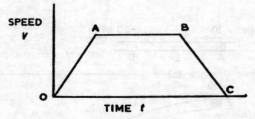

SPEED v

TIME t

Fig. 4.3

As a reminder of work which the student should have already covered we now state the formulae for motion with uniform acceleration.

4.6 Summary of formulae for uniform acceleration

$$v = u + at$$

$$v^2 = u^2 + 2as$$

$$s = ut + \tfrac{1}{2}at^2$$

$$s = v_{av}t$$

$$v_{av} = \frac{u + v}{2}$$

When the speed is increasing the acceleration is positive.

When the speed is decreasing the acceleration is negative, i.e. a retardation or deceleration.

Example   *A train accelerates uniformly from rest to reach 54 km/h in 200 s after which the speed remains constant for 300 s. At the end of this time the train decelerates to rest in 150 s. Find the total distance travelled.*

SOLUTION

$$54 \text{ km/h} = \frac{54}{3.6} = 15 \text{ m/s}$$

Referring to Fig. 4.3, the speed at A is 15 m/s. The time taken for the train to travel from O to A is 200 s, from A to B, 300 s, and from B to C, 150 s. The total distance travelled is given by the area OABC, hence

$$\begin{aligned} \text{distance travelled} &= (\tfrac{1}{2} \times 15 \times 200) + (15 \times 300) \\ &\quad + (\tfrac{1}{2} \times 15 \times 150) \\ &= 7{,}125 \text{ m} \\ &= \textbf{7.125 km} \end{aligned}$$

Problems

1. The cutting stroke of a planing machine is 600 mm and it is completed in 1.2 s. For the first and last quarters of the stroke the table is uniformly accelerated and retarded, the speed remaining constant during the remainder of the stroke. Using a speed–time graph, or otherwise, determine the maximum cutting speed.

(0.75 m/s)

2. A diesel train accelerates uniformly from rest to reach 60 km/h in 6 min, after which the speed is kept constant. Calculate the total time taken to travel 6 kilometres.

(9 min)

3. A train travelling at 30 km/h is slowed by a 'distant' signal at A, and comes uniformly to rest between A and B to stop at B, 300 m from A. After 1 min at rest the 'stop' signal at B allows the train to accelerate uniformly to C, 500 m from B, where it is again travelling at 30 km/h/ Calculate the total time lost between A and C due to signals.

(156 s)

4. The driver of a train shuts off the power and the train is then uniformly retarded. In the first 30 s the train covers 110 m, and it then comes to rest in a further 30 s. Determine (*a*) the initial speed of the train before power is cut off, (*b*) the total distance travelled in coming to rest.

(4.89 m/s; 147 m)

5. A car A starts from rest with a uniform acceleration of 0.6 m/s². A second car B starts from the same point, 4 s later and follows the same path with an acceleration of 0.9 m/s². How far will the cars have travelled when B passes A?

(142.5 m)

4.7 Freely falling bodies

The reader will recall the experiments attributed to Galileo which proved that all bodies dropped at the same place fell to earth with the same acceleration g, due to the force of gravity. Thus if a body falls to earth from a height which is small compared with the radius of the earth, it is found to increase its velocity by an equal amount each second, i.e. its acceleration downwards is *uniform*. Similarly if the body is projected upwards its deceleration upwards is uniform, and equal to g. The value of g may be taken as 9.8 m/s² for practical engineering purposes (*see* page 97), and it is found to be independent of the weight, size and shape of the body provided that air resistance is neglected. The formulae for uniformly accelerated motion in a straight line apply to the motion of freely falling bodies provided that the acceleration g replaces the acceleration a.

For motion downwards $a = g$ (acceleration)
For motion upwards $a = -g$ (deceleration)

The acceleration g may be regarded as a *vector* directed towards the centre of the earth, i.e. vertically 'downwards'.

Example *A small rocket is launched vertically from rest and reaches an altitude* h. *The acceleration is constant at 11.5 m/s² until the fuel burns out after 5.8 s. Assuming that from the point of burn-out onwards the rocket is travelling freely vertically upwards with constant deceleration* g = 9.8 m/s² *find the value of* h.

SOLUTION
The initial velocity u of the rocket is zero. The speed v after 5.8 s is given by

$$v = u + at$$
$$= 0 + 11.5 \times 5.8$$
$$= 66.7 \text{ m/s}$$

The height achieved s_1 after 5.8 s is obtained from

$$s_1 = ut + \tfrac{1}{2}at^2$$
$$= 0 + \tfrac{1}{2} \times 11.5 \times 5.8^2$$
$$= 193.4 \text{ m}$$

The remaining part of the flight is with deceleration $g = 9.8 \text{ m/s}^2$. The final velocity at the top of the flight is zero, and if s_2 is the further distance travelled, then from the formula $v^2 = u^2 - 2as$, where the initial velocity is now 66.7 m/s,

$$0 = 66.7^2 - 2 \times 9.8 \times s_2$$

therefore

$$s_2 = 227 \text{ m}$$

therefore

$$h = 193.4 + 227 = \textbf{420 m}$$

4.8 Relative velocity; velocity diagram

In dealing with speed and velocity it has been assumed so far that the earth's surface has been 'fixed'. Yet it is known that the earth rotates around its axis and that the earth's centre is in motion around the sun. In fact, therefore, there is no point completely at rest and all velocities have been measured *relative to the surface of the earth*. In a similar way any moving point A may be regarded as 'fixed' and the velocity of any other point B measured relative to the point A; that is, the velocity of B is obtained as it would appear to an observer moving with point A. The velocity of B relative to the earth, v_B (or just 'velocity of B') is then made up of two parts:
1. v_{BA}, the velocity of B relative to A (as if A were at rest).
2. v_A, the velocity of A relative to earth, i.e. the 'velocity of A'.

Account must be taken of *direction* as well as magnitude, i.e. speed. The velocities v_A, v_B, and v_{BA}, are each vectors and are added and subtracted in the same way as other vectors such as force vectors are added or subtracted. Thus the velocity of B is the *vector sum* or the velocity of B relative to A and the velocity of A relative to earth.

Consider the simple case when two bodies are moving in the same straight line. Assume in the first instance that they are moving in the same direction. Then to find the velocity of B relative to that of A the procedure is as follows:

Choose a point **o** to represent a point at rest relative to earth, Fig. 4.4 (*a*).

From **o** draw a line **oa** to represent v_A in magnitude and direction.

Similarly from the same point **o** draw a line **ob** to represent v_B.

Then the line **ab**, taken in the sense **a** to **b**, represents v_{BA}, the velocity of B relative to A, i.e. it is as if A were the fixed point and **a** represented a point at rest.

Similarly **ba**, taken in the sense **b** to **a**, represents v_{AB}, the velocity of A relative to B, in both magnitude and direction.

If the bodies are moving in opposite directions in the same straight line then the vector diagram is shown in Fig. 4.4 (*b*), and $v_{BA} = $ **ab**, $v_{AB} = $ **ba**, as before.

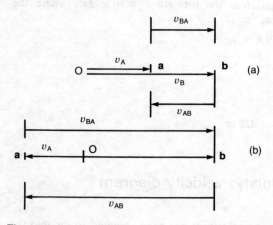

Fig. 4.4

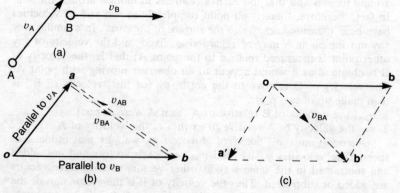

Fig. 4.5

The general case when A and B are not moving in the same straight line is shown in Fig. 4.5 (*a*). The procedure is exactly as before. Point **o** is the 'earth point' and vectors **oa** and **ob** represent velocities v_A and v_B respectively. Assume A brought to rest by giving it an equal and opposite velocity of magnitude v_A, vector **oa**′, Fig. 4.5 (*c*). The same velocity has to be added to B. The velocity of B relative to A therefore, is the resultant of **ob** and **oa**′, i.e. **ob**′, or the closing line **ab** of the vector diagram Fig. 4.5 (*b*). Similarly, the closing line **ba** is the velocity of A relative to B, i.e. as if B were at rest.

The vector diagram **oab** is called the *velocity diagram*.

Note that an absolute velocity, i.e. a velocity relative to earth, is always measured *from* point **o** in the velocity diagram.

Example *A tool is traversed across a lathe bed at 1 mm/s relative to the slide. The slide is traversed at 2.5 mm/s along the lathe. What is the velocity of the tool?*

SOLUTION
Draw **oa** horizontal to represent the velocity of the slide, 2.5 mm/s (Fig. 4.6).

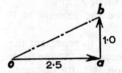

Fig. 4.6

From **a** draw **ab** perpendicular to **oa** to represent the velocity of the tool across the bed, relative to the slide, 1 mm/s.

Then **ob** represents the velocity of the tool.

By measurement from the velocity diagram:

speed of tool = length of **ob**

= **2.7 mm/s**

Its direction of motion is that corresponding to **ob** which makes an angle of **21° 48′** with **oa**, i.e. with the axis of the lathe.

Example *Two ships are steadily steaming towards each other. When 1,000 m apart ship B takes avoiding action by turning through 30° to port. The speed of ship A is 20 m/s and that of B is 30 m/s. Calculate their nearest distance apart and how long before this distance is reached after B takes avoiding action. Neglect the time taken to alter course.*

SOLUTION
Fig. 4.7 (*a*) shows a diagram to scale of the paths of the two ships. The motion of B is at 30° to the line BA and AB = 1,000 m. The velocity diagram is drawn as follows (Fig. 4.7 (*b*)):

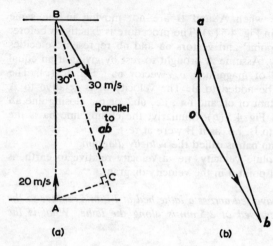

Fig. 4.7

Draw **oa** parallel to AB to represent the velocity of A, 20 m/s.

Draw **ob**, 30° to BA, to represent the velocity of B, 30 m/s.

Then **ab** represents in magnitude and direction the velocity of B relative to A, i.e. the motion of B as seen from A.

In Fig. 4.7 (*a*) BC is drawn from B parallel to **ab** to represent the *path* of B relative to A. The shortest distance between A and B is found by drawing the perpendicular AC from A on BC. By measurement:

nearest approach = AC = **311 m**

distance travelled by B on path BC = length BC

= 950 m

velocity of B relative to A on this path = **ab**

= 48.5 m/s

Problems

1. A rocket leaves its pad vertically with an acceleration of 7.5 m/s² which remains constant until the fuel is exhausted after 8 s. It then continues to travel freely in the vertical direction. Find the total time taken for the rocket to reach its highest point and the altitude achieved.

(14.1 s; 424 m)

2. An aircraft carrier is steaming at 26 knots into a head wind of 28 knots when an aircraft with an *air speed* of 110 knots lands on the deck against the wind. The aircraft is brought to rest by the arrester gear in 3 s. What is the distance travelled by the plane along the deck after striking the gear? *Air speed* is the speed of the plane relative to the air. 1 knot = 0.514 m/s.

(43 m)

3. An aircraft travelling due west at 600 km/h just passes over another aircraft travelling due north at the same speed. What is the velocity and direction of the first aircraft relative to the second?

(849 km/h, SW)

4. To a destroyer steaming due east at 30 knots a cruiser whose speed is 24 knots appears to be steaming NW. What are the two possible directions in which the cruiser may be moving?

(17° N. of E., 73° N. of E.)

5. An aircraft is flying due north at 1,000 km/h while another, 400 km to the east, is travelling NW, at 1,800 km/h. What is their closest distance of approach?

(84 km)

6. Two trains pass each other on parallel tracks. The first is 180 m long and travels at 60 km/h, the second is 120 m long and travels at 40 km/h. Calculate the total time taken to pass each other completely (*a*) if travelling in the same direction, (*b*) if travelling in opposite directions.

((*a*) 54 s; (*b*) 10.8 s)

7. Two aircraft leave airports A and B 100 km apart at the same instant. The first flies directly from A to B at 200 km/h, the second flies on a course inclined at 60° to the line joining B to A at 300 km/h. Find (*a*) the nearest distance of approach of the two aircraft, (*b*) the time taken to reach the nearest distance. Neglect the curvature of the earth.

(60 km, 10 min 50 s)

4.9 Angular velocity of a line

Let a line AB of fixed length move in the plane of the paper in any manner whatsoever (Fig. 4.8). After a small interval of time d*t* let the line AB move to A′B′ and let AB make with A′B′ a small angle dθ rad. Then the *angular velocity* ω of the line AB is defined:

$$\omega = \frac{d\theta}{dt}$$

and is measured in radians per second (rad/s).

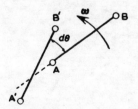

Fig. 4.8

4.10 Motion of a body in a plane

A rigid body forming part of a mechanism will always be of fixed length whatever its motion. Any two points A and B on the body will therefore remain at a fixed distance apart. Since there is no stretch along the line AB, there will be no velocity of B relative to A along AB, Fig. 4.9. However, since the body may rotate, B may have a velocity relative to A by rotation about A, as if A were fixed. The motion of B relative to A *can occur only in a direction perpendicular to the line* AB. This must be so whatever the motion of A. If B rotates about A with angular velocity ω then, after a small time dt, let AB′ be the position of AB, where $\angle$BAB′ = dθ (Fig. 4.9).

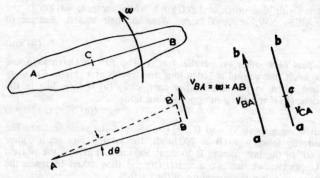

Fig. 4.9

    Distance travelled by B = BB′

therefore

    velocity of B normal to AB = v_{BA} = $\dfrac{BB'}{dt}$

since dθ is small,

    BB′ = AB × dθ

therefore

$$v_{BA} = AB \times \frac{d\theta}{dt}$$

$$= AB \times \omega, \text{ since } \omega = \frac{d\theta}{dt}$$

hence

$$\omega = \frac{v_{BA}}{AB}$$

$$= \frac{\text{velocity of B relative to A}}{\text{length AB}}$$

The relative velocity v_{BA} would be represented by a vector **ab**, of length ωAB, drawn perpendicular to AB, the sense of the vector corresponding to the motion of B relative to A (Fig. 4.9). In the same way for any point C on the line AB we may write:

$$v_{CA} = \omega.AC$$

thus $\dfrac{v_{CA}}{v_{BA}} = \dfrac{\omega AC}{\omega AB}$

i.e. $\dfrac{\textbf{ac}}{\textbf{ab}} = \dfrac{AC}{AB}$

Therefore, if point **c** is located on **ab** such that

$$\frac{\textbf{ac}}{\textbf{ab}} = \frac{AC}{AB}$$

then the velocity of C relative to A is given by **ac**.

Vector **ab** is called the *velocity image* of the line AB. When the velocity image of a link has been obtained, therefore, the relative velocity between any two points on the link (or on an extension of the link) is given by the length between the corresponding points on the image. This is a most useful fact to remember when dealing with mechanisms.

4.11 Velocity triangle for a rigid link. Application to mechanisms

Suppose now that AB represents a link in a mechanism and that v_A, the velocity of A, is known completely whereas v_B the velocity of B is known only in direction (Fig. 4.10). The problem is to find the

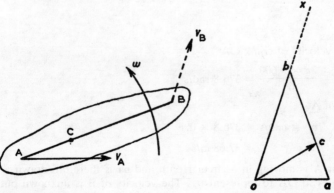

Fig. 4.10

magnitude of v_B and the angular velocity of the link. To do this we draw a *velocity triangle* for the link, thus:

Draw **oa** to represent v_A in magnitude direction and sense.
Through **o** draw a line **ox** parallel to the given direction of v_B.
Through **a** draw a line *perpendicular to the link* to cut **ox** in **b**.
Then **ob** represents v_B and **ab** is the velocity image of AB.
oab is the velocity triangle for link AB.

To find the velocity of any point C on AB first locate point **c** on the velocity image **ab** such that

$$\frac{ac}{ab} = \frac{AC}{AB}$$

then

$$v_C = oc$$

The angular velocity of the link is

$$\omega = \frac{v_{BA}}{AB} = \frac{ab}{AB}$$

Note that (*a*) **oac** is the velocity triangle for link AC and

$$\omega = \frac{v_{CA}}{AC} = \frac{ac}{AC}$$

and (*b*) all absolute velocities are measured from **o**.

The velocity triangle is most useful when dealing with the problem of finding the velocities of points in mechanisms. Each link is taken in turn and the velocity diagram obtained before proceeding to the next link in the chain. The method is shown in the following examples.

Example   *The crank OA of the engine mechanism shown (Fig. 4.11) rotates at 3,600 rev/min anticlockwise. OA = 100 mm, and the connecting-rod AB is 200 mm long. Find (a) the piston velocity; (b) the angular velocity of AB; (c) the velocity of point C on the rod 50 mm from A.*

SOLUTION
Angular velocity of crank OA

$$\omega = \frac{2\pi \times 3,600}{60} = 376.8 \text{ rad/s}$$

velocity of A

$$v_A = \omega OA = 376.8 \times 0.1$$
$$= 37.68 \text{ m/s}$$

Point A is moving in a circular path and **oa** is therefore drawn at right angles to OA to represent v_A. The velocity of B is unknown but

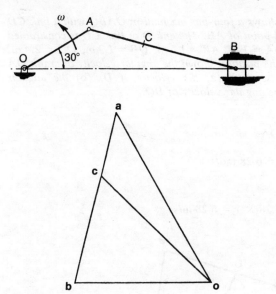

Fig. 4.11

its *direction* is horizontal. Hence in the velocity diagram a line of indefinite length is drawn horizontally through **o**. The velocity of B *relative* to A must be *perpendicular* to AB; therefore the velocity triangle for the link AB is completed by drawing a line **ab** perpendicular to AB. The lines **ab**, **ob** cut at point **b** which determines the magnitude of the velocity v_B of the piston B. From the diagram

$$v_B = \mathbf{ob} = 27 \text{ m/s}$$

$$v_{BA} = \mathbf{ab} = 33.6 \text{ m/s}$$

$$\text{Angular velocity of AB} = \frac{\text{velocity of B relative to A}}{\text{length of AB}}$$

$$= \frac{33.6}{0.2}$$

$$= \mathbf{168 \text{ rad/s}}$$

To find velocity of C: since C is on AB mark off **ac** on the velocity image **ab** such that

$$\frac{\mathbf{ac}}{\mathbf{ab}} = \frac{AC}{AB}$$

Then **oc** represents in magnitude and direction the velocity of C, the sense being from **o** to **c**. Thus

$$v_C = \mathbf{oc} = \mathbf{32.4 \text{ m/s}}$$

Example   *Fig. 4.12 shows a four-bar mechanism OABQ with a link CD attached to C the mid-point of AB. The end D of link CD is constrained to move vertically. OA = 1 m, AB = 1.6 m, QB = 1.2 m, OQ = 2.4 m, and CD = 2 m. For the position shown the angular velocity of crank OA is 60 rev/min clockwise; find (a) the velocity of D; (b) the angular velocity of CD; (c) the angular velocity of BQ.*

SOLUTION
Angular velocity of OA

$$\omega = \frac{2\pi \times 60}{60} = 6.28 \text{ rad/s}$$

velocity of A

$$v_A = \omega OA = 6.28 \times 1 = 6.28 \text{ m/s}$$

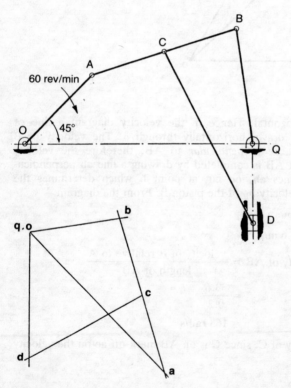

Fig. 4.12

Velocity diagram
Points O and Q are at rest. Points A and B are moving in circular paths, thus the directions of their velocities are known.

Draw **oa** normal to OA to represent v_A, 6.28 m/s.

Draw through **q** (coincident with **o**) normal to QB a line **qb** of indefinite length to represent v_B, the magnitude of which is unknown.

Draw through **a** a line perpendicular to AB to represent v_{BA}, the velocity of B relative to A; thus point **b** is located.

Since C is the mid-point of AB, **c** is the mid-point of the velocity image **ab** in the velocity diagram.

The velocity of D is vertical, therefore draw **od** vertically through **o**, this line being of indefinite length.

The velocity of D relative to C is normal to DC, therefore draw **cd** from **c** perpendicular to CD.

The intersection **d**, of **od** and **cd**, completes the diagram.

From the diagram,

$v_D = $ **od** $= $ **4 m/s** in direction **o** to **d**

$$\text{Angular velocity of CD} = \frac{\text{velocity of D relative to C } (\mathbf{cd})}{\text{length of CD } (CD)}$$

$$= \frac{4.21}{2}$$

$$= \mathbf{2.1 \ rad/s}$$

$$\text{Angular velocity of BQ} = \frac{\text{velocity of B relative to Q } (\mathbf{qb})}{BQ}$$

$$= \frac{3.02}{1.2}$$

$$= \mathbf{2.51 \ rad/s}$$

Problems

1. The engine mechanism of Fig. 4.13 has a crank 200 mm long and connecting rod 500 mm long. If the crank speed is 50 rev/s clockwise, find for the position shown: (*a*) the piston velocity; (*b*) the angular velocity of the connecting rod; (*c*) the velocity of a point C on the rod 200 mm from the crankpin.

(66 m/s; 66 rad/s anticlockwise; 61.8 m/s)

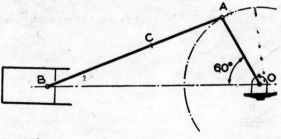

Fig. 4.13

2. The crank OA in the mechanism shown, Fig. 4.14 rotates anticlockwise at 5 rev/s and is 300 mm long. The link AB is 600 mm long and the end B moves in horizontal guides. Find for the position shown: (*a*) the velocity of B; (*b*) the velocity of point C, the mid-point of AB; (*c*) the angular velocity of AB.

(6.7 m/s; 7.5 m/s; 11.1 rad/s clockwise)

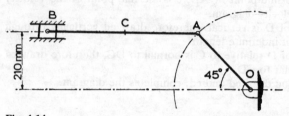

Fig. 4.14

3. In the mechanism shown, Fig. 4.15 crank OA rotates at 100 rev/min clockwise. Links AB and AC are pin-jointed at A and the pin ends B and C are attached to blocks sliding in horizontal and vertical guides, respectively. For the position shown when C is vertically below A find the velocity of B and C and the angular velocity of links AB and AC. OA = AB = AC = 150 mm.

(B, 1.57 m/s; C, 1.36 m/s; AB, 10.48 rad/s anticlockwise; AC, 5.24 rad/s clockwise)

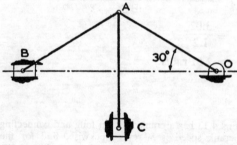

Fig. 4.15

4. The ends A, B of a link 1.5 m long are constrained to move in vertical and horizontal guides (Fig. 4.16). When A is 0.9 m above O it is moving at 1.5 m/s upwards. What is the velocity of B at this instant and the angular velocity of the link?

(1.13 m/s, 1.25 rad/s clockwise)

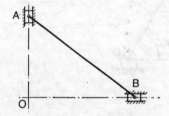

Fig. 4.16

5. The crank OA of an engine rotates at 1,800 rev/min clockwise and is 300 mm long. There are two connecting rods AB, AC each 450 mm long, connected to the single crankpin (Fig. 4.17). The cylinders are arranged to form a 60° 'vee'. Find for the configuration shown: (*a*) the velocity of each piston; (*b*) the angular velocity of connecting rod AC.

(B, 67.2 m/s; C, 24 m/s; 124 rad/s anticlockwise)

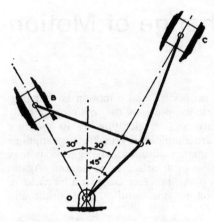

Fig. 4.17

6. Fig. 4.18 shows a crank press for impact extrusion. Crank OA rotates at 1 rev/s clockwise. OA = 100 mm, AB = 400 mm, CB = 240 mm, BD = 300 mm. Find the velocity of the plunger D when the crank makes an angle of 30° with the horizontal.

(170 mm/s)

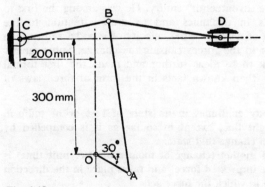

Fig. 4.18

CHAPTER 5
Inertia and Change of Motion

Dynamics is the study of forces on bodies whose motion is changing. This necessitates being able to describe precisely the motion of a body using information on its velocity and acceleration as its position changes. Problems in dynamics are simplified by making assumptions and approximations to produce a mathematical model, e.g. a body may be assumed to be rigid or weightless or a surface to be smooth. Again, a rocket with attached satellite in flight, however large, is a 'particle' in relation to space, but in relation to one another, the rocket and satellite are large rigid bodies.

5.1 Newton's laws of motion

The fundamental facts concerning the science of dynamics were discovered by Galileo in the seventeenth century. He was among the first to carry out experiments in dynamics and to draw attention to the importance of the *rate of change* of the velocity of a body, rather than its velocity, in relation to the forces causing the change. Subsequently Sir Isaac Newton took these ideas further and made the first formal presentation of all the then known facts in the form of three laws of motion;

> **Law 1. Every body continues in its state of rest or of uniform motion in a straight line, except in so far as it is compelled by impressed forces to change that state.**
>
> **Law 2. Change of motion (change of momentum per unit time) is proportional to the impressed force, and takes place in the direction of the straight line in which the force acts.**
>
> **Law 3. To every action there is an equal and opposite reaction.**

These three laws form the fundamental statements on which the study of dynamics is based and we shall refer back to them in the following paragraphs.

5.2 Inertia and mass

A change in motion of a body (treated simply as a particle, without rotation) can occur both by a change in speed and in direction, i.e. by a change in its velocity. The rate of change of velocity is its acceleration and is determined by circumstances external to the body itself. Consider, for example, two spring connected bodies A and B, Fig. 5.1. Imagine A and B to be drawn apart and then allowed to move *freely* in a horizontal plane. At an instant when A is moving from left to right with acceleration a_A, B is moving from right to left with acceleration a_B. Neither a_A nor a_B will remain constant but we are concerned only with their values at any instant.

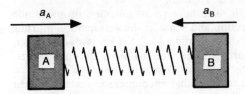

Fig. 5.1

One body is said to act upon or influence the motion of the other. Further, it is a fundamental idea or proposition, and might be verified experimentally, that the ratio of the accelerations a_A/a_B is constant for these two bodies throughout the motion. Moreover, the ratio a_A/a_B is independent of the type of 'connexion' between them. For example, they may influence each other by virtue of gravitational or magnetic attraction, electrically and so on. The ratio of the accelerations depends solely upon the bodies themselves; upon an inherent property which determines the ratio a_A/a_B. This property we call *inertia*. Each body has this property, which is measured by a quantity called its *mass*.

Let the body A have mass m_A and the body B have mass m_B. Then the *ratio of the masses* is defined by the proportion

$$\frac{m_A}{m_B} = \frac{a_B}{a_A}$$

or $m_A a_A = m_B a_B$.

If B be a standard *unit mass*, then $m_B = 1$, and we may define the mass of A by the relation

$$m_A = \frac{a_B}{a_A}$$

It is unnecessary to define mass as a *quantity of matter*, or inertia as a *reluctance to accelerate*, nevertheless these are sometimes useful and more familiar terms. The ideas of mass and inertia are important whenever changes of motion are considered.

5.3 Force

Since

$$m_A a_A = m_B a_B$$

it can be seen that the product **mass × acceleration** for the body A is of the same magnitude as the corresponding product for body B. This product is called a *force*, and it should be noted that it may vary from instant to instant. A force F is therefore defined by the product

$$F = ma \qquad \text{(Newton's second law of motion)}$$

and it is regarded as that which changes the motion of a particular body. We say *that force causes acceleration*.

A force F is further defined to be a *vector quantity*, of magnitude ma, which has the same direction and sense as the acceleration a. Thus the force on body A is from left to right and that on body B is from right to left. The action, or force exerted by body B upon A, is, therefore, equal and opposite to that of A upon B (Newton's third law of motion). This force is the same as that encountered in statics. When the acceleration a of a body is zero then force F must be zero; conversely, we say that if there is no force acting on a body it will have no acceleration and its motion will remain unchanged (Newton's first law of motion). If more than one force acts on a body the acceleration is in the direction of the *resultant* force and proportional to the magnitude of the resultant force.

5.4 Weight

Consider a body A falling 'freely' near the earth's surface (Fig. 5.2). The acceleration of A is $a_A = g$, vertically downwards towards the centre of the earth, where g is the acceleration due to gravity. As before, if suffix B denotes the earth, the product *mass × acceleration* for both the body and the earth gives the force between them, i.e.

$$\text{force} = m_A a_A = m_B a_B$$

or $\quad \text{force} = m_A g = m_B a_B$

and $\quad a_B = \dfrac{m_A g}{m_B}$

But the mass m_B of the earth is very large indeed compared with falling bodies, so that the acceleration a_B of the earth towards the moving body is negligibly small. Nevertheless, the force $m_A g$ is a finite quantity, known as the weight W of the mass m_A. Thus

$$W = m_A g$$

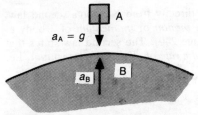

Fig. 5.2

This quantity W is the same 'weight' as measured in statics. For example, to prevent the downward acceleration g of a falling body A we may imagine an equal and opposite upward acceleration $-g$ to be added to the motion of A. That is, the net acceleration $= g - g = 0$, and the body is then at rest or in uniform motion. This would require a force $F = m_A g$ upwards. Thus, if A were at rest on a table (Fig. 5.3), then $W = m_A g$, is the force exerted by the table on the body. Or, if A is a body suspended from a spring, $F = W$ is the upward force exerted by the spring on A.

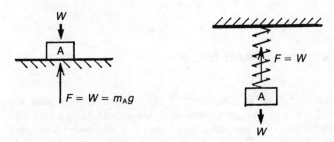

Fig. 5.3

5.5 The equation of motion

Since a weight W will produce an acceleration g downwards, any other force F as we have defined it, will give the *same* body an acceleration a according to the proportion:

$$\frac{F}{a} = \frac{W}{g} = m$$

or $F = ma$

$$= \frac{W}{g} a$$

This is the *equation of motion* of a body of mass m or weight W acted upon by a force F.

This equation can be derived directly from Newton's second law. To do this we must first define the *motion* or *momentum* of a body as the product *mass × velocity* (*see* page 240). The second law states that *the applied force is proportional to the rate of change of momentum*, i.e.

force α rate of change of momentum

α rate of change of 'mass × velocity'

and if the mass m is constant, then

force α mass m × rate of change of velocity

therefore

F = a constant × m × rate of change of velocity

The rate of change of velocity is the acceleration a, therefore

F = a constant × ma

The units of force, mass, velocity and time are chosen so as to make the value of the constant equal to unity, hence

$F = ma$

5.6 Units of mass and force

We have seen that mass is a measure of the reluctance of a body to accelerate and we can say that mass is a magnitude (a quantity of inertia) characteristic of a particular body. Mass is also defined simply as a 'quantity of matter'. The SI unit of mass is the *kilogram* (**kg**) and the international prototype of this particular quantity of matter is in the custody of the Bureau International des Poids et Mesures at Sèvres near Paris. A smaller unit used is the *gram* (**g**) which is 10^{-3} kg, and for large masses, the *megagram* (**Mg**) or *tonne* (**t**) which is 10^3 kg. It should be noted that the prefix *mega*, which means 10^6, refers to the unit 'gram' and not to the basic unit 'kilogram'.

The SI unit of force is the *newton* (**N**) defined as *that force which, when applied to a body having a mass of one kilogram, gives it an acceleration of one metre per second squared* (Fig. 5.4). Thus from the equation of motion:

$F = ma$

1 (newton) = 1 (kilogram) × 1 (metre per second squared)

1 (N) = 1 (kg) × 1 (m/s²)

A newton is a small quantity and in mechanics we often use the multiple forms *kilonewton* (**kN**) = 10^3 N, *meganewton* (**MN**) = 10^6 N, and *giganewton* (**GN**) = 10^9 N. Note that a force of 1 kN applied to a body of mass 1 Mg will give it an acceleration of 1 m/s².

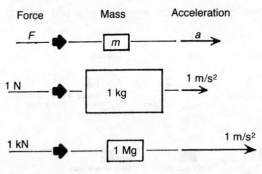

Fig. 5.4

Mass and weight

Mass is the amount of matter in a body. Weight is the earth's gravitational force on a body. These are different physical quantities and in the SI system the units are different. However, although the units indicate clearly whether mass or weight is intended, care should be taken in the use of the words 'mass' and 'weight' since colloquially 'weight' is used as a synonym for 'mass'. It is wrong to say that a body 'weighs 1 tonne'; either the body has a mass of 1 tonne or it weighs 9.8 kN. The term *load* however may be used in connexion with a mass or a force (weight). Statements such as 'a load of 100 kg' or 'a load of 1 MN' should not cause difficulty since the units indicate whether a mass or force is intended.

A beam balance compares the weights of two bodies and hence their masses since the weight of a body is proportional to its mass ($W/m = g = $ constant). A spring balance measures force and when used to compare the weights of bodies and hence their masses it is being used as a substitute for a beam balance. A spring balance may therefore be calibrated in newtons for measurement of force (weight) or kilograms for measurement of mass. The reading in kilograms is only true for the particular value of g, the acceleration due to gravity, used in calibrating the spring. The error involved in assuming that the reading is valid everywhere is very small since the value of g varies only by about 0.5 per cent over the earth's surface. For practical engineering purposes the value of 9.8 m/s^2 is satisfactory and the slightly more accurate value 9.81 m/s^2 may be used where required. For certain purposes a standard international value is required and this is 9.80665 m/s^2.

The relation between the weight W of a body and its mass m is

$$W = mg$$

thus W (N) $= m$ (kg) $\times$ 9.8 (m/s^2)

or W (kN) $= m$ (Mg or tonne) $\times$ 9.8 (m/s^2)

Fig. 5.5 shows the relationship between the weight W of a body and its mass m.

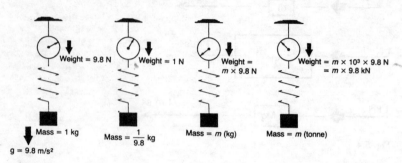

Fig. 5.5

5.7 Inertia force

The ideas of force met with in dynamics are similar to those used in statics. Hence it would be useful if we could use the methods of statics to solve problems in dynamics and obtain a 'free-body diagram' showing all the forces acting on it as if the body is 'at rest'. To do this we imagine a body existing free from all actions or influences from other bodies. We say that no force acts upon it, hence its motion is unchanged; in particular if it is 'at rest' it will remain at rest. Let a force F act upon an otherwise free body of mass m (Fig. 5.6). We have seen that there is a direct connexion between the ideas of force, mass and acceleration, such that the body will accelerate in the direction of the force with an acceleration a given by the equation of motion:

$$F = ma$$

In order to treat this problem as one of statics, it is useful to think of the body as being in 'equilibrium', even though accelerated, i.e. it is necessary to consider the accelerating force F as being balanced by an equal and opposite force of magnitude ma, Fig. 5.6. This force is known as an *inertia force* and always 'acts' to balance the resultant force on the body, i.e. in a direction opposite to that of the acceleration a.

It is useful to consider the inertia force as a resistance to change of motion, or a reluctance to being accelerated. Similarly we sometimes think of a tractive resistance or friction force as a resistance to be overcome in order to *maintain* steady motion. Such resisting forces are of a similar nature to the accelerating force F but act in a direction opposite to that of the velocity, so as to tend to slow down or decelerate the body. Thus if R is the resistance to steady motion without acceleration, and E the *applied* force, i.e. the *effort* required to overcome the resistance *and* produce the acceleration, then the forces acting on the body are as shown in the free-body diagram, Fig. 5.7. For 'static' balance in accelerated motion we may equate forces, thus:

$$E = ma + R$$

Note that the *accelerating* force is

$$F = E - R = ma$$

The nature of tractive resistance will be investigated below.

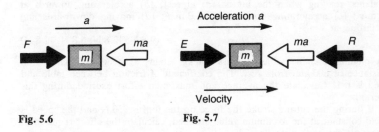

Fig. 5.6 **Fig. 5.7**

5.8 Active and reactive forces

A useful distinction is often made between active and reactive forces. An *active force* is one which can itself cause or tend to cause a change in motion; for example a push, a weight or the tractive effort of a vehicle. A *reactive force* is called into play by the action of an active force, and cannot of itself cause appreciable change of motion. Examples of reactive forces include the upward reaction of a support upon which a load rests, a bearing reaction, an inertia force, and some friction forces.

Example *A planing machine table of mass 450 kg attains a speed of 0.6 m/s in a distance of 500 mm from rest, with uniform acceleration. The coefficient of friction between table and bed is 0.1. Calculate the maximum effort required to drive the table.*

SOLUTION
From $v^2 = u^2 + 2as$

$$0.6^2 = 0 + 2 \times a \times 0.5$$

hence $a = 0.36$ m/s^2

The effort E to drive the table must overcome the friction force and the inertia force. For 'static' equilibrium

E = friction force + inertia force

$= \mu W + ma$

$= 0.1 \times 450 \times 9.8 + 450 \times 0.36$

$= \mathbf{603\ N}$

Problems

1. A mass of 1 kg is hung from a spring balance in a lift. What is the spring balance reading when the lift is: (*a*) at rest; (*b*) accelerating upwards at 3 m/s²; (*c*) accelerating downwards at 3 m/s²; (*d*) moving downwards and retarding at 3 m/s²?

(9.8 N; 12.8 N; 6.8 N; 12.8 N)

2. A planing machine table of mass 450 kg attains a speed of 0.6 m/s at a distance of 600 mm from rest. The coefficient of friction between table and bed is 0.1. Calculate the average and maximum effort exerted during this period.

If during the cutting stroke the force on the tool is 950 N and the speed is held constant at the maximum value attained, calculate the effort required to maintain the cutting stroke.

(288 N; 576 N; 1390 N)

3. The tool force on a shaping machine during the cutting stroke is 180 N and the reciprocating parts are equivalent to a moving mass of 45 kg. If power is suddenly shut off what would be the further distance cut by the tool if the cutting speed were initially 1.2 m/s? Assume the cutting force to be independent of the speed.

(180 mm)

5.9 Variable forces

A force may vary in a number of ways, e.g. with time, distance moved or velocity. When a spring is compressed the force varies with the amount of compression. When a plunger is moved up and down in oil its motion is damped by the oil and the damping force at low speeds is proportional to the velocity of the plunger; at high speeds it is proportional to the square of the velocity. In ascertaining the motions of bodies subject to variable forces the use of 'equation of motion' and 'inertia force' involves complex equations and it is often more convenient to employ the *principle of conservation of momentum* or the *work-kinetic energy equation*. These concepts will be developed later.

5.10 Tractive resistance

On a level track the resistances to steady motion of a vehicle are due to:
(a) rolling resistance
(b) air resistance

The *rolling resistance* arises mainly from the deformation of the tyres or of the track, but includes also axle friction. In a car it is increased by too-low a pressure in the tyres. In a rail wagon further effects arise from the sliding friction in bearing journals, friction at wheel flanges–particularly in rounding a curve–and friction due to misalignment of axles or inadvertent rubbing of brake blocks on tyre rims. The rolling resistance R_r is measured by the force required just to move a vehicle at a given steady speed on the level. The resistance increases with load and is affected by the running temperatures of wheels and tyres. Speed has a relatively small effect under normal driving conditions but high speeds produce a sharp increase in the resistance. To allow for the speed of the vehicle the resistance is expressed in the form $R_r = c_1 + c_2 v$, where v is the road speed, and c_1 and c_2 are constants.

The resistance of air to the motion of a vehicle is dependent upon the shape of the vehicle and its projected frontal area as well as the density of the air. This resistance is called *drag* (*see* page 270) and is of two main forms:

(i) *form drag* which is the resistance of the air to being pushed out of the way; the flatter the frontal shape the greater the drag force, hence the need for streamlining; aerofoils, projecting lips, and spoilers affect this form of resistance.

(ii) *skin or friction drag* which is the resistance due to the disturbance of the air flow in the layers of air at the skin of the car, not only at the front but also at the underside and rear.

The drag force D for a vehicle is roughly proportional to the *square* of the road speed, particularly at higher speeds, and proportional to the projected frontal area A, as well as the density of the air, ρ, i.e.

$$D \propto \rho A v^2$$
$$= \text{constant} \times \rho A v^2$$

The 'dynamic pressure' of the airstream flowing over a vehicle can be shown to be $\frac{1}{2}\rho v^2$, and it is convenient in various fields of work to express D in terms of this pressure, i.e.

$$D = \text{constant} \times (\tfrac{1}{2}\rho v^2) \times A \times 2$$
$$= C_d \times \tfrac{1}{2}\rho v^2 A$$

where C_d is a dimensionless *coefficient of drag* for a particular vehicle–and is used to compare the aerodynamic performance of cars. At sea-level $\rho = 1.23$ kg/m^3, and if A is in m^2, and v in km/h, the formula for D in newtons becomes

$$D = C_d \times \tfrac{1}{2} \times 1.23 \times \left(\frac{v}{3.6}\right)^2 \times A$$

$$= \frac{C_d A v^2}{21} \text{ N}$$

A modern saloon car, well designed aerodynamically, has a coefficient of drag of about 0.32.

Wind resistance depends upon both vehicle and wind speed: the effect of wind may either assist or retard the vehicle. The *total resistance R* at a given speed may be expressed as so many newtons per tonne mass of vehicle, particularly for trains. For example, for a train of coaches in good condition the total resistance may be about 50 N/tonne. For goods wagons the figure might be about 100 N/tonne depending upon the speed.

5.11 Tractive effort

The *tractive effort E* required to propel a road vehicle or train can be taken as equivalent to the pull in a tow-rope or coupling. In practice vehicles are driven by an engine torque transmitted to the driving axles. The tractive effort is the *driving* force at the road surface, the force with which the ground *pushes on the vehicle*. Similar problems arise with aircraft but the term *thrust* is used instead of tractive effort (*see* page 272).

Vehicle moving at constant speed   When a vehicle travels at *constant* speed on the level, Fig. 5.8, there is no unbalanced force, and the effort E equals the resistance R (rolling and air resistance), i.e. tractive effort = total resistance

i.e.

$$E = R$$

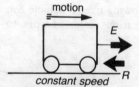

Fig. 5.8

When the vehicle is at its *maximum* speed it is not accelerating and again the effort is equal to the resistance. On ascending a gradient the tractive effort must be increased to overcome the component of the weight. On descending the component of the weight assists the motion and reduces the effort required.

Vehicle accelerating   When the effort is increased above that required to maintain constant speed the vehicle accelerates and the effort must overcome the resistance and provide the accelerating force, i.e. balance the inertia force. For a vehicle of mass *m* accelerating *up* an incline the free-body diagram is shown in Fig. 5.9. The effort is

E = resistance + resolved part of weight + inertia force

i.e.

$$E = R + mg \sin \theta + ma$$

Note that for accelerated motion *down* the slope, the resistance and inertia forces both act upwards, and the component of the weight assists the effort.

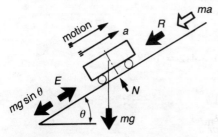

Fig. 5.9

Example   *A car of mass 1.1 tonne is driven at constant speed up an incline of 10°. The rolling resistance is 170 N and the air resistance is 100 N. Find the tractive effort. What is the effort when the car has a uniform acceleration of 0.6 m/s²?*

SOLUTION
Rolling resistance $R_r = 170$ N
Since the speed is constant there is no inertia force, hence

tractive effort = component of weight down the slope

+ *total* resistance

= $1.1 \times 10^3 \times 9.8 \sin 10° + (100 + 170)$

= **2140 N**

When the car accelerates, the inertia force is *ma* and the tractive effort has to be increased by this amount, i.e.

$$\text{tractive effort} = 2140 + 1.1 \times 10^3 \times 0.6$$
$$= 2800 \text{ N}$$
$$= \mathbf{2.8 \text{ kN}}$$

Example   *On test, a motor car just reaches 130 km/h on the level in still air. Experiments show that the total resistance to motion due to windage and road drag is given in newtons by (132 + 0.6v + 0.1v²), where v is the road speed in km/h. Find the tractive effort required. If the car has an all-up mass of 1200 kg and the tractive effort is assumed constant what would be the acceleration of the car at 50 km/h?*

SOLUTION
At a steady speed of 130 km/h,

$$\text{tractive effort} = \text{resistance}$$
$$= 132 + 0.6 \times 130 + 0.1 \times 130^2$$
$$= \mathbf{1900 \text{ N}}$$

At 50 km/h,   resistance $= 132 + 0.6 \times 50 + 0.1 \times 50^2$
$$= 412 \text{ N}$$

The tractive effort is 1900 N, hence

accelerating force $F = 1900 - 412 = 1488$ N

and $\qquad\qquad\qquad F = ma$

i.e. $\qquad\qquad\quad 1488 = 1200\, a$

hence $\qquad\qquad\qquad a = \mathbf{1.24 \text{ m/s}^2}$

Example   *Coal wagons are lowered 30 m from rest down an incline of 1 in 10 with uniform acceleration, by means of a cable attached to a rope brake. The total mass of the wagons is 48 tonnes and the resistance to motion is 130 N/tonne. At the end of the incline the wagons are travelling at 3 m/s. Calculate the pull in the cable.*

SOLUTION
Initial speed of wagons $u = 0$, and final speed $v = 3$ m/s

From $v^2 = u^2 + 2.a.s.$

$$3^2 = 0 + 2.a.30$$

therefore $\quad a = 0.15$ m/s²

Inertia force $= ma = 48 \times 10^3 \times 0.15 = 7200$ N
Component of weight down the slope,

$$W \sin \theta = 48 \times 10^3 \times 9.8 \times \tfrac{1}{10}$$
$$= 47040 \text{ N}$$

Resistance, $R = 130 \times 48 = 6240$ N

Fig. 5.10 shows the free-body diagram for the wagon. For 'static' equilibrium, the pull E in the cable is given by

$E = 47040 - 7200 - 6240$

$= 33,600$ N

$= \mathbf{33.6 \ kN}$

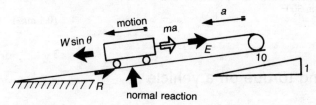

Fig. 5.10

Problems

1. A small road vehicle of mass 500 kg runs freely down a hill against a rolling resistance of 200 N and air resistance 50 N. If the hill has an inclination to the horizontal of $\sin^{-1}0.2$, find the acceleration of the vehicle.

   (1.46 m/s²)

2. Determine the tractive effort required to accelerate a car at 0.1 m/s² down an incline of 1 in 100. The car has a mass of 1.5 tonne and the resistance to motion is 200 N.

   (203 N)

3. A 65 tonne locomotive pulls a 285 tonne train of coaches on a down gradient of 1 in 400 against a track resistance of 50 N/tonne of total mass. Find the time taken to reduce speed from 80 km/h to 44 km/h with the engine stopped, if the braking force is 70 kN.

   (44.4 s)

4. A train of total mass 500 tonne is hauled by two locomotives up an incline of 1 in 75, increasing its speed from 15 to 45 km/h in 100 s. The tractive resistance to motion is 45 N/tonne and the leading locomotive develops a tractive effort of 60 kN at the maximum speed attained. Calculate the effort exerted by the second locomotive at this speed.

   (69.5 kN)

5. An electric locomotive together with its train has a mass of 200 tonne. The average tractive resistance to motion is 55 N/tonne while starting from rest up a gradient of 1 in 100. Calculate the time taken to travel the first 1 km if the average tractive effort exerted by the locomotive is 35 kN.

   (5 min 2 s)

6. To maintain a uniform speed of 108 km/h on the level a car requires a tractive effort of 550 N. The total resistance to motion is given by $R = 200 + kv^2$ where v is the speed in km/h. What is the value of the constant k?

   (0.03)

7. In a test a car attained a maximum speed of 96 km/h against a head wind of 10 km/h. It is calculated that the total air drag force amounts to 0.5 kN. The projected frontal area is 2.2 m² and the air drag in newtons is given by $C_dAv^2/20$ where the projected area A is in m², and v is the air speed in km/h relative to the vehicle. Find the value of the drag coefficient C_d.

(0.41)

8. A 200 tonne train travels on a level track against a resistance given by the expression $(40 + 0.012v^2)$ N/t mass of train where v is the speed in km/h. If the tractive effort is constant at 200 N/tonne mass of train find the acceleration at an instant when the speed is 48 km/h and the train is climbing a gradient of 1 in 300.

(0.1 m/s²)

5.12 Driving torque on a vehicle

A vehicle is usually driven by an engine *torque* at the driving axle. A wheeled vehicle will be driven forward by this torque only if there is a friction force F at the road surface, Fig. 5.11. This friction force does *not* form a resistance to motion and only prevents slipping of the wheel.

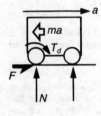

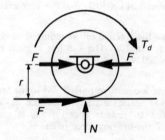

Fig. 5.11   **Fig. 5.12**

The force F is equivalent to a couple $F \times r$, where r is the wheel radius, together with a force F at the axle bearing, Fig. 5.12. Then the driving torque T_d must balance the couple Fr, i.e.

$$T_d = Fr$$

(this is strictly true only if the rotational inertia of the wheels is neglected). Also, the force F acting on the bearing from left to right, Fig. 5.12, is the effort which drives the vehicle; thus:

$$F = E$$

(again, this is strictly true only if the wheel inertia is negligible).

If the linear speed of the vehicle is v and the radius of the driving wheels is r, then the angular velocity of the driving wheels is

$$\omega_d = \frac{v}{r}$$

The driving wheels rotate much slower than the engine, and the ratio of the engine to the driving axle speed is the gear reduction n, thus

$$n = \frac{\text{angular speed of engine } (\omega_e)}{\text{angular speed of axle } (\omega_d)}$$

i.e. $\omega_e = n\omega_d$

The work done in unit time at the engine is the same as the work done in unit time at the axle, *neglecting all losses*, so

$$T_e\omega_e = T_d\omega_d$$

where T_e is the engine torque. Therefore

$$T_d = nT_e$$

5.13 Maximum possible tractive effort

Previously we have been concerned with the tractive effort *required* to drive a vehicle. We now consider the limitations on *obtaining* the tractive effort. It may be recalled that in pure rolling motion no slip occurs between wheel and road. Nevertheless on a perfectly smooth surface the wheels would always slip and no effort could be exerted at the driving wheels. The maximum possible tractive effort is limited by the greatest force F which can be exerted at the track surface. This limiting force depends on the load N normal to the track *at the driving wheels* and on the *limiting coefficient of adhesion* μ_a. Thus:

$$F = \mu_a N$$

The coefficient of adhesion μ_a may be thought of as a limiting coefficient of friction. It may also take account of wheel flattening. The normal reaction N and hence the force F available depends on whether the vehicle is driven at the rear, front or all four wheels. When a car is accelerating or hill-climbing the effect of the inertia force and weight component down the slope is to increase the normal load on the rear wheels and decrease that on the front wheels. Thus for rear-wheel drives wheel grip is increased under these conditions whereas for front-wheel drives the grip is reduced. In the case of four-wheel drives the arrangement of the drive usually ensures that equal torques are applied to rear and front axles so that the tractive forces at rear and front are equal. Similar arguments apply to the braking of vehicles except that on road vehicles the brakes are normally always applied to all four wheels. For these reasons the problem of a road vehicle driven at one axle only, which is usually the case, is complex, and we consider only problems in which a vehicle is driven or braked at all wheels. Thus for a vehicle of weight W driven on all its wheels up a slope inclined at an angle θ to the horizontal:

$$N = W \cos \theta$$

therefore

maximum tractive effort $E = \mu_a W \cos \theta$

and on the level

maximum tractive effort $E = \mu_a W$

Example   *A truck of total mass 20 tonne is driven along a level track against a track resistance of 200 N/tonne. The engine develops an engine torque of 240 N-m at a maximum speed of 2,000 rev/min. The gear reduction from an engine to driving axle is $9:1$ and the wheel diameter is 800 mm. Find the maximum linear speed of the vehicle in km/h and the time taken to reach this speed on the level.*

SOLUTION
Maximum angular velocity of engine,

$$\omega_e = \frac{2\pi \times 2,000}{60}$$

$$= 209.4 \text{ rad/s}$$

maximum angular velocity of wheels $= \dfrac{209.4}{9} = 23.3$ rad/s

hence

$$\frac{v}{r} = 23.3$$

$$v = 23.3 \times 0.4 = 9.32 \text{ m/s}$$

therefore,

maximum linear speed of truck $= 9.32 \times 3.6 = \textbf{33.6 km/h}$

$$\text{Axle torque} = \text{engine torque} \times \text{gear ratio}$$

$$= 240 \times 9 = 2,160 \text{ N-m}$$

$$\text{tractive force} = \frac{\text{axle torque}}{\text{wheel radius}} = \frac{2,160}{0.4} = 5,400 \text{ N}$$

$$\text{resistance to motion} = 20 \times 200 = 4,000 \text{ N}$$

$$\text{net tractive force} = 5,400 - 4,000 = 1,400 \text{ N}$$

From $F = ma$

$$1,400 = 20 \times 1,000 \times a$$

$$a = 0.07 \text{ m/s}^2$$

From $v = at$

$$t = \frac{9.32}{0.07}$$

$$= 133 \text{ s}$$

Example *A motor car has a total mass of 1,000 kg and has road wheels of 600 mm effective diameter. The engine torque is 130 N-m and the tractive resistance is constant at 450 N. The gear ratio between engine and road wheels is 13.5 to 1. Neglecting friction, rotational inertias, and transmission losses find the acceleration rate when the car climbs a gradient of 1 in 20.*

SOLUTION

$$\text{Tractive effort} = \frac{\text{axle torque}}{\text{wheel radius}}$$

$$= \frac{13.5 \times 130}{0.3}$$

$$= 5850 \text{ N}$$

Accelerating force

$F =$ tractive effort $-$ resistance
   $-$ weight component down the slope

$$= 5850 - 450 - 1000 \times 9.8 \times \frac{1}{20}$$

$$= 4910 \text{ N}$$

and $F = ma$

therefore

$$a = \frac{4910}{1000} = 4.91 \text{ m/s}^2$$

Example *A locomotive and train have masses 100 and 500 tonne, respectively. The coefficient of adhesion between wheel and track is 0.5. If 80 per cent of the weight of the locomotive is carried by the driving wheels and the resistance to motion is 100 N/t find the maximum possible starting acceleration.*

SOLUTION

Load on *driving* wheels $= 0.8 \times 100 \times 9.8 = 784$ kN.
Maximum tractive effort

E = coefficient of adhesion × load on *driving* wheels

$$= 0.5 \times 784$$

$$= 392 \text{ kN}$$

Resistance

$$R = 100 \times (100 + 500)$$

$$= 60,000 \text{ N} = 60 \text{ kN}$$

The tractive effort must balance both the resistance and inertia force. Therefore

$$E = R + ma$$

i.e. $392 = 60 + 600\,a$

since the total mass of the train being accelerated is 600 tonne. Hence

$$a = \mathbf{0.553 \text{ m/s}^2}$$

Example   *A car of mass 1,000 kg is driven on all four wheels. The coefficient of friction between wheels and road surface is 0.65. Calculate the maximum tractive effort and corresponding acceleration when ascending an incline of 10°.*

SOLUTION

Maximum tractive effort = maximum friction force

$$= \mu N$$

$$= \mu W \cos \theta$$

$$= 0.65 \times 9,800 \times \cos 10°$$

$$= \mathbf{6,270 \text{ N}} = \mathbf{6.27 \text{ kN}}$$

For balance of forces,

tractive effort = inertia force + component of weight down slope

therefore

$$6,270 = ma + W \sin \theta$$

$$= 1,000 \times a + 9,800 \times \sin 10°$$

hence

$$a = \mathbf{4.6 \text{ m/s}^2}$$

Problems

1. The total mass of a small diesel locomotive is 50 tonne and the whole of the weight is carried on the driving wheels. The limiting coefficient of adhesion

between wheels and rails is 0.1, and the tractive resistance to motion is 90 N/tonne. The locomotive pulls a train of wagons totalling 300 tonne having a tractive resistance of 45 N/tonne. Calculate (*a*) the maximum tractive effort exerted by the locomotive; (*b*) the starting acceleration on the level.

(49 kN; 0.0886 m/s²)

2. A diesel locomotive of mass 60 tonne hauls a train of wagons of mass 200 tonne up an incline of 1 in 200. The tractive resistance is 110 N/t mass. Calculate the maximum retardation (*a*) if the brakes are applied at the locomotive only; (*b*) if the brakes are applied at every wagon and the locomotive. The coefficient of adhesion is 0.65.

(1.63 m/s²; 6.53 m/s²)

3. A racing car of mass 1 tonne is driven on all four wheels and has a limiting coefficient of adhesion between tyres and road of 0.6. (*a*) Calculate its maximum starting acceleration; (*b*) what is the time and distance required to come to rest from 100 km/h with the brakes locked? It may be assumed that the car does not skid sideways.

(5.88 m/s²; 4.73 s; 65.6 m)

4. A car of mass 1,800 kg travels down an incline of 1 in 4. Calculate the *maximum* braking torque which could be applied to all wheels before slipping occurs. The coefficient of static friction is 0.55 and the wheel diameter 750 mm.

(3.52 kN-m)

5. To maintain a *uniform* speed of 72 km/h on the level a car of mass 1.5 t requires an axle torque of 115 N-m. The wheel diameter is 500 mm. Find the tractive resistance.

If while running at 72 km/h the car climbs a gradient of 1 in 15 find the time taken for the speed to fall to 40 km/h assuming that the tractive resistance increases by 50 per cent and that the torque applied to the wheels is maintained unchanged.

(460 N; 11 s)

6. A car of total mass 1,300 kg travels down an incline of 1 in 13. The wheels are 600 mm in diameter and the tractive resistance is 300 N. The engine torque is 65 N-m and the gear reduction ratio between engine and axle is 14 : 1. Find the acceleration of the car, neglecting rotational inertias, friction and transmission losses.

(2.86 m/s²)

5.14 Application of inertia force to connected bodies

Consider a mass m_2, being pulled along a level plane by a cord passing over a light frictionless pulley, and attached to the freely hanging mass m_1, shown in Fig. 5.13. If we assume no resistance to motion, m_1 will accelerate downwards with an acceleration a and, provided there is no stretch in the cord, m_2 will have the same acceleration horizontally.

Let T be the tension in the cord. If there is no friction at the pulley this will produce an upward force T on m_1 and a horizontal force T from left to right on m_2. The acceleration a and the tension T will be calculated by reducing the problem to a 'static' one by the use of the idea of an inertia force. The free-body diagrams for m_1 and m_2 are shown in Fig. 5.13. The forces acting on m_1 are its weight m_1g downwards and the tension T upwards. For balance an inertia force m_1a must be added in a direction opposite to that of a, i.e. vertically upwards. For 'static' balance of the vertical forces on m_1

$$m_1g = T + m_1a$$

The horizontal forces on m_2 are the tension T from left to right and the inertia force m_2a in the direction opposite to that of a, i.e. from right to left.

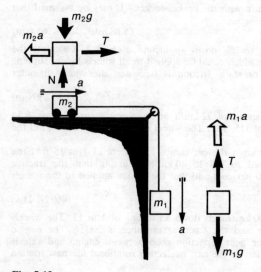

Fig. 5.13

There is a vertical reaction N on m_2 supporting the weight m_2g, which does not enter into the problem, in the absence of friction. Hence

$$T = m_2a$$

From these two equations we find

$$m_1g = m_1a + m_2a$$
$$= (m_1 + m_2)\, a$$

This corresponds to

$$F = ma$$

where $F = m_1g$ is the accelerating force on the *two* bodies and

$m = m_1 + m_2$, the total mass of the two bodies. From the above equations T and a can be found,

It should be noted that if there is friction at the pulley the tension in the cord is greater in the vertical portion than in the horizontal portion.

5.15 The simple hoist

Consider a mass m_1 and a lighter mass m_2 hanging either side of a *light* frictionless pulley system and connected by a light cord. The acceleration of m_1 will be downward and of m_2 upwards, as shown in Fig. 5.14. The tension T in the cord will be the same on both sides of the pulley. There is therefore a force T upwards on both masses.

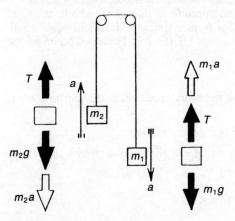

Fig. 5.14

For mass m_1 the downward force due to the weight is balanced by the tension, together with an inertia force acting in the opposite direction to that of a i.e. upwards. For balance, therefore

$$m_1g = T + m_1a$$

or $\qquad T = m_1g - m_1a$

For mass m_2 the tension acting upwards is balanced by the weight together with the inertia force both acting downwards. Thus:

$$T = m_2g + m_2a$$

Example *A car of mass 1 tonne hauls a trailer of mass 0.5 tonne with a common acceleration of 0.15 m/s². Calculate the pull in the horizontal tow rope and the tractive effort required.*

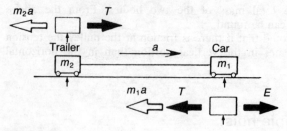

Fig. 5.15

SOLUTION

Let E be the tractive effort and T the tension in the tow rope. The external forces acting on the car alone in the horizontal direction, are E and T, Fig. 5.15.

The *inertia force* on the car $= m_1a = 1 \times 1,000 \times 0.15 = 150$ N.

This force acts in the direction opposite to that of the acceleration. E, T and the inertia force together form a system of forces in equilibrium, as shown in the free-body diagram for the car. For equilibrium of the car,

$$E = T + 150$$

Similarly, since the trailer has the same acceleration the inertia force is m_2a, i.e.

$$0.5 \times 1,000 \times 0.15 = 75 \text{ N}$$

The horizontal forces on the trailer are the tension T and the inertia force, hence for balance

$$T = \textbf{75 N}$$

thus $E = 75 + 150$

$$= \textbf{225 N}$$

Alternatively, the effort is the force required to accelerate the car and trailer *together*. Since the total mass is 1.5 tonne this force is

$$1.5 \times 1,000 \times 0.15 = 225 \text{ N}$$

Example *A 100-tonne locomotive pulls a train of ten 32-tonne coaches on the level. The tractive effort at 72 km/h is 44 kN and the track resistance is 70 N/tonne of total mass. Find the acceleration produced and the draw-bar pull on the leading coach under these conditions.*

SOLUTION

Total mass of train,

$$m = 100 + 10 \times 32 = 420 \text{ tonne}$$

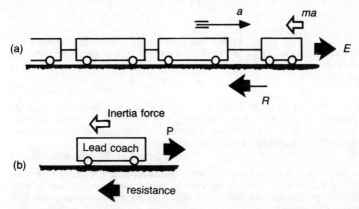

Fig. 5.16

Resistance,

$$R = 420 \times 70 = 29400 \text{ N}$$

The forces in the horizontal direction on the locomotive and train *together*, Fig. 5.16 (a) are the effort, *E*, resistance, *R*, and the inertia force *ma*, acting opposite to the direction of the acceleration, i.e. to the motion of the train. Then

accelerating force

$$F = E - R$$
$$= 44000 - 29400$$
$$= 14600 \text{ N}$$

and $F = ma$

i.e. $14600 = 420 \times 10^3 \times a$

therefore,

$$a = 0.035 \text{ m/s}^2$$

The draw-bar is between the locomotive and the leading coach. To find the draw-bar pull *P* on the coaches consider the forces acting on the train alone, Fig. 5.16 (b). These are, *P*, the track resistance on the train (320×70 N), and the inertia force due to the accelerated mass of the train, therefore

draw-bar pull P = resistance + inertia force

$$= 320 \times 70 + 320 \times 10^3 \times 0.035$$
$$= 33600 \text{ N}$$
$$= \textbf{33.6 kN}$$

Problems

1. Two loads, each of mass 2 kg are tied together by a light inextensible cord. They are accelerated along the level by a pull of 18 N at one load. Find the acceleration of the system and the tension in the cord. Resistance to motion may be neglected.

$$(4.5 \text{ m/s}^2; 9 \text{ N})$$

2. A locomotive of mass 80 tonne pulls a train of mass 200 tonne with an acceleration of 0.15 m/s² along the level. The resistance to motion of both locomotive and train is 45 N/tonne. Calculate (a) the tractive effort required, (b) the pull in the coupling hook at the locomotive.

$$(54.6 \text{ kN}; 39 \text{ kN})$$

3. In an experiment a load of mass 4.5 kg is pulled along a level track by a mass of 0.5 kg attached to it by a light inextensible cord passing over a light frictionless pulley and hanging vertically. Calculate the distance travelled from rest in 2 seconds.

$$(1.96 \text{ m})$$

4. A mine cage of mass 500 kg is returned to the surface by a wire cable passing over a loose pulley at the pit head. The cable is fastened to a counterweight of mass 600 kg. Find the acceleration of the empty cage if allowed to move freely.

$$(0.89 \text{ m/s}^2)$$

5. A motor car develops a tractive effort of 1.8 kN on the level, when towing another exactly similar car whose engine is out of action. Find the tension in the tow rope and the acceleration. The resistance to motion is 650 N on each car. The mass of each car is 1 tonne.

$$(900 \text{ N}; 0.25 \text{ m/s}^2)$$

6. A mass of 15 kg is supported by a light rope which passes over a light smooth pulley, and carries at its other end a mass of 6 kg. The 6-kg mass is held fast by a pawl. If the pawl is released find the tension in the rope and the time taken for the 15-kg mass to reach the level of the pawl. The 15-kg mass is initially 2 m above the level of the pawl.

$$(84 \text{ N}; 0.975 \text{ s})$$

7. An 80-tonne locomotive develops a tractive effort of 70 kN when starting to pull a train of coaches of total mass 300 tonne on the level against a track resistance of 90 N/tonne for the locomotive and 60 N/tonne for the train. Find the starting acceleration. What is the draw-bar pull on the leading coach?

$$(0.12 \text{ m/s}^2; 54 \text{ kN})$$

CHAPTER 6

Motion in a Circle

When a body moves in a circular path it is accelerated even although its speed round the circle may be constant. This is because its velocity is changing since the direction of motion is changing. This vector change in velocity gives rise to a special acceleration of the body called **centripetal acceleration**, directed radially inwards towards the centre of motion. A further acceleration occurs if the speed is not constant, i.e. a *linear acceleration* tangential to the circle of motion. This acceleration can be related to the motion of a radial line joining the body to the centre of motion. The angle swept out in unit time by this radial line is the *angular velocity* of the line and a change in this angular velocity is the **angular acceleration** of the line. These accelerations, the relationship between them, and the forces involved will now be dealt with in detail.

6.1 Centripetal acceleration

A linear acceleration can be caused by a change in direction without a change in speed, that is, by a *vector* change in velocity.

Consider a point A moving in a circular path of radius r with constant angular velocity ω about a fixed point O (Fig. 6.1(a)). Let the line OA move to OA' in a small time dt, and let angle AOA' $= d\theta$ rad. The initial vector velocity of A

$$v = \mathbf{oa}$$

$$= \omega \times OA$$

$$= \omega r$$

perpendicular to the line OA (Fig. 6.1(b)). After time dt the vector velocity of A

$$v = \mathbf{oa}'$$

$$= \omega r$$

perpendicular to OA'.

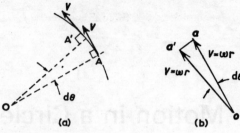

Fig. 6.1

The vector *change* in velocity of A in time dt

$$= \mathbf{aa}'$$

$$= \omega r \times d\theta, \text{ since } d\theta \text{ is small}$$

Therefore the acceleration of A,

$$a = \text{rate of vector change in velocity}$$

$$= \frac{\mathbf{aa}'}{dt}$$

$$= \frac{\omega r \, d\theta}{dt}$$

$$= \omega^2 r, \text{ since } \frac{d\theta}{dt} = \omega$$

Also since

$$\omega = \frac{v}{r}$$

then

$$a = \left(\frac{v}{r}\right)^2 \times r$$

$$= \frac{v^2}{r}$$

The direction of the change **aa**′ is in the sense **a** to **a**′, that is, along the line AO. Thus the acceleration of A due to its rotation is directed radially inward from A to O and is of amount

$$a = \omega^2 r = \frac{v^2}{r}$$

This acceleration is called the *centripetal acceleration*.

6.2 Centripetal force

Consider now a body of mass m at A, rotating about O. The centripetal acceleration a can only take place if there is a force acting in the direction A to O. The magnitude of this force F is given by

$$F = ma$$
$$= m \times \omega^2 r$$
$$= m \times \frac{v^2}{r}$$

and its direction is that of a, i.e. radially *inwards*. It is an active force since it is the force causing the body to move in a circular path and is known as the *centripetal force*. For example a body whirled in a horizontal circle at the end of a light cord is maintained in its circular path by the tension in the cord acting radially inwards at its connexion with the body at A, Fig. 6.2 (a).

6.3 The inertia force in rotation

For balance of forces at the body A the centripetal force F may be considered as being in equilibrium with an equal and opposite inertia force of magnitude $m\omega^2 r$, Fig. 6.2 (c). This inertia force is a reactive force, since it cannot of itself cause motion. If the cord is cut the active force F disappears and the body A moves in a straight line tangential to the circular path, Fig. 6.2 (b). It does not fly radially outwards.

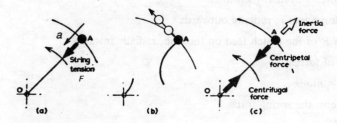

Fig. 6.2

6.4 Centrifugal force

Now consider the tension in the cord at O. This is equal and opposite to the centripetal force at A and therefore acts radially outward. This force at O is called the *centrifugal force* and may be thought of as due to the cord tension required to provide the motion in a circle, or the action of A upon the point O due to the rotation of A.

Example *A centrifugal clutch is shown at the rest position with its axis mounted vertically, Fig. 6.3. The rotating bobs are each of mass 225 g* *and the spring strength is 7.5 kN/m. The centre of mass of each bob is at 150 mm radius in the rest position. Calculate the radial force on the clutch face at 720 rev/min.*

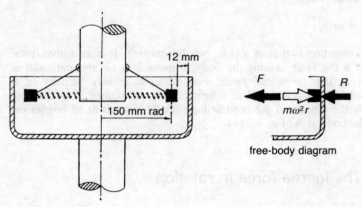

Fig. 6.3

SOLUTION

As the shaft speed rises the bobs fly out until they engage the inside face of the clutch cylinder. The forces on each rotating mass are then as shown on the free-body diagram for a bob:

    spring force F, radially inwards

    inertia force $m\omega^2 r$, radially outwards

    reaction R of the clutch face on the bob, radially inwards

For balance of forces:

    $F + R = m\omega^2 r$

At engagement, the spring force

    F = stiffness × extension

      = 7.5 × 1,000 × 0.012

      = 90 N

radius of rotation

    $r = 150 + 12 = 162$ mm = 0.162 m

    Inertia force = $m\omega^2 r = 0.225 \left(\dfrac{2\pi \times 720}{60} \right)^2 \times 0.162$

              = 207 N

hence

    $90 + R = 207$

therefore

$$R = 117 \text{ N}$$

The force *on* the clutch face is equal and opposite to this.

Note: As the speed rises from rest there is a particular speed at which engagement just commences. At this speed the spring force is equal to the inertia force. As the speed rises above the engagement speed the spring force remains constant but the inertia force increases in value, thus increasing R. The value of R governs the friction force between bob and rim surfaces and hence determines the power transmitted.

Example *A body of mass 0.5 kg slides in smooth guides when whirled in a horizontal circle at the end of a spring. The spring stiffness is S = 1.5 kN/m. The natural unstretched length of the spring is l = 150 mm. Calculate the radius of rotation of the weight and the stretch in the spring when the speed of rotation is 360 rev/min.*

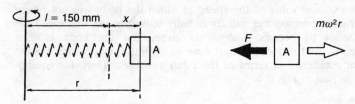

Fig. 6.4

SOLUTION

The radial forces acting on the body A are: the spring force F inwards and the inertia force $m\omega^2 r$ outwards. Hence

$$F = m\omega^2 r$$

$$= 0.5 \left(\frac{2\pi \times 360}{60}\right)^2 \times r$$

$$= 711r \text{ N}$$

where r is in metres.

Let spring extension equal x m, then spring force

$$F = S \times x$$

$$= 1.5 \times 1{,}000 \times x \text{ N}$$

but $x = r -$ initial length of spring

$$= r - 0.15 \text{ m}$$

therefore

$$F = 1{,}500(r - 0.15) \text{ N}$$

i.e. $F = 711 \, r = 1,500(r - 0.15)$

hence

$\quad r = 0.285 \text{ m} = 285 \text{ mm}$

thus extension

$\quad x = 285 - 150 = \mathbf{135 \text{ mm}}$

6.5 Dynamic instability

When the body A in the above example is displaced along the spring axis it will tend to return to its original position providing the spring is not 'overstretched'. The example showed there will be a definite radius of rotation corresponding to the particular speed of rotation. However, there is a critical value of the speed at which the body will not return to its original position but will fly radially outwards. At this speed the radius tends to become *indefinitely* large, and the body is then *unstable*. To find this *critical speed* we must find an expression for the radius of rotation r in terms of the other variables. Thus, for equilibrium, referring to Fig. 6.4

$\quad F = m\omega^2 r$

and $F = S \times x$

$\quad\quad = S(r - l)$

therefore

$\quad S(r - l) = m\omega^2 r$

hence

$\quad r = \dfrac{Sl}{S - m\omega^2}$

Now r tends to infinity as the denominator tends to zero, i.e. the condition for instability is

$\quad S - m\omega^2 = 0$

i.e. $\omega^2 = \dfrac{S}{m}$

using the figures from the above example

$\quad \omega^2 = \dfrac{1.5 \times 1,000}{0.5}$

hence the critical speed is
$\quad \omega = 54.8 \text{ rad/s}$

$\quad\quad = 523 \text{ rev/min}$

Such problems of dynamic instability are of great importance in rotating and other machinery in motion. A machine running at a critical speed may suffer considerable damage and an effort must be made to avoid such speeds by correct design.

Problems

1. What is the maximum speed at which a car may travel over a humpbacked bridge of radius 15 m without leaving the ground?

(43.7 km/h)

2. A 2-kg mass is attached at the end of a cord 1 m long and whirled in the vertical plane. Find the greatest speed at which the tension in the cord just disappears. What would be the maximum tension in the cord at a speed of 3 rev/s?

(0.498 rev/s; 732 N)

3. A trolley travels at 30 km/h round the inside of a vertical track. Calculate the maximum force on the track if the trolley's mass is 14 kg and the track radius is 2.4 m. What is the least velocity the trolley must have in order not to fall at the highest point?

(543 N; 17.5 km/h)

4. A rotor in a gyro instrument consists essentially of a flat disk of 100 mm diameter and 25 mm thick. It is mounted accurately on a spindle so that the central axis of the spindle coincides with the centre of the disk. What is the maximum out-of-balance force on the spindle at a gyro speed of 12,000 rev/min if the centre of mass of the disk is 0.025 mm out of alignment? Steel has a density of 7.8 Mg/m³.

(60.6 N)

5. A centrifugal clutch is designed just to engage when the centres of gravity of the rotating weights are 200 mm from the centre of rotation. The speed at engagement is 1,440 rev/min. If the revolving masses are each 0.5 kg calculate the spring stiffness required for an extension of 20 mm when the masses move from the rest to the engaged position.

(113.5 kN/m)

6. The centrifugal clutch shown at rest in Fig. 6.5 has springs of stiffness 15 kN/m and is designed to just engage at 10 rev/s. Calculate the required value of the revolving mass A. What is the force on the clutch rim at 16 rev/s?

(0.345 kg; 234 N)

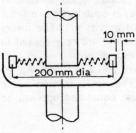

10 mm

200 mm dia

Fig. 6.5

7. A control mechanism is to be actuated by a mass of 4.5 kg rotating at the end of a spring at 300 rev/min. Calculate the minimum stiffness of spring required if the control is to be just stable at this speed.

(4.45 kN/m)

8. A body of mass 1.8 kg is whirled round at the end of a spring of stiffness 1.8 kN/m extension at a speed of 60 rev/min. Calculate the radius of the path of the body if the unstretched length of the spring is 150 mm. At what speed would the radius tend to be indefinitely great?

(156.2 mm; 302 rev/min)

6.6 Vehicle rounding a curve

Fig. 6.6 shows a two-wheeled vehicle (e.g. a cycle) rounding a curve of radius r at constant speed v. The cycle has mass m; A and B are the points of contact of the front and rear wheels, respectively, with the ground; O is the centre of rotation. For simplicity assume the rolling resistance to motion negligible. Hence there is no friction or other resisting force at A or B tangent to the path. Radial forces are necessary in order that the vehicle shall move in a curved path and not a straight line, and these are provided by radially-inward friction forces F_1 and F_2 at A and B, respectively.

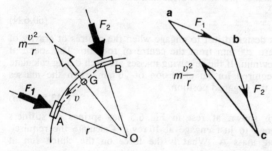

Fig. 6.6

Owing to rotation, every particle of mass dm of which the vehicle is composed has an inertia force $dm(v^2/r)$ acting upon it radially outward. The net radial effect of all these forces is equivalent to a force acting at the centre of mass (i.e. the centre of gravity G). The forces F_1, F_2 and the inertia force are in equilibrium and are represented in the force diagram, Fig 6.6, by **ab**, **bc** and **ca**, respectively.

In practice the angle ∠AOB is often small and the forces F_1, F_2 approximately in the same straight line. The total friction force $F = F_1 + F_2$ is then, for an unbanked flat track, equal and opposite to the inertia force, or

$$F = \frac{mv^2}{r}$$

The friction force is here an active force in that it causes the vehicle to deviate from a straight line. A vehicle on a perfectly smooth sheet of ice, for example, could not move except in a straight line. When the limiting friction force is insufficient to provide the centripetal acceleration (radially inward) the vehicle tends to move in a straight line. It then appears to be skidding 'outwards'.

6.7 Superelevation of tracks: elimination of side-thrust

The superelevation of a railway track is the amount by which the outer rail is raised above the level of the inner rail. The wheels of a train are flanged, the flanges being on the inside of the rails. As the train rounds a curved track, the centripetal force required to provide the circular motion is provided by the inward thrust of the outer rail. To reduce the magnitude of this lateral load, a second rail may sometimes be provided on the inside curve so that the inner wheel flange is contained between two rails. This second rail then takes some of the side thrust. More generally, the side thrust may be eliminated completely at a particular speed by suitable banking of the track. The amount of banking or *cant* depends on the tightness of the curve and the speed of the trains using the track. In practice the amount of superelevation is limited to about 150 mm i.e. about 6° of cant since 25 mm of superelevation on a standard gauge line is equal to 1° of cant. The speed chosen is the average speed at which a train (usually a freight train) may be expected to take the curve. At any speed higher than the one suitable for that angle of banking there will be a side thrust on the outer rail, so that fast passenger trains have some lateral force; at lower speeds than the design value there will be a side thrust in all cases on the inner rail. The amount of *extra* banking needed at a given speed to remove this side thrust altogether is called the 'cant deficiency' and this is normally limited to about 110 mm of superelevation.

The banking of a car race track serves a similar purpose to the superelevation of a rail track, i.e. to eliminate side thrust on the tyres. To serve its purpose for cars of different speed the gradient of the banking is increased towards the outside of the curve. There is a correct angle of banking for any particular speed and this angle is independent of the weight of the vehicle (*see* example page 127). Most racing tracks are now unbanked so that means have had to be found to provide an increase in side thrust to allow high speeds round corners. Large, wide tyres give extra adhesive force but in order to increase the downward force on the vehicle without affecting the weight, aerodynamic devices are used. One method is to fit aerofoils, called 'wings', above the rear of the racing car and to either side of the pointed nose. Another solution is to shape the floor of the car with 'venturi-tunnels' and combined with 'skirts' attached to the side-pods, the suction ground-effect produces an enormous increase in download. Not all of these arrangements however fall within the regulations governing the various racing categories.

Example   *A car of mass 2 tonne rounds an unbanked curve of 60 m radius at 72 km/h. Calculate the side thrust on the tyres.*

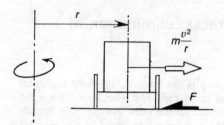

Fig. 6.7

SOLUTION

The radial forces acting on the car are: (*a*) the outward inertia force, $\dfrac{mv^2}{r}$; (*b*) the inward force F exerted by the road on the tyres, i.e. the *side thrust* (Fig. 6.7). These two forces are in balance, therefore

$$F = \frac{mv^2}{r}$$

$$= 2 \times 1,000 \times \frac{20^2}{60} \text{ since 72 km/h} = 20 \text{ m/s}$$

$$= \mathbf{13{,}330 \text{ N}}$$

(Note that F is *not* equal to the limiting friction force which is μmg.)

Example   *A racing car travels at 180 km/h on a track banked at 30° to the horizontal. The limiting coefficient of friction between tyres and track is 0.7. Calculate the minimum radius of curvature of the track if the car is not to slide 'outwards'.*

SOLUTION

The *total* reaction R of the track on the car acts at an angle ϕ to the normal, where $\tan \phi = \mu = 0.7$, *see* paragraph 3.2. This reaction is in balance with the inertia force and the weight of the car. These three forces act at a point and can be represented by the triangle of forces, **oab**, Fig. 6.8. From the triangle of forces

$$\tan (30° + \phi) = \frac{mv^2}{r} \div mg = \frac{v^2}{gr}$$

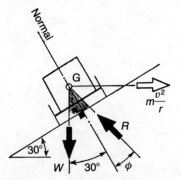

Fig. 6.8

But $\tan \phi = 0.7$, thus $\phi = 35°$

and $v = 180$ km/h $= 50$ m/s

hence

$$\tan (30° + 35°) = \frac{50^2}{9.8 \times r}$$

therefore

$$r = \textbf{119 m}$$

Example   *Calculate the angle of banking on a bend of 100 m radius so that vehicles can travel round the bend at 50 km/h without side thrust on the tyres. For this angle of banking what would be the value of the coefficient of friction if skidding outwards commences for a car travelling at 120 km/h?*

SOLUTION

If the angle of banking is θ, Fig. 6.9, then the forces on the vehicle parallel to the slope are (i) the component of the weight $mg \sin \theta$ acting inwards; (ii) the component of the inertia force $\frac{mv^2}{r} \cos \theta$ acting outwards; (iii) the side thrust F acting inwards.

The equation of forces parallel to the slope is therefore

$$F + mg \sin \theta = \frac{mv^2}{r} \cos \theta$$

If there is to be no side thrust, $F = 0$, hence

$$\tan \theta = \frac{v^2}{gr} = \frac{(50/3.6)^2}{9.8 \times 100} = 0.198$$

and $\qquad \theta = \textbf{11° 12'}$

Note that the angle of banking is independent of the weight of the car.

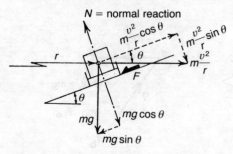

Fig. 6.9

When skidding outwards commences limiting conditions exist so that

$$F = \mu \times N$$

$$= \mu \left(\frac{mv^2}{r} \sin \theta + mg \cos \theta \right)$$

where $v = 120$ km/h $= 33.3$ m/s. The equation of forces parallel to the slope becomes

$$\mu \left(\frac{mv^2}{r} \sin \theta + mg \cos \theta \right) + mg \sin \theta = \frac{mv^2}{r} \cos \theta$$

hence

$$\mu = \frac{(v^2/gr) - \tan \theta}{(v^2/gr) \tan \theta + 1}$$

$$\frac{v^2}{gr} = \frac{(120/3.6)^2}{9.8 \times 100} = 1.135$$

therefore

$$\mu = \frac{1.135 - 0.198}{1.135 \times 0.198 + 1} = \mathbf{0.765}$$

The student should rework this problem using the method of the previous example.

Example   *Calculate the superelevation of the outside rail of a curved track if a train is to traverse the curve without side thrust on the rails at 50 km/h. The radius of the curve is 180 m and the track gauge is 1,440 mm.*

SOLUTION

If the superelevation is h and the track width x, the angle of banking θ is given by:

$$\sin \theta = \frac{h}{x}$$

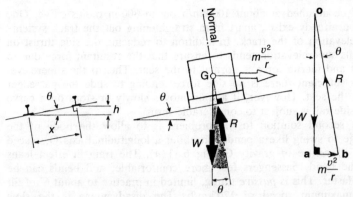

Fig. 6.10

The forces acting on the vehicle are shown in the free-body diagram, Fig. 6.10, i.e.: the weight W, the total reaction R and the inertia force mv^2/r. Since there is no side thrust the reaction R exerted by the track on the vehicle must be normal to the track incline. From the triangle of forces (Fig. 6.10):

$$\tan\theta = \frac{\mathbf{ab}}{\mathbf{oa}} \text{ where } \mathbf{oa} = W = mg$$

$$\mathbf{ab} = \frac{mv^2}{r} = m \times \frac{(50/3.6)^2}{180} = 1.075 \, m$$

Hence

$$\tan\theta = \frac{m \times 1.075}{m \times 9.8} = 0.1095$$

But since $\tan\theta$ is small, $\tan\theta = \sin\theta$, approximately

therefore

$$\tan\theta = \frac{h}{1.44}$$

thus $h = 1.44 \tan\theta = 1.44 \times 0.1095 = 0.158 \, \text{m} = \mathbf{158 \, mm}$

6.8 Passenger comfort – the pendulum car

Anyone who has been thrown outwards while in a vehicle travelling round a curve at speed will be familiar with the reality of the radial inertia force. Of course, what is experienced by a passenger is the tendency to move in a straight line while the vehicle turns. Experiments have shown that an uncompensated radial acceleration (v^2/r) in excess of about $0.1\,g$ ($0.98 \, \text{m/s}^2$) is definitely unpleasant. This acceleration would be attained at about 100 km/h on an 800-m radius curve. One solution already exists, apart from straightening out the track system–superelevation of the track. In addition to reducing the side thrust on the rails superelevation tends to ensure that the resultant force due to weight and inertia force is normal to the seat. Then if the superelevation is sufficient there is no side force tending to slide the passenger across the seat. However, if a train moves slowly or stops on a curve the inside rail is subject to considerable thrust.

A second solution to the problem is to allow the body of the carriage to swing like a pendulum about a longitudinal axis O, placed above its centre of gravity G, Fig. 6.11 (*a*). The train in effect leans into the bend, passengers feel more comfortable, and bends can be taken faster. This is *passive* tilting, limited in practice to about 6° of tilt at a maximum speed of 120 km/h. The disadvantage is the slow response time when entering and leaving a bend. In an *active* system the carriage is tilted by hydraulic jacks, electronically controlled. In practice there is a combination of banked track and tilting carriage. Speeds above 200 km/h are possible with a pendulum car subject to restrictions for other reasons such as braking requirements.

The forces acting on the swinging carriage are: its weight W, the inertia force mv^2/r radially outward, and the reaction R at the pivot. The three forces are in equilibrium, hence all three forces pass through G. The line of action of R is therefore from G to O. The resultant force is always normal to the carriage floor; a similar argument applies also to any passenger seated in the swinging carriage.

The triangle of forces **abc** is shown in Fig. 6.11 (*b*). If there is no superelevation of the track the angle between the vertical and the reaction R is the angle α through which the carriage swings about O. From the triangle of forces:

$$\tan \alpha = \frac{\mathbf{bc}}{\mathbf{ab}}$$

$$= \frac{mv^2}{r} \div W$$

$$= \frac{v^2}{gr} \text{ since } W = mg$$

For example, at 120 km/h on a 1,200-m radius curve, $v = 33.3$ m/s, and

$$\tan \alpha = \frac{33.3^2}{9.8 \times 1,200} = 0.095$$

thus $\alpha = 5.4°$

and the pendulum car swings outwards nearly 6°. It should be noted, however, that allowing the carriage to pivot does not affect the side thrust on the track. In this case, since there is no superelevation the side thrust would be equal to the inertia force.

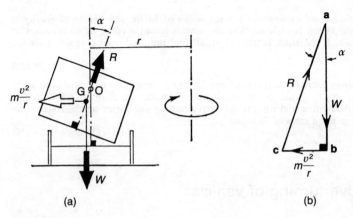

Fig. 6.11

Problems

1. Calculate the minimum limiting coefficient of friction between tyres and road in order that a car shall negotiate an unbanked curve of 120 m radius at 100 km/h.

(0.66)

2. Calculate the minimum radius of unbanked track which a motor cycle may traverse at 130 km/h without skidding outwards if the coefficient of sliding friction is 0.6.

(222 m)

3. A vehicle travels at 72 km/h round a track banked at 20° to the horizontal. The coefficient of sliding friction between tyres and road is 0.5. Calculate the radius at which skidding would occur.

(38.7 m)

4. A race track is to be banked so that at 120 km/h a car can traverse a 180-m radius curve without side thrust on the tyres. Calculate the angle of banking required.

(32° 13′)

5. A car travels round a curve of 60 m radius which is banked at 10° to the horizontal, the slope being *away* from the inside of the curve. If the coefficient of friction between tyres and road is 0.7 calculate the maximum speed at which the curve can be traversed without skidding.

(60 km/h)

6. A vehicle traverses a banked track of radius 90 m and angle of banking 60°. If its speed is 12 m/s, calculate the minimum coefficient of friction between tyres and track if the vehicle is not to slip *down* the track.

(1.22)

7. A road curve of 75 m radius is banked so that the resultant reaction for any vehicle is normal to the road surface at 72 km/h. Calculate the angle of banking and the value of the coefficient of friction if skidding outwards commences for a car travelling at 120 km/h.

(28° 30′; 0.53)

8. A car of mass 1 tonne has a track width of 1.5 m. The centre of gravity is 600 mm above road level. The car travels round a curve of 60 m radius at 72 km/h. If the track is banked at 30° find the total normal reaction on the outer wheels.

(6.25 kN)

9. A railway carriage built on the pendulum-car principle has a maximum angle of tilt of 9°. What is the maximum allowable speed on a 600-m radius unbanked curve and what is the corresponding side thrust on the track if the carriage has a mass of 50 tonne?

(110 km/h; 77.6 kN)

6.9 Overturning of vehicles

The tendency of a vehicle to slide 'outwards' when rounding a curve has been shown to result in a limiting speed at which sliding just occurs. In addition there is also a limiting speed at which the vehicle will overturn. This is an example of dynamic instability. To maintain a vehicle in a circular path requires a centripetal force at the centre of mass, but this force can only be supplied by friction at the road surface or the side thrust of a rail track. This lateral force is equivalent to the same force at the centre of mass together with a couple tending to rotate the vehicle about the centre of mass (Fig. 6.12). The two forces forming this couple are the lateral force at the track surface and the inertia force at the centre of mass. Alternatively, the inertia force acting radially outwards may be thought of as tending to tip the vehicle about its offside wheels. The weight provides the balancing couple tending to maintain stable equilibrium and prevent tipping.

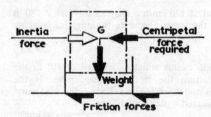

Fig. 6.12

Example *A vehicle has a track width of 1.4 m and its centre of gravity is 650 mm above the road surface in the centre plane. If the limiting coefficient of friction between tyres and road is 0.6 determine whether the vehicle will first overturn or sideslip when rounding a curve of 120 m radius at speed on a level track. State the maximum permissible speed on the curve.*

SOLUTION
Fig. 6.13 shows the free-body diagram for the vehicle. Let v be speed of vehicle in metres per second and r = radius of curve (120 m). For sideslip to occur the inertia force must be just equal to or greater than the total limiting inward friction force, i.e.

$$\frac{mv^2}{r} = \mu W = \mu mg$$

or $$\frac{v^2}{120} = 0.6 \times 9.8$$

i.e. $$v = 6.6 \text{ m/s or } 95.6 \text{ km/h}$$

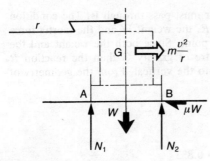

Fig. 6.13

For overturning to occur the ground reaction N_1 at the *inner* wheels must be zero, i.e. the tilting moment due to the inertia force must be greater than the stabilizing moment due to the dead weight. Taking moments about the outer wheel track B thereby eliminating N_2 and μW, and assuming the car just about to overturn:

$$\frac{mv^2}{r} \times 0.65 = W \times 0.7$$

i.e. $$\frac{mv^2 \times 0.65}{120} = m \times 9.8 \times 0.7$$

i.e. $$v = 35.6 \text{ m/s or } 128 \text{ km/h}$$

Sideslip takes place first at the lower speed of **95.6 km/h** and this therefore, is the maximum speed permissible on the curve.

Example *Calculate the maximum speed at which a car may traverse a banked track of 30 m radius without overturning if the centre of gravity of the car is 0.9 m above ground level and the track width of the wheels is 1.5 m. The track is banked at 30° to the horizontal.*

SOLUTION
When overturning starts the inner wheels at A (Fig. 6.14) just lift and the car starts to rotate about the outer wheels at B. Hence the total

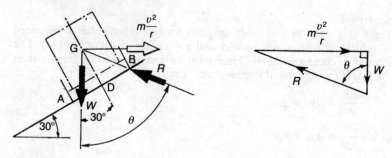

Fig. 6.14

reaction R of the track on the car must pass through B. The condition for (unstable) equilibrium is that R, the weight W and the inertia force mv^2/r shall all pass through one point. Since both the weight and the inertia force act through the centre of gravity G then the reaction R must act along BG at an angle θ to the vertical. From the geometry of the car

$$\theta = 30° + \angle DGB$$

and, since

$$\tan \angle DGB = \frac{DB}{GD} = \frac{0.75}{0.9} = 0.833$$

then
$$\angle DGB = 39°\, 48'$$

hence
$$\theta = 30° + 39°\, 48'$$
$$= 69°\, 48'$$

From the triangle of forces:

$$\tan \theta = \frac{mv^2}{r} \div W$$

or
$$\tan 69°\, 48' = \frac{v^2}{9.8 \times 30} \text{ since } W = mg$$

thus
$$v^2 = 9.8 \times 30 \times \tan 69°\, 48'$$

and
$$v = 28.2 \text{ m/s or } \mathbf{101.5 \text{ km/h}}$$

The student should rework this example, resolving forces parallel and perpendicular to the slope, and taking moments about B.

Problems

1. Calculate the maximum speed at which a car can traverse an unbanked curve of 24 m radius if its wheels are 1.8 m apart and its centre of gravity is 1.2 m above the road. Assume that slipping does not occur.

    (13.3 m/s or 47.8 km/h)

2. Calculate the smallest radius unbanked curve which a racing car can traverse at 200 km/h without overturning if its wheels are 1.5 m apart and its centre of gravity is 650 mm above the road.

(273 m)

3. A sports car is to be built capable of rounding a 100-m curve at 120 km/h and its wheels are to be 1.5 m apart. Calculate the maximum allowable height of its centre of gravity above ground level.

(0.66 m)

4. Calculate the maximum speed at which a car can traverse a 30-m radius track banked at 20° to the horizontal. Its centre of gravity is 0.9 m above the ground and its wheels are 1,350 mm apart.

(76.3 km/h)

5. A double-deck bus whose wheels are 2.4 m apart and whose centre of gravity is 1.5 m above the ground is to round a road banked at 10° to the horizontal at 50 km/h. Calculate the minimum radius of the curve if overturning is not to occur.

(17.4 m)

6. A four-wheeled vehicle turns a corner of radius 10 m on a level track. Its centre of gravity is 0.9 m above road level and its wheels are 1.5 m apart. Calculate (a) the fastest speed at which it may traverse the bend without the inner wheels leaving the ground; (b) the fastest speed at which it may travel round on two outer wheels without tipping more than 30°.

(9.03 m/s; 4.13 m/s)

7. Determine the speed at which overturning will occur for a vehicle whose wheels are 1.5 m apart and whose centre of gravity is 0.9 m above the ground, when travelling round a banked track of 60 m radius: the banking is 10° away from the inside of the curve. Calculate the angle of banking which would tip the vehicle when at rest.

(26.3 m/s; 39° 48′)

8. A car of mass 1,120 kg is travelling along a curved unbanked road of radius 60 m. Its wheel track is 1.35 m and its centre of gravity is 0.9 m above the ground, and mid-way between front and rear axles. The coefficient of friction is 0.5. Find (a) the vertical ground reaction at each wheel when the car is travelling at 15 m/s; (b) the maximum speed without overturning; (c) the skidding speed.

(1,340 N; 4,140 N; 21.1 m/s; 17.15 m/s)

6.10 Banking of an aircraft

When an aircraft turns in a circle a centripetal force is required to maintain the circular motion and this is supplied by the plane banking, Fig. 6.15. In straight level flight there is a lifting force due to the air acting vertically upwards on the wings to balance the weight of the aircraft. This lifting force is explained in paragraph 12.6. When the plane banks at an angle to the horizontal the *lift* continues to act normal to the lateral axis of the plane, Fig. 6.15, providing a vertical component to oppose the force of gravity and a horizontal component for the required centripetal force. Thus for *correct* banking, with no sideslip, the radially inwards component of the lifting force must equal the centripetal force. The angle of banking depends on the plane's speed and on the radius of the turning circle, but is independent of the mass of the plane. This is shown in the following example.

Example   *An aircraft turns in a horizontal circle of radius r = 180 m when cruising at a speed of 185 km/h. Fig. 6.15 shows the forces acting on the plane as it banks without slip at an angle θ to the horizontal. The weight of the plane is W, and the upwards lifting force of the air is L, acting normal to the axis of the plane. Find the correct angle of banking required. If the plane makes a 180° turn in one minute at the same speed and its mass is 12 tonnes what is the angle of banking required and the lifting force?*

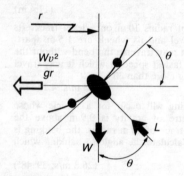

Fig. 6.15

SOLUTION

Resolving forces vertically,

$$L \cos \theta = W$$

i.e. $$L = W/\cos \theta$$

The centripetal force is $\dfrac{Wv^2}{gr}$, and this is provided by the horizontal component of L, therefore

$$L \sin \theta = \frac{Wv^2}{gr}$$

i.e. $$\frac{W}{\cos \theta} \cdot \sin \theta = \frac{Wv^2}{gr}$$

i.e. $$\tan \theta = \frac{v^2}{gr}$$

$$= \frac{(185/3.6)^2}{9.8 \times 180}$$

$$= 1.5$$

i.e. $\theta = \mathbf{56°}$

Note that the correct angle of banking is independent of the weight of the plane.

When the plane turns through 180° in one minute the distance travelled is πr, hence

$$\pi r = \text{speed} \times \text{time}$$

$$= (185/3.6) \times 60$$

i.e. $r = 981.5 \text{ m}$

and $\tan \theta = \dfrac{v^2}{gr} = \dfrac{(185/3.6)^2}{9.8 \times 981.5} = 0.2745$

i.e. $\theta = \mathbf{15.4°}$

$L \cos 15.4° = W$

$= 12 \times 9.8 \text{ kN}$

i.e. $L = \mathbf{122 \text{ kN}}$

Problems

1. An aircraft of mass 300 tonne makes a steady horizontal turn at 585 km/h. The angle of banking to the horizontal is 35°. Find the centripetal acceleration, the radius of the turning circle, and the lifting force.

   (6.86 m/s²; 3850 m; 3.59 MN)

2. An aircraft makes a turn in a circle of radius 1.3 km at a speed of 450 km/h. Find the correct angle of banking. If the mass of the plane is 4000 kg what is the total lift?

   (50.8°; 62 kN)

3. An aircraft makes a horizontal turn of 360° in one minute at 555 km/h. Find the radius of the turning circle and the correct angle of banking. If the aircraft weighs 20 kN what is the lifting force?

   (1.47 km; 58.7°; 38.5 kN)

4. An aircraft has a mass of 170 tonne and cruises in level flight at 925 km/h. Find the radius of the turning circle and the angle of banking when the plane makes a 180° turn in one minute. If the wing area is 350 m² what is the wing loading during the turn?

   (4.91 km; 53.9°; 8.08 kN/m²)

5. A helicopter of mass 22 tonne makes a level turn at a steady speed of 160 km/h. The angle of banking is 50° (i.e. the lateral axis of the rotor is at 50° to the horizontal). Find the radius of the turning circle, the total lifting force and the centripetal force.

   (*see* paragraphs 12.6 and 12.8; 169 m; 335.4 kN; 257 kN)

CHAPTER 7

Balancing

7.1 Static balance – two masses in a plane

Consider a light arm pivoted freely at the fulcrum O (Fig. 7.1) and carrying masses m_1, m_2 at distances r_1, r_2 from O, respectively. In general the arm will rotate about O and the system is said to be out of balance. For equilibrium there must be balance of moments about O, i.e.

$$m_1 g \times r_1 = m_2 g \times r_2$$

or

$$m_1 r_1 = m_2 r_2$$

When in balance the arm may be set in any position and it will remain at rest in that position. The weights are said to be in *static balance* and the centre of gravity of the system is located at O.

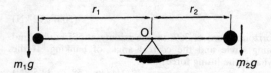

Fig. 7.1

7.2 Dynamic balance – two masses in a plane

Now consider two light arms fixed to a shaft at bearing O and rotating with angular velocity ω, Fig. 7.2 (a). The arms are in the same plane and carry masses m_1, m_2 at radii r_1, r_2 respectively. Owing to the rotation each mass exerts an inertia force radially outward on the bearing O.

The force due to m_1 is $m_1 \omega^2 r_1$ (**oa** in the force diagram, Fig. 7.2 (b)).

The force due to m_2 is $m_2 \omega^2 r_2$ (**ab** in the force diagram).

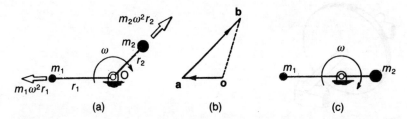

Fig. 7.2

The resultant out of balance force on the bearing is given by **ob** in the force diagram.

When the dynamic load on the bearing is zero the rotating system is said to be in *dynamic balance*. The condition for no load at O is that the two inertia forces shall: (*a*) act along the same straight line but with opposite sense; (*b*) be equal in magnitude.

The relative positions of the masses are as in Fig. 7.2(*c*); the condition for equal inertia forces is:

$$m_1\omega^2 r_1 = m_2\omega^2 r_2$$

Thus, since ω^2 is the same for both masses

$$m_1 r_1 = m_2 r_2$$

This is also the condition for static balance. Hence, if two bodies *in the same plane* are in static balance when pivoted about a given axis they will be in dynamic balance at any speed when rotating about the same axis.

7.3 Method of balancing rotors

It was shown above that for a two mass system to be in static balance the *mr* product for each mass had to be the same. This is also the condition for the masses to be balanced when rotating, and suggests a method for ensuring balance for rotating rotors such as turbine disks or car-wheel assemblies.

Fig. 7.3 shows a turbine rotor idealized in the form of a disk, mounted on a shaft placed on a pair of parallel knife-edges. The rotor may be allowed to rotate freely on the knife-edges and, if not uniform, the heavier section will rotate to the lowest point. The point is marked with chalk and a small balance mass attached at a point diametrically opposite. The rotor is turned through 90° and again allowed to rotate freely and again the heavier section will rotate to the bottom. The balance mass is then increased or decreased accordingly and the process repeated until the rotor remains at rest in any position. It is then in static balance and remains balanced when rotating at speed.

Fig. 7.3

In practice it is usually possible to balance a rotor to an accuracy of 0.001 m-kg, i.e. the amount of residual unbalance is equivalent to a mass of 1 kg at 1 mm radius. For a rotor of mass 10 tonne this is equivalent to a displacement x of the centre of gravity from the axis of rotation given by

$$10 \times 1{,}000 \times x = 0.001$$

$$x = 10^{-7} \text{ m or } 0.1 \ \mu\text{m}$$

The corresponding out-of-balance centrifugal force when running at 3,600 rev/min is

$$m\omega^2 r = 10 \times 1{,}000 \times \left[\frac{2\pi \ 3{,}600}{60}\right]^2 \times 10^{-7} \text{ kg-m/s}^2$$

$$= 142 \text{ N}$$

It is usual to limit the out-of-balance force to be not greater than 1 per cent of the rotor weight.

7.4 Static balance – several masses in one plane

We now consider the static balance of several masses in the same plane of magnitudes m_1, m_2, . . ., at radii r_1, r_2, . . . from a common pivot O (Fig. 7.4). If the system is to be in static balance the shaft at O must remain at rest in all positions, i.e. there must be no resultant moment about O. When the centre of gravity of the system is at the centre of the shaft the resultant vertical force due to the dead weight of all the attached bodies passes through O and the shaft is then in static balance.

The static moment of mass m_1 about O

$$= m_1 g \times \text{OX} \quad (\text{Fig. 7.4}(b))$$

$$= m_1 g \times r_1 \cos \alpha_1$$

Similarly for m_2 and m_3. For static balance the sum of all such moments must be zero, i.e. since g is constant, $\Sigma mr \cos \alpha$ must be zero.

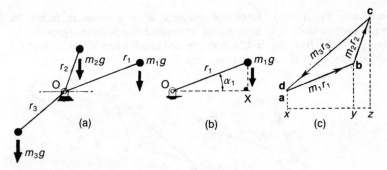

Fig. 7.4

Now construct a polygon in the following way:

> Draw **ab**, magnitude $m_1 r_1$, parallel to radius r_1 (Fig. 7.4(c)).
> Draw **bc**, magnitude $m_2 r_2$, parallel to radius r_2.
> Draw **cd**, magnitude $m_3 r_3$, parallel to radius r_3.

If the polygon closes, **d** coincides with **a**. Hence from Fig. 7.4(c), if **x**, **y**, **z** are the feet of the perpendiculars from **a**, **b**, **c**, respectively, on to a horizontal line

$$\mathbf{xy} + \mathbf{yz} + \mathbf{zx} = 0 \text{ for a closed polygon, as shown}$$

But
$$\mathbf{xy} = m_1 r_1 \cos \alpha_1$$

$$= \text{moment of } m_1 \text{ about O} \div g$$

Similarly for m_2, m_3. Hence

$$\mathbf{xy} + \mathbf{yz} + \mathbf{zx} = \Sigma mr \cos \alpha$$

$$= 0 \text{ for static balance}$$

But this is also the condition for the polygon to close. Hence the condition for static balance is that the vector polygon formed by the *mr* values must close. This result is unchanged for all angular positions of the shaft; thus if the polygon closes for one position it closes for all positions.

7.5 Dynamic balance of several masses in one plane

Suppose the shaft and attached masses shown in Fig. 7.5(a) to be rotating with angular velocity ω. Owing to the rotation there will be inertia forces of magnitude $m_1 \omega^2 r_1$, $m_2 \omega^2 r_2$, ..., acting radially outward at each mass. These forces can be represented by a force polygon which must close if there is no resultant unbalanced force, as shown in Fig. 7.5 (*b*). Since the quantity ω^2 is a common factor for each inertia force it is convenient to replace the force polygon by an *mr* polygon. But this is the polygon obtained when investigating the static balance of the same system, Fig. 7.4.

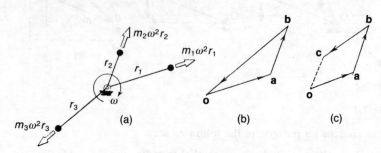

Fig. 7.5

Thus the conditions for static and dynamic balance are identical, i.e. that the *mr* polygon must close.

If the inertia forces are not in balance the *mr* polygon will not close, Fig. 7.5 (*c*). The closing line **co** taken in the sense **c** to **o** represents the *mr* value required to produce balance. The line **oc** taken in the sense **o** to **c** represents the resultant unbalanced *mr* effect. To obtain the actual magnitude of the unbalanced forces multiply by ω^2. Thus the resultant unbalanced force on the shaft is:

oc $\times \omega^2$, in direction **o** to **c**

and the required *balancing* force (equilibrant) is:

co $\times \omega^2$, in direction **c** to **o**

An unbalanced inertia force produces a load on the shaft bearing which, in a machine, results in increased wear and leads to early failure of the bearing metal.

Example   *A shaft carries two rotating masses of 1.5 kg and 0.5 kg, attached at radii 0.6 m and 1.2 m, respectively, from the axis of rotation. The angular positions of the masses are shown in Fig. 7.6 (a). Find the required angular position and radius of rotation r of a balance mass of 1 kg.*

If no balance mass is used what is the out-of-balance force on the shaft bearing at 120 rev/min?

SOLUTION
The *mr* values are 0.9 kg-m for A and 0.6 kg-m for B and these are represented by **oa** and **ab**, respectively, in the 'force' polygon, Fig. 7.6(*b*). The resultant out-of-balance *mr* value is given by **ob** in direction **o** to **b**. From a scale drawing:

ob = 1.31 kg-m

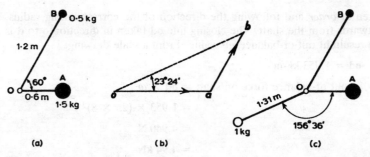

Fig. 7.6

The equilibrant is equal and opposite to the out-of-balance force, and since the *mr* value for the balance mass is:

$$1 \times r \text{ kg-m}$$

therefore, for balance

$$1 \times r = 1.31$$

and $\quad r = 1.31 \text{ m}$

thus the radius of rotation of the balance mass is **1.31 m**.

The balance mass must be positioned so that its inertia force is acting in direction **b** to **o**, i.e. at an angle of 156° 36′ to the radius of mass A, as shown in Fig. 7.6 (*c*). If no balance mass is used

$$\text{out-of-balance force} = \mathbf{ob} \times \omega^2$$

$$= 1.31 \times \left[\frac{2\pi \times 120}{60} \right]^2 \text{ kg-m/s}^2$$

$$= \textbf{207 N}$$

Example *Four bodies A, B, C and D are rigidly attached to a shaft which rotates at 8 rev/s. The bodies are all in the same plane and the masses and radii of rotation, together with their relative angular positions, are: A, 1 kg, 1.2 m, 0°; B, 2 kg, 0.6 m, 30°; C, 3 kg, 0.3 m, 120°; D, 4 kg, 0.15 m, 165°. Find the resultant out-of-balance force on the shaft and hence determine the magnitude and position of the balance mass required at 0.6 m radius.*

SOLUTION

Fig. 7.7 (*a*) shows the relative angular positions of the masses.

For mass A, $\quad mr = 1 \times 1.2 \quad = 1.2 \text{ kg-m}$

$\qquad\qquad$ B, $\quad mr = 2 \times 0.6 \quad = 1.2 \text{ kg-m}$

$\qquad\qquad$ C, $\quad mr = 3 \times 0.3 \quad = 0.9 \text{ kg-m}$

$\qquad\qquad$ D, $\quad mr = 4 \times 0.15 = 0.6 \text{ kg-m}$

The *mr* polygon is shown in Fig. 7.7 (*b*), each *mr* value being taken *in order* and following the direction of the corresponding radius, outwards from the shaft. The closing line **od** taken in direction **o** to **d** is the resultant out-of-balance *mr* value. From a scale drawing:

$$\textbf{od} = 1.953 \text{ kg-m}$$

$$\therefore \text{ out-of-balance force on shaft} = \textbf{od} \times \omega^2$$

$$= 1.953 \times (2\pi \times 8)^2$$

$$= 4,940 \text{ N}$$

$$= \textbf{4.94 kN}$$

od makes an angle of 51° 47′ with **oa**. The equilibrant is equal and opposite to the resultant force, i.e. in direction **d** to **o**. The *mr* value for the balance mass is $m \times 0.6$ kg-m. Therefore for balance

$$m \times 0.6 = 1.953$$

i.e. $$m = \textbf{3.26 kg}$$

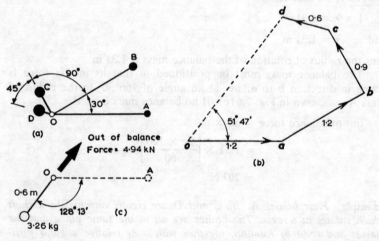

Fig. 7.7

The balance mass must be positioned so that its inertia force acts in the direction **d** to **o**. Its position relative to A is shown in Fig. 7.7 (*c*).

7.6 Dynamic forces at bearings

If a shaft carries rotating masses *not in the same plane* it may be in static balance and therefore balanced as regards inertia forces, but it may yet be subject to an unbalanced couple. Whether the masses are in the same plane or not the inertia forces act radially outwards and may therefore balance; but since the lines of action of the inertia forces act in different planes each force produces a different moment about any given plane of the shaft. Thus an unbalanced moment may arise.

For a shaft to be in complete dynamic balance there must be no unbalanced force or couple.

An unbalanced couple cannot, of course, exist in practice, but must be resisted by reactions at the bearings. As the shaft rotates so does the direction of the unbalanced couple and the bearings are therefore subject to rotating radial forces. Also, since the dead weight reactions are constant and upwards it means that the bearing reactions are constantly changing in direction and magnitude as the shaft rotates. This condition if allowed to persist sets up undesirable vibrations. The following examples, restricted to one mass, or two masses in the same axial plane, will be used to show the existence of unbalanced couples and to bring out the main points in the methods of calculating the bearing reactions. The balancing of several masses in different planes of rotation cannot be dealt with completely at this stage.

Example *A shaft is supported in bearings at A and B, 2 m apart, Fig. 7.8. A rotor of total mass 30 kg is mounted at a point 0.6 m from bearing A. Owing to faulty mounting the centre of gravity of the rotor is offset 3 mm from the axis of rotation O. Calculate the dynamic loads on the bearings when the shaft rotates at 600 rev/min. Calculate also the maximum and minimum loads on the bearings.*

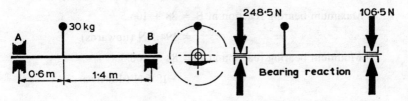

Fig. 7.8

SOLUTION

Inertia force $= m\omega^2 r$

$$= 30 \times \left(\frac{2\pi\ 600}{60}\right)^2 \times 0.003$$

$$= 355 \text{ N}$$

Taking moments about bearing B:

$$R_A \times 2 = 355 \times 1.4$$

$$R_A = 248.5 \text{ N}$$

Taking moments about bearing A:

$$R_B \times 2 = 355 \times 0.6$$

$$R_B = 106.5 \text{ N}$$

(or $R_A + R_B = 355$, i.e. $R_B = 355 - 248.5 = 106.5$ N).

Therefore the dynamic loads on the bearings are **248.5 N** at A and **106.5 N** at B. Both reactions oppose the unbalanced inertia forces as shown.

When at rest, taking moments about B:

$$R_A' \times 2 = 30 \times 9.8 \times 1.4$$

$$R_A' = 206 \text{ N}$$

and $R_A' + R_B' = 30 \times 9.8 = 294$ N

thus $R_B' = 88$ N

The dead weight reactions are therefore **206 N** at A and **88 N** at B, both vertically upwards.

As the shaft rotates the dynamic load on each bearing rotates. The maximum and minimum bearing reactions therefore occur when the lines of action of the dynamic and dead weight reactions coincide.

Maximum bearing reaction at A = 206 + 248.5

= 454.5 N (upwards).

Minimum bearing reaction at A = 206 − 248.5

= −42.5 N (downwards).

Maximum bearing reaction at B = 88 + 106.5

= 194.5 N (upwards).

Minimum bearing reaction at B = 88 − 106.5

= −18.5 N (downwards).

Example   *The shaft shown in Fig. 7.9 (a) carries two masses at C and D in the same axial plane but diametrically opposite to one another. Calculate the dynamic loads on the bearings when the shaft rotates at 144 rev/min. Each mass is 3 kg at 600 mm radius.*

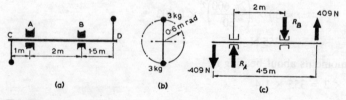

(a)   (b)   (c)

Fig. 7.9

SOLUTION

Fig. 7.9 (*b*) shows the end view when the masses are in the vertical plane; the shaft is evidently in static balance since the *mr* values are equal and opposite. When rotating, therefore, the inertia forces are in balance. Nevertheless the two rotating masses exert a pure couple anticlockwise due to the two equal inertia forces acting at distance CD apart.

$$\text{Moment of couple} = m\omega^2 r \times \text{CD}$$

This couple is in equilibrium with reactions R_A and R_B at bearings A and B, respectively, acting as shown in Fig. 7.9 (*c*) to produce a clockwise couple. For each rotating mass

$$\text{inertia force} = m\omega^2 r$$

$$= 3 \times \left[\frac{2\pi \ 144}{60} \right]^2 \times 0.6$$

$$= 409 \text{ N}$$

To calculate the reactions take moments about each bearing in turn. Moments about A:

$$R_B \times 2 = 409 \times 3.5 + 409 \times 1$$

$$R_B = \textbf{920.3 N}$$

Moments about B:

$$R_A \times 2 = 409 \times 3 + 409 \times 1.5$$

$$R_A = \textbf{920.3 N}$$

The two reactions are equal since they together form a pure couple exerted by the bearings, in opposition to that exerted by the inertia forces of the rotating masses. The reactions in this case could have been found by simply equating the two couples, i.e.

$$R \times 2 = \text{inertia force} \times \text{CD}$$

$$= 409 \times 4.5$$

$$R = \textbf{920.3 N}$$

However, when the two masses have unequal inertia forces, i.e. when the shaft is not in static balance, the bearing reactions are generally not equal. The reactions must then be found by taking moments about each bearing in turn.

7.7 Car wheel balancing

Present-day motor cars require accurately balanced wheels. If a wheel assembly is out of balance, forces come into play which may affect steering and tyre wear and cause rough running. Even when a wheel and tyre is in static balance it may be dynamically out of balance owing to a heavy spot on one side of the wheel. This dynamic unbalance is apparent in wobble of the wheel when rotated at speed.

A car wheel may be statically balanced as a rotor, except that when the balance mass has been determined it is usual to split it into two parts, placing one part on each side of the wheel. Dynamic balance is achieved by running the wheel at speed in a suitable machine to check the wobble. Balance masses to correct for dynamic unbalance must be placed on opposite sides of the wheel, 180° apart, in order to provide the necessary couple, Fig. 7.10.

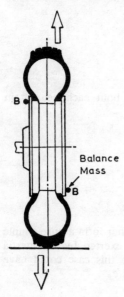

Balance Mass

Fig. 7.10

Problems

1. Two masses revolve together in the same plane at an angular distance 45° apart. The first is a 3-kg mass at a radius of 225 mm, the second 5 kg at 175 mm radius. Calculate the out-of-balance force at 2 rev/s and the position of a 10-kg balance mass required to reduce this force to zero.

(226 N; balance mass at 143 mm radius and 160° 33′ to 5 kg mass)

2. A casting is bolted to the face plate of a lathe. It is equivalent to 2 kg at 50 mm from the axis of rotation, another 1 kg at 75 mm radius and 4 kg at 25 mm radius. The angular positions are, respectively, 0°, 30°, 75°. Find the balance mass required at 150 mm radius to eliminate the out-of-balance force. State the angular position of the balance mass.

(1.56 kg; 215° from 2-kg mass)

3. A turbine casing is placed on a rotating table mounted on a vertical axis. The casing is symmetrical except for a projecting lug of mass 15 kg at a radius of 1.2 m and a cast pad of mass 25 kg at 0.9 m radius. The lug and the pad are positioned at right angles to one another. The casing is bolted down symmetrically with respect to the axis of rotation. Find the magnitude and position of the balance mass required at a radius of 1.5 m.

(19.2 kg at 141° 41' to pad)

4. Two equal holes are drilled in a uniform circular disk at a radius of 400 mm from the axis. The mass of material removed is 187.5 g. Calculate the resultant out-of-balance force if the holes are spaced at 90° to each other and the speed of rotation is 1,000 rev/min.

Where should a mass be placed at a radius of 250 mm in order to balance the disk, and what should be its magnitude?

(582 N; 0.211 kg at 45° to a drilled hole)

5. Three masses are bolted to a face plate as follows: 5 kg at 125 mm radius, 10 kg at 75 mm radius, and 7.5 kg at 100 mm radius. The masses must be arranged so that the face plate is in balance. Find the angular position of the masses relative to the 5-kg mass.

(Each at 114° 36' to 5-kg mass)

6. Four masses, A, B, C and D, rotate together in a plane about a common axis O. The masses and radii of rotation are as follows: A, 2 kg, 0.6 m; B, 3 kg, 0.9 m; C, 4 kg, 1.2 m; D, 5 kg, 1.5 m. The angles between the masses are: ∠AOB = 30°, ∠BOC = 60°, ∠COD = 120°. Find the resultant out-of-balance force at 12 rev/s and the radius of rotation and angular position of a 10-kg mass required for balance.

(21.6 kN; 380 mm, 39° 7' to OA)

7. A cast-steel rotor is 40 mm thick and has four holes drilled at radii 50, 100, 125 and 150 mm, respectively. The holes are 25 mm in diameter and located as shown in Fig. 7.11. Find the magnitude of the total unbalanced force on the spindle bearings at 12 rev/s. Where should a hole of 12 mm diameter be drilled for balance? Steel has a density of 7.8 Mg/m³.

(74 N; at 364 mm radius, 112° 32' clockwise to hole A)

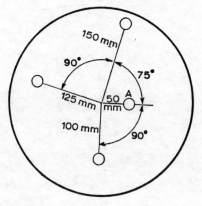

Fig. 7.11

8. A casting of mass 25 kg is bolted to the faceplate of a lathe rotating at 2 rev/s. If the centre of gravity of the casting is offset 25 mm from the axis of rotation and 75 mm in front of the spindle bearing, calculate the out-of-balance couple carried by the bearing.

(7.42 N-m)

9. A mass of 25 kg revolves at a radius of 600 mm. It is mounted on a shaft which runs in two bearings distant 600 and 900 mm, respectively, on either side of its plane of rotation. Calculate the dynamic force on each bearing at a speed of 144 rev/min. What are the *maximum* and *minimum* forces on each bearing?

(1,365 N, 2,050 N; 2,197 N, 1,903 N; 1,463 N, 1,267 N)

10. A shaft rotates in bearings 2 m apart. A mass of 20 kg rotates with the shaft at a radius of 100 mm in a plane 600 mm from the left-hand bearing. A second mass of 12 kg at a radius of 250 mm is in the central plane of the shaft. Both masses are in the same axial plane when viewed from the end of the shaft and on the same side of the shaft. Calculate the dynamic force on each bearing at 5 rev/s.

(2,860 N; 2,071 N)

CHAPTER 8

Periodic Motion

8.1 Periodic motion

When a body moves to and fro so that every part of its motion recurs regularly it is said to have *periodic motion*. For example, in the engine mechanism of Fig. 8.1 (*a*) when crank OC rotates uniformly the piston X moves back and forth between the two limiting points A and B, the motion being repeated at regular intervals of time. It is important to note that neither the velocity nor the acceleration of the piston is uniform and the methods and formulae for uniformly accelerated motion do not apply.

Now consider the motion of piston X more carefully. As X moves

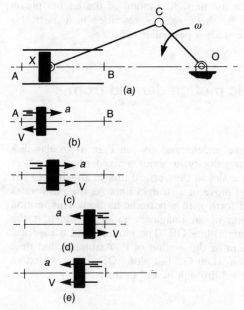

(a)

(b)

(c)

(d)

(e)

Fig. 8.1

towards A its velocity v is from right to left, Fig. 8.1 (*b*). At A it comes instantaneously to rest and reverses direction. Before reaching A it must be slowing down or retarding, i.e. the acceleration *a* of X is from left to right, in the opposite sense to the velocity. After reversing direction X accelerates from rest, both *v* and *a* are from left to right, Fig. 8.1 (*c*). At B the piston comes again instantaneously to rest, hence near B it is retarded and the acceleration *a* is from right to left, Fig. 8.1 (*d*). After reversal of the motion at B both *v* and *a* are from right to left, Fig. 8.1 (*e*). As X reaches A for the second time the whole sequence repeats itself.

The periodic reciprocating motion of the engine piston is complex but is approximately the same as an important but simpler periodic motion termed *simple harmonic motion*. This latter type of motion will now be dealt with in detail.

8.2 Simple harmonic motion

We define simple harmonic motion (s.h.m.) as a periodic motion in which:

1. The acceleration is always directed towards a fixed point in its path.
2. The acceleration is proportional to its distance from the fixed point.

The motion is similar to the periodic motion of the engine piston except that the acceleration has been exactly described in a particular way. In order to fix ideas we study a particular case.

8.3 Simple harmonic motion derived from a circular motion

Fig. 8.2 shows a 'Scotch yoke' mechanism. A pin P in a circular disk rotates at a uniform angular velocity ω about a fixed point O where OP = r. The pin engages in a slot in the vertical link E attached to bar F; the latter is constrained to move in a straight line. As the pin rotates bar F reciprocates back and forth with a periodic motion. This motion corresponds exactly with that of an imaginary point X which is the projection of P on the horizontal line OB. The motion of X is identical with the horizontal component of the motion of P. Assuming that time is measured from the position when OP lies along OB, i.e. $t = 0$ when P is at B, then the angle turned through by OP in time t is

$$\theta = \omega t$$

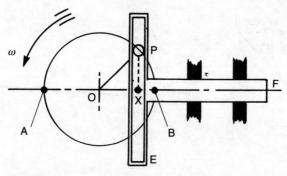

Fig. 8.2

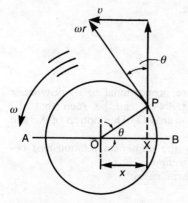

Fig. 8.3

and the displacement of X *measured from the midposition* at time t is OX given by

$$x = r \cos \theta$$

$$= r \cos \omega t$$

The velocity of P is tangent at P to the circle of rotation and its magnitude is ωr. The velocity of X is the horizontal component of the velocity of P (Fig. 8.3), i.e.

$$v = \omega r \sin \theta$$

$$= \omega r \sin \omega t$$

$$= \omega \sqrt{(r^2 - x^2)} \text{ since } \sin \theta = \frac{\sqrt{(r^2 - x^2)}}{r}$$

The centripetal acceleration of P is ωr^2, and is directed radially inwards from P to O, Fig. 8.4. The acceleration of X is the horizontal

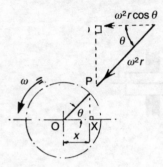

Fig. 8.4

component of the acceleration of P, i.e.

$$a = \omega^2 r \cos \theta$$

$$= \omega^2 r \cos \omega t$$

$$= \omega^2 x \text{ since } x = r \cos \theta$$

The acceleration of X is therefore proportional to its distance x from the fixed point O. From Fig. 8.4 it can be seen that the acceleration of X is *always* directed towards O. The motion of X (and of bar F) is therefore simple harmonic.

The above formulae give merely the numerical relationships between x, v and a, *without regard to direction*.

Note particularly the following special cases:

Extreme positions of s.h.m.
The velocity of X is zero at A and B, Fig. 8.5. At these points the velocity of P is vertical and therefore has no component along AB. The acceleration of X at these two points is the centripetal acceleration of P, Fig. 8.6, i.e.

$$a = \omega^2 r$$

and this is the *maximum* acceleration of the s.h.m.

Thus $v = 0$ and $a = \omega^2 r$ when $\theta = 0°$ and $180°$.

Mid-position of s.h.m.
When P is at C or D, X coincides with O, the mid-point of its path. The velocity of X is that of P, i.e. its velocity reaches its maximum value ωr, Fig. 8.5 i.e. $v = \omega r$ when $\theta = 90°$ or $270°$. The acceleration of P is vertical and since there is no horizontal component the acceleration of X is zero, Fig. 8.6, i.e. $a = 0$, when $\theta = 90°$ or $270°$.

Two important facts are

1. *When the acceleration of X is zero the velocity is a maximum.*
2. *When the velocity of X is zero the acceleration is a maximum.*

This latter statement should not be surprising since s.h.m. requires that the acceleration is proportional to the distance from O, and therefore reaches its maximum value at its greatest distance from O.

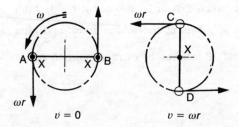

Fig. 8.5 Velocity of X

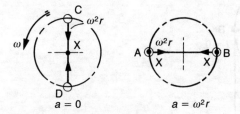

Fig. 8.6 Acceleration of X

8.4 Periodic time

The *periodic time* or *period* t_p of s.h.m. is the time taken for point X to complete one to-and-fro oscillation, i.e. to pass through any point twice in the same direction. This is also the time for the rotating arm OP to sweep out an angle 2π rad at ω rad/s Fig. 8.7. The line OP is said to generate the s.h.m. and ω is sometimes called the *circular frequency*. Thus:

$$\text{period, } t_p = \frac{2\pi}{\omega} \text{ seconds}$$

Also, since $a_x = \omega^2 x$

$$\omega = \sqrt{\left(\frac{a_x}{x}\right)}$$

hence

$$t_p = 2\pi \sqrt{\left(\frac{x}{a_x}\right)}$$

therefore

$$t_p = 2\pi \sqrt{\left(\frac{\text{displacement}}{\text{acceleration}}\right)} \text{ s}$$

In Fig. 8.7 it can be seen that when P is at B, $\theta = 0$ and $t = 0$. After time $t = \dfrac{\pi}{2\omega} = \dfrac{t_p}{4}$, when $\theta = 90°$, P is at the mid-position, and after time $t = \dfrac{\pi}{\omega} = \dfrac{t_p}{2}$, when $\theta = 180°$, P reaches the extreme position A.

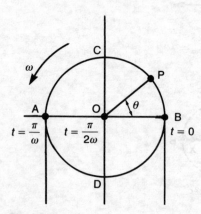

Fig. 8.7

8.5 Frequency

The *frequency* (n) of oscillation is the number of complete cycles, back and forth, made in unit time. The frequency n is therefore the reciprocal of the period t_p. The unit of frequency is the *hertz* (**Hz**) which is one cycle per second. Thus

$$n = \frac{1}{t_p} \text{ cycles per second}$$

$$= \frac{\omega}{2\pi} \text{ Hz}$$

therefore

$$n = \frac{1}{2\pi} \sqrt{\left(\frac{\text{acceleration}}{\text{displacement}}\right)} \text{ Hz}$$

8.6 Amplitude

The distance r through which the point X moves on either side of the fixed point O is called the *amplitude* of the motion. The total distance $2r$ is called the *stroke* or *travel*.

Note: The above results were obtained for the motion of a point X reciprocating to-and-fro due to the rotation of a point P. These results, however, apply to any body performing s.h.m.

Example   *A body moving with s.h.m. has a velocity of 3 m/s when 375 mm from the mid-position and an acceleration of 1 m/s² when 250 mm from the mid-position. Calculate the periodic time and the amplitude.*

SOLUTION
To calculate t_p we must first find ω.

$$a = \omega^2 x$$

therefore

$$\omega = \sqrt{\left(\frac{a}{x}\right)}$$

$$= \sqrt{\left(\frac{1}{0.25}\right)}$$

since

$$a = 1 \text{ m/s}^2 \text{ when } x = 0.25 \text{ m}$$

i.e. $\omega = 2$ rad/s
therefore

$$t_p = \frac{2\pi}{\omega} = \frac{2\pi}{2} = \textbf{3.142 s}$$

To calculate the amplitude r, refer to Fig. 8.3. When $x = 0.375$ m

$$v = 3 \text{ m/s}$$

and $v = \omega\sqrt{(r^2 - x^2)}$

thus $3 = 2\sqrt{(r^2 - 0.375^2)}$

hence $r = \textbf{1.55 m}$

Example   *A body performs s.h.m. in a straight line. Its velocity is 12 m/s when the displacement is 50 mm, and 3 m/s when the displacement is 100 mm, the displacement being measured from the mid-position. Calculate the frequency and amplitude of the motion. What is the acceleration when the displacement is 75 mm?*

SOLUTION

First determine the amplitude r. From

$$v = \omega\sqrt{(r^2 - x^2)}$$

when $x = 0.05$ m, $v = 12$ m/s, and

$$12 = \omega\sqrt{(r^2 - 0.05^2)} \qquad \ldots [1]$$

when $x = 0.1$ m, $v = 3$ m/s, and

$$3 = \omega\sqrt{(r^2 - 0.1^2)} \qquad \ldots [2]$$

Divide equation [1] by equation [2]:

$$\frac{12}{3} = \frac{\sqrt{(r^2 - 0.05^2)}}{\sqrt{(r^2 - 0.1^2)}}$$

Squaring both sides

$$16 = \frac{r^2 - 0.0025}{r^2 - 0.01}$$

hence

$$r = 0.1025 \text{ m} = \textbf{102.5 mm}$$

To find ω, from equation [1]:

$$12 = \omega\sqrt{(0.1025^2 - 0.0025)}$$

hence

$$\omega = 134 \text{ rad/s}$$

thus

$$\text{frequency } n = \frac{\omega}{2\pi}$$

$$= \frac{134}{2\pi}$$

$$= \textbf{21.3 Hz}$$

To find the acceleration when $x = 0.075$ m. From

$$a = \omega^2 x$$

$$a = 134^2 \times 0.075$$

$$= \textbf{1,347 m/s}^2$$

Example   *A body oscillates along a straight line with s.h.m. The frequency is 0.4 Hz and the amplitude 300 mm. Find the displacement of the body 0.3 s after leaving the position of maximum displacement.*

SOLUTION

$$\omega = 2\pi n = 2\pi \times 0.4 = 2.51 \text{ rad/s}$$

Fig. 8.7 shows the geometrical representation of the motion. The displacement x is measured from the midposition. Time is measured from the position of maximum displacement (B) where $t = 0$. When $t = 0.3$ s, the angle θ turned through is

$$\theta = \omega t = 2.51 \times 0.3 = 0.753 \text{ rad}$$
$$= 43.14°$$

hence

$$x = r \cos \theta$$
$$= 300 \cos 43.14°$$
$$= \textbf{219 mm}$$

Example   *A piston is driven by a crank and connecting rod as shown in Fig. 8.8. The crank is 75 mm long and the rod 450 mm. Assuming the acceleration of the piston to be simple harmonic find its velocity and acceleration in the position shown when the crank speed is 360 rev/min clockwise. What is the maximum acceleration of the piston and where does it occur?*

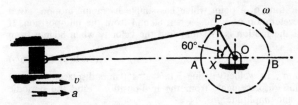

Fig. 8.8

SOLUTION
It can be seen that the motion of the piston is approximately the same as that of point X, the projection of crank-pin P on the line of stroke. The greater the length of the rod compared with that of the crank the more closely does the motion of the piston agree with that of X and therefore of s.h.m.

$$\text{amplitude } r = 0.075 \text{ m}$$
$$\omega = \frac{2\pi \times 360}{60} = 37.7 \text{ rad/s}$$

For the position shown

$$x = \text{OX} = 0.075 \times \cos 60° = 0.0375 \text{ m}$$

velocity of piston,

$$v = \omega\sqrt{(r^2 - x^2)}$$
$$= 37.7\sqrt{(0.075^2 - 0.0375^2)}$$
$$= \textbf{2.45 m/s} \text{ (inwards towards O)}$$

Acceleration of piston,

$$a = \omega^2 x$$
$$= 37.7^2 \times 0.0375$$
$$= \mathbf{53.3\ m/s^2}\ \text{(inwards towards O)}$$

Maximum acceleration of piston,

$$= \omega^2 r$$
$$= 37.7^2 \times 0.075$$
$$= \mathbf{106.6\ m/s^2}$$

The maximum acceleration occurs when X coincides with A or B, i.e. when the piston is at the top or bottom dead centre position. To be exact there is a difference in the accelerations at A and B.

Problems

1. Find the periodic time of a point which has simple harmonic motion, given that it has an acceleration of 12 m/s^2 when 80 mm from the mid-position. If the amplitude of the motion is 110 mm, find the velocity when 80 mm from the mid-position.

(0.513 s; 0.925 m/s)

2. A body has s.h.m., its velocity being 3 m/s at 150 mm displacement, and 2.4 m/s at 225 mm displacement, from the mid-position. Find the periodic time, frequency, and amplitude.

(0.585 s; 1.71 Hz; 316 mm)

3. A body moves with s.h.m. and completes twenty oscillations per second. Its speed at a distance of 25 mm from the centre of oscillation is one-half the maximum speed. Find the amplitude and the maximum acceleration of the body.

(22.8 mm; 454 m/s^2)

4. A body oscillates along a straight line with s.h.m. The amplitude is 300 mm and when the body is at point A, 150 mm from the centre of oscillation, it is moving with a speed of 3 m/s. Calculate the shortest time taken from A to reach a point 260 mm from the centre of oscillation.

(0.0454 s)

5. In a simple crank and connecting rod mechanism the crank is 50 mm long and the connecting rod is 350 mm long. When the crank is 30° from the top dead centre position find the velocity and acceleration of the piston at 10 rev/s. Assume the motion of the piston to be simple harmonic. What is the maximum velocity and acceleration of the piston?

(1.57 m/s; 170.5 m/s^2; 3.14 m/s; 197.4 m/s^2)

6. A body moves with s.h.m. making 20 oscillations per minute. The stroke is 300 mm. Find the velocity and acceleration 0.4 s after passing through the midposition.

(0.21 m/s; 0.49 m/s^2)

7. The maximum velocity of a body moving with s.h.m. is 2.5 m/s and its velocity when passing through a point 300 mm from the midposition is

1.5 m/s. Find the number of oscillations per minute, the amplitude and the maximum acceleration.

(63.7; 375 mm; 16.7 m/s²)

8.7 Dynamics of simple harmonic motion

A simple harmonic motion, or close approximation to it, occurs in many important mechanical problems, e.g. a mass oscillating at the end of a spring, the simple pendulum, the motion of a piston on a connecting rod which is long compared to its crank. In every case it is necessary to *show* that the motion is simple harmonic or to find the degree of approximation involved. In order to do this the procedure is:

1. Write down an 'equation of motion,' i.e. balance the inertia and applied forces.
2. Check if the acceleration of the body is proportional to its displacement from a fixed point or mid-point of its motion.
3. Check if the acceleration is *always* directed towards the fixed point.

Having checked the conditions for s.h.m. we may conclude that the motion is simple harmonic or, alternatively, note the approximation involved in the equation of motion in order to consider the motion as simple harmonic. For example, the motion of a Scotch yoke is exactly s.h.m. whereas that of an engine piston is only very approximately s.h.m. unless the connecting rod is at least six times the crank length.

8.8 The mass and spring

(a) Horizontal motion

Consider a body A of mass m and weight $W = mg$, attached to a light spring of stiffness S, which is anchored at B, Fig. 8.9. The body is constrained to move in horizontal guides, assumed frictionless, and in the rest position it is at point O.

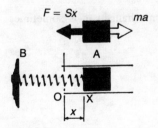

Fig. 8.9

Imagine the body to be pulled to the right a distance r and then released. The pull F in the spring will initially cause the body to move from rest towards the left. When it is a distance x from O the pull F of the spring is from right to left and, since x is also the extension of the spring at this instant:

$$F = Sx$$

This pull is 'balanced' by the inertia force ma, where a is the acceleration of the body. This inertia force must be from left to right and therefore a is from right to left as expected. For equilibrium

$$F = ma$$

or $Sx = ma$

thus

$$a = \frac{S}{m} \times x$$

$$= (\text{constant}) \times x$$

The acceleration of the body is therefore proportional to the distance x from the fixed point O and directed from right to left, i.e. towards O.

When the spring is compressed and the body is to the left of O, then both the spring force and the inertia force are reversed in direction so that acceleration a is still directed towards O. Thus the acceleration is *always* directed towards O. Hence the motion of the body is simple harmonic and the mass is said to move with *free* or *natural* oscillations.

Now compare the expressions:

$$a = \omega^2 x \text{ for s.h.m.}$$

and $a = \dfrac{S}{m} x$ for body A

Evidently

$$\omega^2 = \frac{S}{m}$$

or $\omega = \sqrt{\left(\dfrac{S}{m}\right)}$

By comparison with the Scotch yoke mechanism ω is sometimes called the *equivalent angular velocity*.

The frequency of oscillation is

$$n = \frac{\omega}{2\pi}$$

i.e. $n = \dfrac{1}{2\pi} \sqrt{\left(\dfrac{S}{m}\right)}$

and the period is

$$t_p = \frac{2\pi}{\omega}$$

i.e. $t_p = 2\pi\sqrt{\left(\dfrac{m}{S}\right)}$

(b) Vertical motion

Let the body A be supported vertically by the spring, Fig. 8.10. At rest, the force in the spring is W. Hence the deflexion at rest or *static deflexion* d, is given by:

$$W = S \times d$$

or $\quad d = \dfrac{W}{S} = \dfrac{mg}{S}$

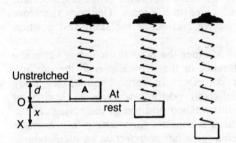
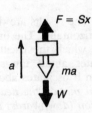

Fig. 8.10

Let the body be pulled a distance r below the static position O and then released. We should expect the body to move upwards from this position with an acceleration towards O due to the upward pull of the spring. At X, a distance x from O, the total extension of the spring is $d + x$. Hence the spring force is

$$F = S(d + x)$$

This force is balanced by the weight W together with the inertia force ma, both acting downwards. Hence

$$F = W + ma$$

or $\quad S(d + x) = W + ma$

but $S \times d = W$, and thus

$$W + Sx = W + ma$$

or $\qquad Sx = ma$

i.e. $\quad a = \dfrac{S}{m}x$

$\qquad = \text{constant} \times x$

The acceleration a is therefore proportional to the distance x from the position of equilibrium O (a fixed point) and always directed towards O. The motion of the body A is therefore simple harmonic.

The period is the same as for horizontal motion, i.e.

$$t_p = 2\pi\sqrt{\frac{m}{S}}$$

but in this case

$$\frac{mg}{S} = d$$

the static deflexion, so that the period may be written:

$$t_p = 2\pi\sqrt{\left(\frac{d}{g}\right)}$$

The amplitude of the motion is the maximum value of the displacement x, i.e. the initial displacement r given to the mass. The body therefore oscillates an equal distance r above and below the static rest position O.

Note: The results are the same whether the mass is oscillating vertically or horizontally. The only difference in the two cases is in the position about which the oscillation takes place. In vertical motion the body oscillates about the static deflected position whereas in horizontal motion it oscillates about the unstretched position of the spring. *In the vertical motion the dead weight is a constant force acting in a constant direction (downwards) and such a force has no effect on an oscillation.*

Example   *A load is suspended from a vertically mounted spring. At rest it deflects the spring 12 mm. Calculate the number of complete oscillations per second.*

If the load's mass is 3 kg what is the maximum force in the spring when it is displaced a further 25 mm below the rest position and then released?

SOLUTION

$$\text{Frequency } n = \frac{1}{2\pi}\sqrt{\frac{g}{d}}$$

$$= \frac{1}{2\pi}\sqrt{\frac{9.8}{0.012}}$$

$$= \textbf{4.54 Hz}$$

If the load is pulled down 0.025 m the amplitude of oscillation, $r = 0.025$ m.

maximum acceleration

$$a_{max} = \omega^2 r$$

$$= (2\pi \times 4.54)^2 \times 0.025$$

$$= \textbf{20.4 m/s}^2$$

Maximum inertia force

$$= ma_{max}$$
$$= 3 \times 20.4$$
$$= \mathbf{61.2\,N}$$

The maximum force in the spring occurs when the mass is at its lowest position. Then the spring force must balance *both* the weight and the inertia force. Hence maximum spring force is:

$$(3 \times 9.8) + 61.2 = \mathbf{90.6\,N}$$

Example *A composite spring has two close-coiled springs A and B in series. A has a stiffness $S_1 = 2000\ N/m$, B, $S_2 = 800\ N/m$. The composite spring carries a mass of weight 50 N and oscillates freely in the vertical direction. Find the frequency of the oscillation.*

SOLUTION
The deflection d of the mass is the sum of the extensions of the two springs, d_1 and d_2, under the action of the same force, i.e. the weight W. Fig. 8.11. Since extension = force/stiffness, therefore

$$d_1 = \frac{W}{S_1} = \frac{50}{2000} = 0.025 \text{ m}$$

$$d_2 = \frac{W}{S_2} = \frac{50}{800} = 0.0625 \text{ m}$$

and $d = d_1 + d_2 = 0.025 + 0.0625 = 0.0875$ m

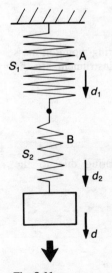

Fig. 8.11

Frequency,

$$n = \frac{1}{2\pi}\sqrt{\frac{g}{d}} = \frac{1}{2\pi}\sqrt{\frac{9.8}{0.0875}} = \textbf{1.68 Hz}$$

Example *An instrument of mass 10 kg is attached to a spring mounted horizontally. The periodic time was observed to be 1.3 s. Find the stiffness of the spring.*

The assembly is now placed in a vehicle with the axis of the spring horizontal and parallel to the longitudinal axis of the vehicle. Fig. 8.12. Find the resulting extension of the spring when the vehicle accelerates smoothly at 4 m/s² and the instrument does not vibrate. Neglect friction.

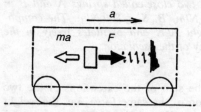

Fig. 8.12

SOLUTION

$$t_p = 2\pi\sqrt{\left(\frac{m}{S}\right)}$$

thus $1.3 = 2\pi\sqrt{\left(\frac{10}{S}\right)}$

and $S = \frac{(2\pi)^2 \times 10}{1.3^2}$

$$= \textbf{234 N/m}$$

The accelerating vehicle will in turn accelerate the spring-mounted mass through the force in the spring. The forces on the instrument are the spring force F, and the inertia force ma. For balance

$$F = ma$$

$$= 10 \times 4$$

$$= 40 \text{ N}$$

but $F = Sx$, where x is the extension of the spring due to the acceleration of the vehicle. Therefore

$$x = \frac{F}{S}$$

$$= \frac{40}{234}$$

$$= 0.171 \text{ m} = \textbf{171 mm}$$

Problems

1. A helical spring, of stiffness 18 kN/m supports a body of mass 25 kg. The body is given a free vibration of amplitude 12 mm. Determine (a) the period of the motion, (b) the maximum acceleration, (c) the maximum velocity.

    (0.234 s; 8.67 m/s²; 0.323 m/s)

2. A mass of 2.5 kg is hung vertically from a spring of stiffness 5.4 kN/m. Calculate the maximum amplitude of vibration if the mass is not to jump from the hook.

    (4.53 mm)

3. A mass of 10 kg is hung from the end of a thin wire. It is found that the static stretch in the wire is 2.6 mm. Calculate the frequency of free vibration when a mass of 14 kg is on the wire.

    (8.26 Hz)

4. An instrument is spring mounted to the body of a rocket, the spring axis lying along the rocket axis. The natural frequency of the instrument upon its mount is 40 Hz. If the rocket is accelerated smoothly to an acceleration ten times that of gravity in such a way that the instrument is not set into vibration find the static deflexion of the instrument on its mounting during the acceleration.

    (1.55 mm)

5. A body of mass 30 kg performs s.h.m. in a straight path, the greatest distance from its mid-position being 750 mm. Calculate the force acting on the body, (a) at the beginning of its travel, (b) mid-way between the end of its path and mid-position, if it makes eighty strokes per minute.

    (395 N; 197.5 N)

6. A spring-loaded slide valve of mass 4 kg opens and closes with s.h.m. It has a lift of 12 mm, equal to twice the amplitude of the s.h.m. If the total time to open and close the valve is 0.1 s, find (a) the stiffness of the spring, (b) the maximum accelerating force exerted by the spring.

    (15.8 kN/m; 95 N)

7. A mass of 1.8 kg is hung vertically on the end of a spring. When set in vibration the frequency is 1.6 Hz. Find the stiffness of the spring.
    If the maximum total extension of the spring during the vibration is 150 mm, what is the amplitude of the vibration?

    (182 N/m; 53 mm)

8. A mass of 6 kg is suspended from a vertically mounted spring. The static extension is 6 mm. A further load of 12 kg is hung from the spring, pulled 10 mm below the equilibrium position and released. Find (a) the period of the resulting oscillation, (b) the maximum spring tension.

    (0.27 s; 274 N)

9. A light shaft is supported horizontally in bearings at its ends and carries a disc of mass 45 kg at the midpoint. When disturbed the shaft and disc vibrate freely and the frequency is measured as 35 Hz and the amplitude 0.4 mm. What is the static deflection at the disc? Calculate the maximum velocity and acceleration of the s.h.m. What is the maximum accelerating force at the midpoint of the shaft? Assume s.h.m. for the disc.

    (0.203 mm; 0.088 m/s; 19.34 m/s²; 870 N)

10. A mass of 8 kg is connected to two springs A and B, Fig. 8.13. The surfaces are smooth and the mass is set oscillating with amplitude 50 mm. If the spring A has stiffness 800 N/m what must be the stiffness of spring B for the frequency to be 2 Hz? What is the maximum force in each spring if both

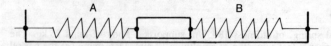

Fig. 8.13

springs are initially unloaded. What is the maximum accelerating force on the mass?

(463 N/m; A, 40 N; B, 23.15 N; 63.15 N)

11. A Scotch yoke mechanism reciprocates with s.h.m. The period is 0.1 s, the amplitude 300 mm, and the mass of the reciprocating parts 30 kg. When the mechanism is 150 mm from the mid-position calculate (*a*) the velocity, (*b*) the acceleration, (*c*) the accelerating force, (*d*) the power delivered to, or returned from, the mechanism at this point. Neglect friction.

(16.3 m/s; 592 m/s^2; 17.76 kN; 290 kW)

8.9 Simple pendulum

A simple pendulum is formed by a concentrated body of mass m and weight W at the end of a light cord of length l suspended at Q, Fig. 8.14. When displaced with the cord taut from the rest position O,

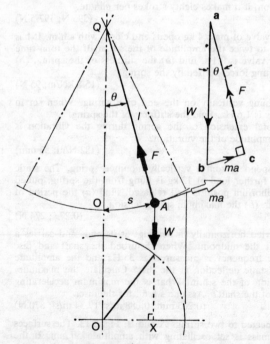

Fig. 8.14 Simple pendulum

the mass moves in an arc about Q. When released from a displaced position A it tends to return to the rest position, i.e. the mass always accelerates towards O from any displaced position along the arc AO. The forces acting on the mass at A are:

1. The weight W, vertically downwards.
2. The tension F in the cord at A, acting from A to Q.
3. The inertia force ma required for balance. This force acts tangentially to the arc at A, i.e. perpendicular to QA.

These three forces are represented in the triangle of forces, **abc**. Thus:

$$\sin \theta = \frac{\mathbf{bc}}{\mathbf{ab}}$$

$$= \frac{ma}{W}$$

$$= \frac{a}{g} \text{ since } W = mg$$

For *small* angular displacements of the cord, we may assume:

$$\sin \theta = \theta \text{ rad}$$

$$= \frac{s}{l}$$

where s = arc OA. Thus

$$\theta = \frac{a}{g}$$

i.e. $a = g\theta$

$$= g \times \frac{s}{l}$$

$$= \frac{g}{l} \times s$$

$$= \text{constant} \times s$$

Thus the acceleration a is directed along the tangent to the arc at A, towards O, and is proportional to the distance s from O, measured along the arc. The motion of the pendulum is therefore *approximately* simple harmonic. It is approximate since we have had to limit θ to small values. However, it turns out to be a very good approximation even when $\theta = 10°$. Other approximations are involved in assuming that the size of the mass is small compared with the length of the cord and that the cord is weightless.

Compare now

$$a = \frac{g}{l}s$$

where a is an acceleration tangent to an arc and s a distance measured along an arc, with

$$a = \omega^2 x$$

where a is directed along a straight line and x is a distance along the straight line. Then

$$\omega^2 = \frac{g}{l}$$

and the period

$$t_p = \frac{2\pi}{\omega}$$

i.e. $t_p = 2\pi\sqrt{\left(\dfrac{l}{g}\right)}$

and frequency $n = \dfrac{1}{2\pi}\sqrt{\left(\dfrac{g}{l}\right)}$

Note that the period is the time for one complete swing, to and fro. It is proportional to the square root of the length of the pendulum and independent of the mass of the suspended body. It does, however, depend on the value of g, the acceleration due to gravity.

8.10 Resonance

If every time an oscillating body reaches the point of maximum displacement it receives an external impulse the amplitude will increase and build up to a maximum value depending on what forces are acting to *damp down* the oscillation. If no damping forces are present the system will finally collapse. The frequency of the external impulse is equal to the frequency of the free oscillation and the effect of the impulse being in unison with the oscillation is called *resonance*. Examples of this phenomenon occur in a wide range of engineering situations including aircraft wings, motor vehicles, machine tools, and in many other fields such as radio and musical instruments. For example, severe vibrations may occur in a drilling machine if it is operated at, or near, the natural frequency of free oscillations of the drill and its holder. Also, if the foundations on which the machine rests vibrates there can be resonance effects on the drill. Rotating machinery has critical speeds which coincide with the natural frequencies of the system. A turbine, for example, starting up from rest may have to pass through one of its natural frequencies before reaching its operating speed, and care must be taken to pass through such speeds as quickly as possible.

Vibrations may be reduced in a number of ways. If the cause is an out-of-balance force this may be eliminated by balancing the machine. Heavy spring mountings to isolate machinery from their foundations and damping devices such as car shock absorbers are employed to reduce the amplitude of oscillations. *Note that the frequency during resonance is the natural frequency of the system.*

Example *A simple pendulum was observed to perform forty oscillations in 100 s, of amplitude 4°. Find (a) the length of the pendulum, (b) the maximum linear acceleration of the pendulum bob, (c) the maximum velocity of the bob, (d) the maximum angular velocity of the pendulum, (e) the velocity of the bob at 2° displacement from the midposition.*

SOLUTION
(*a*) Periodic time

$$t_p = \frac{100}{40} = 2.5 \text{ s}$$

therefore

$$2.5 = 2\pi \sqrt{\left(\frac{l}{9.8}\right)}$$

i.e. $l = \textbf{1.55 m}$

(*b*) Since

$$t_p = \frac{2\pi}{\omega}; \ \omega = \frac{2\pi}{2.5} = 2.51 \text{ rad/s.}$$

Maximum acceleration of the bob occurs at either extreme position when the displacement is a maximum (Fig. 8.15), i.e. when $\theta = 4°$,

$$a_{max} = \omega^2 r$$

where $r = \text{arc OA} = \text{OQ} \times \angle \text{OQA}$

$$= 1.55 \times 4 \times \frac{\pi}{180}$$

$$= 0.1085 \text{ m}$$

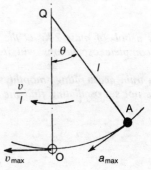

Fig. 8.15

therefore

$$a_{max} = 2.51^2 \times 0.1085$$

$$= \mathbf{0.685 \ m/s^2}$$

(*c*) Maximum linear velocity of the bob occurs when the bob passes the mid-position O, i.e.

$$v_{max} = \omega r$$

$$= 2.51 \times 0.1085$$

$$= \mathbf{0.272 \ m/s}$$

(*d*) The angular velocity of the pendulum is the linear velocity of the bob divided by the length of the pendulum. Therefore the maximum angular velocity occurs when the bob has its maximum linear velocity.

$$\text{Maximum angular velocity} = \frac{v_{max}}{l}$$

$$= \frac{0.272}{1.55}$$

$$= \mathbf{0.175 \ rad/s}$$

(*e*) At 2° displacement from mid-position, $\theta = 2°$, and $x =$ displacement along arc from midposition. Then since θ is small, x is the displacement of the s.h.m., and

$$\frac{x}{l} = 2° \times \frac{\pi}{180} \text{ rad}$$

therefore

$$x = 1.55 \times \frac{2\pi}{180} = 0.054 \text{ m}$$

and $v = \omega \sqrt{(r^2 - x^2)}$

$$= 2.51 \sqrt{(0.1085^2 - 0.054^2)}$$

$$= \mathbf{0.24 \ m/s}$$

Example   *A simple pendulum is formed by a bob of mass 2 kg at the end of a cord 600 mm long. How many complete oscillations will it make per second?*

The same pendulum is suspended inside a train accelerating smoothly along the level at 3 m/s². If the pendulum is not set oscillating find the angle the cord makes with the vertical.

SOLUTION

$$\text{Frequency } n = \frac{1}{2\pi} \sqrt{\left(\frac{g}{l}\right)}$$

$$= \frac{1}{2\pi} \sqrt{\left(\frac{9.8}{0.6}\right)}$$

$$= 0.64 \text{ Hz}$$

When suspended in a smoothly accelerating train the cord makes an angle θ with the vertical, Fig. 8.16. The forces acting on the bob are: the tension F in the cord, the weight $W = 2 \times 9.8 = 19.6 \text{ N}$, and the inertia force

$$ma = 2 \times 3 = 6 \text{ N}$$

The inertia force acts in a direction opposite to that of the acceleration of the train. From the triangle of forces:

$$\tan \theta = \frac{6}{19.6}$$

$$= 0.306$$

therefore

$$\theta = 17°$$

The tension F is given by

$$F^2 = 19.6^2 + 6^2$$

$$= 419$$

hence

$$F = 20.5 \text{ N}$$

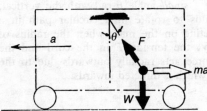
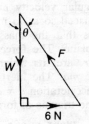

Fig. 8.16

Problems

1. A simple pendulum has a period of 4 s. Find its length.
   If the amplitude is 300 mm find the velocity and acceleration of the bob at 100 mm displacement from the position of equilibrium. What is the maximum velocity and acceleration of the bob?
   (3.97 m; 0.445 m/s; 0.247 m/s^2; 0.47 m/s; 0.742 m/s^2)

2. Calculate the value of g if a simple pendulum of length 2,600 mm makes 100 complete oscillations in 325 s.

(9.72 m/s²)

3. A mass of 1.8 kg suspended from a spring of stiffness 45 N/m is set in oscillation. What length of simple pendulum will have the same frequency of oscillation? What is the frequency?

(392 mm; 0.795 Hz)

4. A 2.25-kg mass hangs from a string of length 900 mm. Calculate the stiffness of spring required to give the same period as the pendulum when carrying the same mass. What is the frequency of oscillation?

The simple pendulum is hung inside a vehicle accelerating smoothly at 2.4 m/s². Calculate the horizontal displacement of the bob if the bob is not set swinging.

(24.5 N/m; 0.525 Hz; 214 mm)

5. A small steel ball runs freely in a groove of radius R in a vertical plane. Show that for small displacements from the equilibrium position the motion of the ball is approximately simple harmonic, of period $T = 2\pi\sqrt{(R/g)}$. Calculate the radius R to give a period of 1 s.

(248 mm)

8.11 Periodic motion of a conical pendulum

Fig. 8.17 shows a body of mass m and weight W suspended by a light arm or cord of length l. If the mass is rotated about a vertical axis at uniform angular velocity ω rad/s and at the same time the cord is slightly displaced so as to make a *small* angle θ rad with the vertical, the result is that the mass tends to rotate in a circular path in a horizontal plane. The forces acting on the mass when the radius of rotation is r, are: its weight W, the tension F in the cord and the inertia force ma. The inertia force acts radially outwards due to the centripetal acceleration $a = \omega^2 r$ which is directed inwards.

From the triangle of forces:

$$\tan \theta = \frac{m\omega^2 r}{W} = \frac{\omega^2 r}{g} \text{ since } W = mg$$

Since θ is small,

$$\tan \theta \cong \theta$$

thus

$$\theta = \frac{\omega^2 r}{g}$$

$$= \frac{\omega^2 l\theta}{g} \text{ since } r = l\theta$$

therefore

$$\omega = \sqrt{\left(\frac{g}{l}\right)}$$

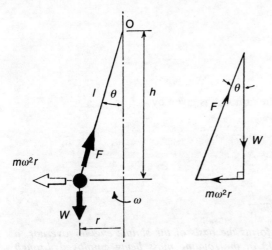

Fig. 8.17 Conical pendulum

This is the *minimum* value of ω at which a small displacement can occur and yet allow the mass to remain in equilibrium. If ω is less than the critical value the cord will remain vertical, if greater than the critical value the bob will start to rise, the cord then sweeping out a conical path.

For a simple pendulum,

$$\omega = \sqrt{\left(\frac{g}{l}\right)}$$

Thus ω for the simple pendulum corresponds exactly with the critical angular velocity of the conical pendulum.

When ω is large, the angle θ is no longer small and the approximation for $\tan \theta$ no longer holds. From the triangle of forces therefore

$$\tan \theta = \frac{\omega^2 r}{g}$$

but $\tan \theta = \frac{r}{h}$

hence $\frac{r}{h} = \frac{\omega^2 r}{g}$

or $h = \frac{g}{\omega^2}$

The distance h of the plane of rotation of the mass from the point of suspension O is known as the 'height' of the pendulum. It is independent of both mass m and length l.

The conical pendulum performs a periodic motion, and its period is

$$t_p = \frac{2\pi}{\omega}$$

but $\omega^2 = \dfrac{g}{h}$

hence

$$t_p = 2\pi\sqrt{\left(\dfrac{h}{g}\right)}$$

The tension F in the arm OA is given by

$$F\cos\theta = W$$

therefore

$$F \times \dfrac{h}{l} = W$$

and $F = W\dfrac{l}{h} = mg\dfrac{l}{h}$

The conical pendulum forms the basis of the simple engine governor, a rise or fall in the height of the rotating mass being employed through suitable linkage to operate a fuel valve.

Example   *A 100-g bob on the end of a light arm forms a simple pendulum of period 1.2 s. The arm is allowed to hang vertically and the bob is then rotated about this vertical axis. At what speed would it start to rise?*

When rotating as a conical pendulum at 20 rad/s what would be the tension in the arm?

SOLUTION
For the simple pendulum

$$\omega = \dfrac{2\pi}{t_p} = \dfrac{2\pi}{1.2} = 5.24 \text{ rad/s} = \textbf{50 rev/min}$$

This is also the speed at which the bob would start to rise in the equivalent conical pendulum. From

$$t_p = 2\pi\sqrt{\left(\dfrac{l}{g}\right)}$$

$$l = \dfrac{gt_p{}^2}{4\pi^2} = \dfrac{9.8 \times 1.2^2}{4\pi^2} = 0.357 \text{ m}$$

When $\omega = 20$ rad/s,

$$\text{height } h = \dfrac{g}{\omega^2} = \dfrac{9.8}{20^2} = 0.0245 \text{ m}$$

Thus tension in cord, Fig. 8.17,

$$F = \dfrac{W}{\cos\theta}$$

$$= \dfrac{mgl}{h}$$

$$= \frac{100 \times 10^{-3} \times 9.8 \times 0.357}{0.0245}$$

$$= \textbf{14 N}$$

Problems

1. A mass hangs from a cord 300 mm long. If the body is rotated about the vertical axis at what speed would it just start to rise?

   When the mass rises 75 mm above the lowest position what is the period of the conical pendulum?

   (54.6 rev/min; 0.953 s)

2. A conical pendulum rotates at 100 rev/min. The cord is 150 mm long and the mass of the bob 1.35 kg. Find (*a*) the amount by which the bob rises above its lowest position, (*b*) the period, (*c*) the tension in the cord.

   (61 mm; 0.6 s; 22.2 N)

3. A simple governor is formed by a pair of balanced rotating spheres, each of mass 3 kg. The link joining each mass to the rotating shaft is 300 mm from the shaft axis to the centre of gravity of the mass. Find (*a*) the change in height when the speed rises from 60 to 70 rev/min; (*b*) the force in the link at 60 rev/min.

   (66 mm; 35.6 N)

4. The arm of a simple conical pendulum is 450 mm long and the rotating mass is 0.9 kg. Find the 'height' of the pendulum and the tension in the arm at 2 rev/s. What *decrease* in speed is necessary to increase the height by 20 per cent? The mass of the arm may be assumed negligible.

   (62 mm; 64 N; 0.17 rev/s)

5. A conical pendulum consists of a 3-kg bob suspended on a string of length 1,500 mm. Find the radius of the circle described by the bob and the maximum speed if the permissible tension in the string is 50 N. What is the acceleration in magnitude and direction at the maximum speed?

   (1,213 mm; 31.8 rev/min; 13.5 m/s^2, radially inwards)

CHAPTER 9
Dynamics of Rotation

9.1 Angular acceleration

If the line OA, Fig. 9.1, rotates about a fixed point O, so that its angular velocity increases from ω_0 in position OA to ω in position OA' in time t then the *average angular acceleration* α of the line is defined as

$$\alpha = \frac{\text{change in angular velocity}}{\text{time taken}}$$

$$= \frac{\omega - \omega_0}{t}$$

The units of α are rad/s^2 if ω is in radians per second and t in seconds.

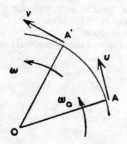

Fig. 9.1

If the angular velocity increases by equal amounts in equal times the acceleration is *uniform*. For uniform acceleration α, we have, by rearranging the above equation:

$$\omega = \omega_0 + \alpha t$$

Now if u and v are the linear velocities of point A when the angular velocities of line OA are ω_0 and ω, respectively, and a is the linear acceleration of point A, i.e. its tangential acceleration, then

$$u = \omega_0 r$$

$$v = \omega r$$

where OA $= r$, and

$$v = u + at$$

thus

$$\omega r = \omega_0 r + at$$

or $\quad \omega = \omega_0 + \dfrac{a}{r} t$

and we have shown that

$$\omega = \omega_0 + \alpha t$$

thus

$$\frac{a}{r} = \alpha$$

or $\quad a = \alpha r$

9.2 Angular velocity–time graph

The relation between angular velocity and time for uniform or constant angular acceleration is given by

$$\omega = \omega_0 + \alpha t$$

and this is represented graphically in the velocity–time graph, Fig. 9.2. For uniform angular velocity the graph is a horizontal line CB, and OC $= \omega_0$.

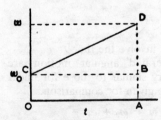

Fig. 9.2

For uniformly accelerated motion the graph is a straight line CD, and AD $= \omega$, OA $= t$. Hence

$$\alpha = \frac{\omega - \omega_0}{t} = \frac{AD - OC}{OA} = \frac{BD}{CB}$$

Thus the angular acceleration α is given by the gradient BD/CB of the graph.

When the acceleration is uniform the *average* angular velocity is simply the mean of the initial and final velocities, or

$$\text{average angular velocity} = \frac{\omega + \omega_0}{2}$$

But $\text{average velocity} = \dfrac{\text{angle turned through}}{\text{time taken}} = \dfrac{\theta}{t}$

hence

$$\frac{\theta}{t} = \frac{\omega + \omega_0}{2}$$

or $\theta = \frac{1}{2}(\omega + \omega_0) \times t$

But $\omega = \omega_0 + \alpha t$, so

$$\theta = \frac{1}{2}[(\omega_0 + \alpha t) + \omega_0] \times t$$

thus $\theta = \omega_0 t + \frac{1}{2}\alpha t^2$

This gives the angle turned through in time t. Again since

$$\theta = \frac{1}{2}(\omega + \omega_0)t$$

and $t = \dfrac{\omega - \omega_0}{\alpha}$

therefore

$$\theta = \frac{1}{2}(\omega + \omega_0) \times \frac{(\omega - \omega_0)}{\alpha}$$

$$= \frac{\omega^2 - \omega_0^2}{2\alpha}$$

and rearranging

$$\omega^2 = \omega_0^2 + 2\alpha\theta$$

This is a useful equation in that it does not involve the time t.

The equations for uniformly accelerated angular motion are summarized below and, since they are very similar to those for linear motion, the equations for linear motion are given for comparison.

$v = u + at$	$\omega = \omega_0 + \alpha t$
$s = \dfrac{u + v}{2}t$	$\theta = \dfrac{\omega_0 + \omega}{2}t$
$s = ut + \frac{1}{2}at^2$	$\theta = \omega_0 t + \frac{1}{2}\alpha t^2$
$v^2 = u^2 + 2as$	$\omega^2 = \omega_0^2 + 2\alpha\theta$

9.3 Use of ω–t graph

Many problems are conveniently solved by making use of the fact that the area under the ω–t graph is equal to the angle turned through. Fig. 9.3(a) shows the graph CD for uniformly accelerated motion:

$$\text{area under } \omega\text{–}t \text{ graph} = \text{area OADC}$$
$$= \text{area OABC} + \text{area CBD}$$
$$= \text{OC} \times \text{OA} + \tfrac{1}{2} \times \text{CB} \times \text{BD}$$
$$= \omega_0 t + \tfrac{1}{2} \times t \times (\omega - \omega_0)$$
$$= \omega_0 t + \tfrac{1}{2} t \times \alpha t, \text{ since } \omega - \omega_0 = \alpha t$$
$$= \omega_0 t + \tfrac{1}{2} \alpha t^2$$
$$= \theta$$
$$= \text{angle turned through in time } t$$

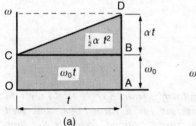 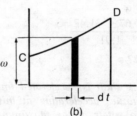

(a)   (b)

Fig. 9.3

In the general case when the motion is not uniformly accelerated suppose CD, Fig. 9.3(b) represents the ω–t graph. Since

$$\frac{d\theta}{dt} = \omega$$

then

$$d\theta = \omega \, dt$$

and

$$\theta = \int \omega \, dt$$

$$= \text{area under } \omega\text{–}t \text{ graph}$$

Example *The speed of a shaft increases from 300 to 360 rev/min while turning through eighteen complete revolutions. Calculate (a) the angular acceleration; (b) the time taken for this change.*

SOLUTION

(a) $\omega_0 = 300 \times \dfrac{2\pi}{60} = 31.42$ rad/s

   $\omega = 360 \times \dfrac{2\pi}{60} = 37.67$ rad/s

   $\theta = 18 \times 2\pi = 113$ rad

Using

   $\omega^2 = \omega_0{}^2 + 2\alpha\theta$

   angular acceleration, $\alpha = \dfrac{\omega^2 - \omega_0{}^2}{2\theta}$

   $= \dfrac{37.67^2 - 31.42^2}{2 \times 113}$

   $= \mathbf{1.91}$ **rad/s**^{2}

(b) $\omega = \omega_0 + \alpha t$

thus

   $t = \dfrac{\omega - \omega_0}{\alpha}$

   $= \dfrac{37.67 - 31.42}{1.91}$

   $= \mathbf{3.27}$ **s**

Example   *A shaft is accelerated uniformly from 8 rev/s to 14 rev/s in 2 s. It continues accelerating at this rate for a further 4 s, and then continues to rotate at the maximum speed attained. What is the time taken to complete the first 200 revolutions?*

SOLUTION

The speed–time graph is shown in Fig. 9.4. ABC represents the uniform acceleration for 6 s; CD represents the motion at constant (maximum) speed for time *t* s. It is convenient in this case to measure the vertical ordinate in revolutions per second, then the area under the graph gives directly the number of revolutions turned through.

Since ABC is a straight line the maximum speed *n* rev/s at C is given by proportion from similar triangles:

   $\dfrac{\text{CF}}{\text{BG}} = \dfrac{\text{AF}}{\text{AG}}$

or $\dfrac{n - 8}{14 - 8} = \dfrac{6}{2}$

thus

   $n = 26$ rev/s

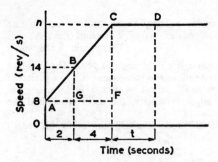

Fig. 9.4

The angle turned through during the first 6 s

= average speed × time taken

$$= \frac{8 + 26}{2} \times 6$$

= 102 rev

Therefore revolutions to be turned through at uniform speed = 200 − 102 = 98. Thus time to turn through 98 rev at maximum speed

$$= \frac{\text{no. of revolutions}}{\text{speed}}$$

$$= \frac{98}{26}$$

= 3.77 s

Total time taken to turn through 200 rev

= 6 + 3.77

= **9.77 s**

Problems

1. The speed of an electric motor rises from 1,430 to 1,490 rev/min in 0.5 s. Find the average angular acceleration and the number of revolutions turned through in this time.

(12.57 rad/s²; 12.2 rev)

2. A flywheel 1.2 m in diameter is uniformly accelerated from rest and revolves completely sixty times in reaching a speed of 120 rev/min. Find (a) the time taken, (b) the angular acceleration, (c) the linear acceleration of a point on the rim.

(60 s; 0.2095 rad/s²; 0.126 m/s²)

3. After the power to drive a shaft is shut off it is seen to describe 120 rev in the first 30 s, and finally comes to rest in a further 30 s. If the retardation is

uniform calculate the initial angular velocity in revolutions per minute (rev/min) and the retardation in radians per second per second (rad/s².)

(320 rev/min; 0.558 rad/s²)

4. A swing bridge has to be turned through a right angle in 140 s. The first 60 s is a period of uniform angular acceleration; the subsequent 40 s is a period of uniform angular velocity and the third period of 40 s a uniform angular retardation. Find the maximum angular velocity, the acceleration and the retardation.

(0.0175 rad/s; 0.00029 rad/s²; 0.000437 rad/s²)

9.4 Dynamics of a rotating particle

In the same way as a change of linear motion requires a force, we shall find that a change of angular motion requires a driving torque.

Let a concentrated body of mass m be attached to the end A of a light arm OA of length r, pivoted at O, Fig. 9.5. OA will rotate freely without the application of a force, provided there is no friction at the pivot. In order to start the rotation, or accelerate the mass, a force F is required at A perpendicular to OA. If a torque is applied to the arm then this force F is provided by the connexion of the arm to the mass.

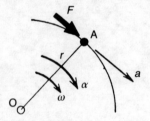

Fig. 9.5

If α is the angular acceleration at the instant considered and a is the linear acceleration of the mass tangent to the circle of motion, then the force required at A is

$$F = ma$$

$$= m\alpha r \text{ since } a = \alpha r$$

The moment of force F about O is the torque T required, i.e.

$$T = F \times r$$

$$= m\alpha r \times r$$

$$= mr^2\alpha$$

The term mr^2 is a most important quantity and is known as the *second moment* of the mass about O, or its *moment of inertia*, denoted by I. The units of I are kg-m^2. The radius r (for a concentrated mass) is called the *radius of gyration* of the mass about O, denoted by k. Thus, for example, if $m = 4$ kg, $r = 2$ m and $\alpha = 3$ rad/s^2, then

$$\text{torque} = mr^2\alpha$$
$$= 4 \times 2^2 \times 3$$
$$= 48 \text{ N-m}$$

Note: The formula $T = mr^2\alpha$ is directly applicable to a thin ring of mean radius r rotating about its axis, since every particle of the ring may be considered as concentrated at the same distance r from the axis of rotation.

Example *A cast-iron pulley is 200 mm wide and 25 mm thick with a mean diameter of 2 m. Considering the pully as a thin ring find the moment of inertia. What torque is required to produce a pulley speed of 5 rev/s in 15 s? Density of cast-iron = 7.2 Mg/m³.*

SOLUTION

Volume of material = mean circumference × thickness × width
$$= (\pi \times 2) \times 0.025 \times 0.2$$
$$= 0.01\pi \text{ m}^3$$

Mass, m = density × volume
$$= 7.2 \times 1,000 \times 0.01\pi$$
$$= 226 \text{ kg}$$

$I = mr^2$
$$= 226 \times 1^2$$
$$= \textbf{226 kg-m}^2$$

$\alpha = \dfrac{2\pi \times 5}{15}$
$$= 2.09 \text{ rad/s}^2$$

$T = I\alpha$
$$= 226 \times 2.09$$
$$= \textbf{472 N-m}$$

Problems

1. A mass of 250 g is mounted on the end of a light arm 240 mm long. The arm is accelerated uniformly from rest to 3,000 rev/min in 20 s. Find the torque required.

(0.226 N-m)

2. A light arm 500 mm long, pivoted at its centre, carries a 10 kg mass at *each* end. If a couple of 3 N-m is applied to the arm calculate the angular acceleration produced.

(2.4 rad/s²)

3. A flywheel is made up of a thin ring 20 mm thick and 150 mm wide, with a mean diameter of 1.5 m. Calculate the time taken to come to rest from 10 rev/s due to a friction couple of 8 N-m. The density of steel is 7.8 Mg/m³ and the effects of the spokes may be neglected.

(488 s)

9.5 Dynamics of a rotating body

Consider a body of total mass m accelerated about a fixed axis O, Fig. 9.6. It is required to find the torque to give the body an angular acceleration α about O. It is necessary to find the torque required to accelerate each particle such as A of mass dm and add these elementary torques to determine the total torque. The force required at A is

$$dF = dm f_A, \text{ normal to OA}$$

where a_A is the linear acceleration of A, tangent to its circular path of motion.

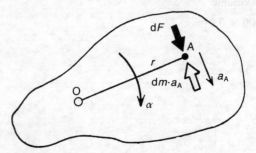

Fig. 9.6

The torque required for the acceleration of A is

$$dT = dF \times r$$

$$= dm.a_A.\times r$$

where r is the radius of rotation of A about O. The total torque T is found by summation of all the elementary torques dT for the whole body. Therefore

$$T = \int dT$$

$$= \int dm \, .a_A.r$$

But a_A differs for every particle, since it depends on radius r, hence we write

$$a_A = \alpha r$$

thus

$$T = \int dm \, \alpha r \times r$$

and since angular acceleration α is the same for every line such as OA, it is therefore a constant at the instant considered, then

$$T = \alpha . \int dm \, r^2$$

or $T = I\alpha$

where

$$I = \int dm \, r^2$$

I is the *second moment* or *moment of inertia* of the whole mass of the body about the axis of rotation O. I is the summation of the quantities (mass) times (radius)^2 for all the particles of the body and depends on the shape and size of the body. Its value depends on the distribution of the mass as well as on the total mass.

It is useful to imagine all the mass of a body concentrated at a particular radius k from the axis O such that the moment of inertia of the concentrated mass is the same as that of the actual body. Thus the *radius of gyration k* is defined by

$$mk^2 = I$$

and the units of I are kg-m^2.

9.6 Inertia couple

Comparing the formulae

$$F = ma$$

and $T = I\alpha$

it is seen that moment of inertia I plays the same part in a change of angular motion as mass m does in change of linear motion. By analogy with the idea of inertia force we may regard the torque T as being balanced by an *inertia couple* $I\alpha$, whose sense is *opposite* to that of the angular acceleration α, Fig. 9.7(a). The problem is then in effect reduced to a static one.

The reality of the effect of an inertia couple will be appreciated by anyone who has tried to accelerate a bicycle wheel rapidly by hand. Although the weight may be carried wholly by the bearings an effort is required to set the wheel spinning. An inertia couple is, of course, *reactive*.

9.7 Accelerated shaft

Consider a shaft (Fig. 9.7(b)) carrying a rotor having a moment of inertia I about the shaft axis. If the bearing friction is equivalent to a couple T_f then, in order to accelerate the shaft and rotor, the driving torque T must balance both the inertia couple $I\alpha$ and the friction couple T_f. Thus:

$$T = I\alpha + T_f$$

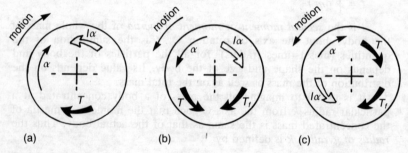

(a)   (b)   (c)

Fig. 9.7

9.8 Shaft being brought to rest

If the shaft is being brought to rest by a braking torque T the friction couple T_f assists the braking action so that T and T_f together must balance the inertia couple $I\alpha$; α is now a retardation, its sense being opposite to that of the motion (Fig. 9.7(c)). Thus:

$$T + T_f = I\alpha$$

If there is no braking torque, the friction couple alone brings the shaft to rest, then

$$T_f = I\alpha$$

Note, in both cases, that

(a) the friction couple T_f opposes the motion, and
(b) the inertia couple $I\alpha$ opposes the *change* of motion.

9.9 Units

Consider the formula

$$T = I\alpha$$

In SI the units of these quantities are: T N-m, I kg-m^2, and α rad/s^2. Thus

$$T(\text{N-m}) = I(\text{kg-m}^2) \times \alpha(\text{rad/s}^2)$$

and since one newton equals one kilogram-metre per second squared, i.e. $1\,\text{N} = 1\,\text{kg-m/s}^2$, and the radian is a number, it can be seen that the units on both sides of the equation agree. The units of T and I may be given in other forms. For example T may be given as kN-m and I as tonne-m^2. In all such cases it is advisable to reduce all quantities to basic units.

9.10 Values of *I* for simple rotors

The derivation of formulae for the mass moment of inertia is left as an exercise in mathematics. Formulae for the moment of inertia I for cylinders and discs about a longitudinal axis are given here without proof.

1. Solid disc or cylinder

For a uniform solid circular cylinder of diameter d, the moment of inertia about the axis O-O, Fig. 9.8(a), is

$$I = m\,\frac{d^2}{8}$$

where m is the mass of the cylinder. By comparison with the formula

$$I = m\,k^2$$

we see that the radius of gyration k for a solid cylinder is given by

$$k^2 = \frac{d^2}{8}$$

$$= \frac{(2r)^2}{8}$$

where r = radius of cylinder; i.e.

$$k = \frac{r}{\sqrt{2}} \text{ or } 0.707r$$

Thus the radius of gyration of a uniform solid circular cylinder is about 0.71 times the radius of the cylinder.

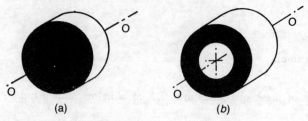

Fig. 9.8

2. Hollow circular cylinder

A hollow circular cylinder is formed by removing from a solid cylinder a concentric cylinder of smaller radius, Fig. 9.8(b). If suffixes 1 and 2 denote outer and inner cylinders, respectively, then for a hollow cylinder the I about longitudinal axis O-O is given by subtraction, thus

$$I = I_1 - I_2$$

$$= m_1 \frac{d_1{}^2}{8} - m_2 \frac{d_2{}^2}{8}$$

where m_1 is the mass of the 'solid' outer cylinder and m_2 that of the inner cylinder.

If ρ is the mass of material per unit volume (density) and l the length of the cylinder, then

$$m_1 = \rho \times \frac{\pi d_1{}^2}{4} \times l$$

$$m_2 = \rho \times \frac{\pi d_2{}^2}{4} \times l$$

and if m is the mass of the actual hollow cylinder, then

$$m = m_1 - m_2$$

$$= \rho \times \frac{\pi(d_1{}^2 - d_2{}^2)}{4} \times l$$

Thus

$$I = \rho \left[\frac{\pi d_1{}^2}{4} l \times \frac{d_1{}^2}{8} \right] - \rho \left[\frac{\pi d_2{}^2}{4} l \times \frac{d_2{}^2}{8} \right]$$

$$= \rho \frac{\pi}{4}(d_1{}^4 - d_2{}^4)\frac{l}{8}$$

$$= \left[\frac{\rho\pi(d_1{}^2 - d_2{}^2)}{4} l \right] \frac{(d_1{}^2 + d_2{}^2)}{8}$$

$$= m \frac{(d_1{}^2 + d_2{}^2)}{8}$$

This formula gives I in terms of the mass m of the actual hollow cylinder. The corresponding radius of gyration k is given by

$$k^2 = \frac{d_1^2 + d_2^2}{8}$$

Note particularly the *positive* sign in the formulae for I and k.

Example *A steel cylinder of 500 mm outside diameter and 200 mm inside diameter is set in rotation about its axis. If the cylinder is 900 mm long, of density 7,800 kg/m³, calculate the torque required to give it an angular acceleration of 0.5 rad/s².*

SOLUTION

Mass of cylinder, m = volume × density

$$= \frac{\pi}{4}(0.5^2 - 0.2^2) \times 0.9 \times 7,800$$

$$= 1,155 \text{ kg}$$

$$I = m\frac{(d_1^2 + d_2^2)}{8}$$

$$= 1,155 \times \frac{(0.5^2 + 0.2^2)}{8}$$

$$= 41.8 \text{ kg-m}^2$$

Torque required, $T = I\alpha$

$$= 41.8 \times 0.5$$

$$= \textbf{20.9 N-m}$$

Example *A flywheel, together with its shaft, has a total mass of 300 kg, and its radius of gyration is 900 mm. If the effect of bearing friction is equivalent to a couple of 70 N-m, calculate the braking torque required to bring the flywheel to rest from a speed of 12 rev/s in 8 s.*

SOLUTION

12 rev/s = $12 \times 2\pi$ = 75.4 rad/s

thus retardation $\alpha = \dfrac{\omega}{t} = \dfrac{75.4}{8} = 9.42 \text{ rad/s}^2$

I of flywheel and shaft = $mk^2 = 300 \times 0.9^2 = 243 \text{ kg-m}^2$

Inertia couple = $I\alpha = 243 \times 9.42 = 2,290 \text{ N-m}$

The braking torque T together with the friction couple of 70 N-m are in equilibrium with the inertia couple, i.e. together they bring the shaft to rest.

$$T + 70 = 2,290$$

thus $\quad T = 2,220 \text{ N-m} = \textbf{2.22 kN-m}$

Problems

1. A flywheel has a moment of inertia of 10 kg-m². Calculate the angular acceleration of the wheel due to a torque of 8 N-m if the bearing friction is equivalent to a couple of 3 N-m.

$$(0.5 \text{ rad/s}^2)$$

2. A light shaft carries a disc 400 mm in diameter, 50 mm thick, of steel (density 7,800 kg/m³). Calculate its moment of inertia about an axis through the centre of the disc and perpendicular to the plane of the disc.

   What torque would be required to accelerate the disc from 60 to 120 rev/min in 1 second, neglecting friction?

   If a friction torque of 1.5 N-m acts, what braking torque would be required to bring the disc to rest from 60 rev/min in 1 second?

$$(0.98 \text{ kg-m}^2; \ 6.16 \text{ N-m}; \ 4.66 \text{ N-m})$$

3. The rotor of an electric motor of mass 200 kg has a radius of gyration of 150 mm. Calculate the torque required to accelerate it from rest to 1,500 rev/min in 6 seconds. Friction resistance may be neglected.

$$(118 \text{ N-m})$$

4. A light shaft carries a turbine rotor of mass 2 tonne and a radius of gyration of 600 mm. The rotor requires a uniform torque of 1.2 kN-m to accelerate it from rest to 6,000 rev/min in 10 min. Find (a) the friction couple, (b) the time taken to come to rest when steam is shut off.

$$(446 \text{ N-m}; \ 16.9 \text{ min})$$

5. A drum rotor is a *thin* cylinder of 1.2 m diameter, 5 mm thick, and 600 mm long. The material is mild steel of density 7.8 Mg/m³. Calculate the moment of inertia of the rotor about the polar axis.

   Find the time taken for the rotor to reach a speed of 3,600 rev/min from rest if the driving torque is 55 N-m and the friction torque is 5 N-m.

$$(31.8 \text{ kg-m}^2; \ 240 \text{ s})$$

6. The flywheel of an engine consists essentially of a thin cast-iron ring of mean diameter 2 m. The cross-section of the ring is 50 mm by 50 mm. Calculate the moment of inertia of the flywheel and find the change in speed of the flywheel if a constant torque of 110 N-m acts on it for 5 seconds. Density of cast iron = 7,850 kg/m³.

$$(123.5 \text{ kg-m}^2; \ 42.4 \text{ rev/min})$$

7. A winding drum of mass 200 tonne has a radius of gyration of 3 m. Find the constant torque required to raise the speed from 40 to 80 rev/min in 60 seconds if the friction torque is 15 kN-m.

   If the wheel is rotating freely at 80 rev/min and a brake is applied bringing it to rest in 120 rev find the brake torque assuming uniform retardation.

$$(140.5 \text{ kN-m}; \ 68.8 \text{ kN-m})$$

8. A shaft carrying a rotor rotates at 3 rev/s. When the driving torque is removed the speed drops to 1.5 rev/s in 6 min due to the braking action of the bearing friction alone. If the rotor's mass is 810 kg and the radius of gyration is 500 mm, find the average value of the friction couple at the bearings.

   If the shaft is of 150 mm diameter and is supported in journal bearings what is the average value of the coefficient of friction at the bearing surface?

$$(5.3 \text{ N-m}; \ 0.0089)$$

9. The rotating table of a vertical boring machine has a mass of 690 kg and a radius of gyration of 700 mm. Find the torque required to accelerate the table to 60 rev/min in three complete revolutions from rest.

$$(354 \text{ N-m})$$

9.11 The hoist

Hoist and pulley problems on connected bodies were dealt with in Chapter 5 but no account was taken there of the rotational inertia of the drum or pulley. We shall now study the effect of combining a hoist drum of moment of inertia I with a hanging load of mass m and weight $W = mg$. Two cases will be considered, according as the load is rising and being accelerated under a driving torque and the load falling and being decelerated by a braking torque. Whatever the case two equations can be written down:

(a) the equation for the balance of couples at the hoist drum;
(b) the equation for the balance of forces at the load.

If the hoist drum radius is r a third equation connecting the angular acceleration α of the drum and the linear acceleration a of the load can be written down, i.e.

$$a = \alpha r$$

The rules for solving problems on hoists are:

1. The friction couple at the drum always opposes the rotation.
2. The inertia couple on the drum always opposes the *change* in rotation.
3. The inertia force on a load always opposes the *change* in linear motion.
4. The direction of the rope tension is always downwards at the drum, upwards on the load, but its effect may be to accelerate or retard.
5. The direction of the weight W is always vertically downwards, but its effect may be to accelerate or retard.

Case 1. Load raised, accelerating upwards, Fig. 9.9(a)
Notes:

1. Acceleration a is upwards, hence the inertia force is downwards.
2. Angular acceleration α is anticlockwise, hence the inertia couple is clockwise.
3. Rotation of the drum is anticlockwise hence the friction couple acts clockwise.

For rotation of the hoist drum the *driving* torque T must balance the friction couple T_f, the inertia couple $I\alpha$, and the torque $E \times r$ due to the tension E in the rope at the drum. Thus

$$T = T_f + I\alpha + E \times r$$

For the load, the tension E in the rope at the load must balance both the dead weight and the inertia force ma, thus

$$E = W + ma$$

Note that if the driving torque T is suddenly withdrawn and no braking torque is applied the load continues to rise accelerating upwards but at

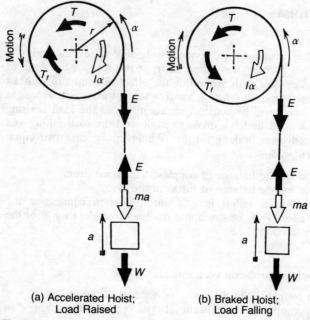

(a) Accelerated Hoist;
Load Raised

(b) Braked Hoist;
Load Falling

Fig. 9.9

a reducing rate as friction acts (until the load hits the drum) to bring
the system to rest. In this case the total retarding torque is $E \times r + T_f$
and for balance at the drum,

$$E \times r + T_f = I\alpha$$

and for balance at the load, since the inertia force now acts upwards

$$E = W - ma$$

Case 2. Load falling and being brought to rest, Fig 9.9(b)
As the load falls and brakes are applied to bring it to rest the drum and
load decelerate. The drum is rotating clockwise therefore the accelerations are as shown. The angular acceleration α of the drum is
anticlockwise, and the linear acceleration a of the load is upwards. The
accelerating torque on the drum is due to the rope tension and the
inertia couple. The braking torque T is assisted by the friction couple.
For balance,

$$T + T_f = Er + I\alpha$$

The retarding force on the load is the rope tension and this is
balanced by the weight and inertia force, thus

$$E = W + ma$$

Note that in the case of a *runaway* hoist, the load falls freely,
accelerating downwards, since no braking torque is applied. The

accelerating torque on the drum is Er due to the rope tension and the total retarding torque is the friction couple together with the inertia couple, so that for balance of couples

$$Er = T_f + I \alpha$$

The retarding force on the load is the rope tension assisted by the inertia force which acts upwards, and the accelerating force is the weight, hence

$$E + ma = W$$

Example   *A hoist drum has a moment of inertia of 85 kg-m² and is used to raise a lift of mass 1 tonne with an upward acceleration of 1.5 m/s². The drum diameter is 1 m. Determine (a) the torque required at the drum; (b) the angular speed of the drum after 3 s from rest.*

SOLUTION

(a) The torque required at the hoist drum is made up of three parts:

1. Torque $I\alpha$, required to accelerate the drum.
2. Torque $W \times r$, required to hold the dead weight of the lift.
3. Torque $ma \times r$, required to accelerate the lift.

$m = 1,000$ kg; $W = mg = 1,000 \times 9.8 = 9,800$ N; $I = 85$ kg-m²

$$\alpha = \frac{a}{r} = \frac{1.5}{0.5}$$

$$= 3 \text{ rad/s}^2$$

Thus,

$$\text{total torque} = I\alpha + W \times r + ma \times r$$
$$= (85 \times 3) + (9,800 \times 0.5) + (1,000 \times 1.5 \times 0.5)$$
$$= \textbf{5,905 N-m}$$

(b) After 3 seconds, the lift speed

$$v = at$$

$$= 1.5 \times 3$$

$$= 4.5 \text{ m/s}$$

and this is the speed of the drum circumference. Therefore angular velocity of the drum

$$\omega = \frac{v}{r} = \frac{4.5}{0.5} = \textbf{9 rad/s}$$

Example   *A wagon of mass 12 tonne is lowered down an incline of 1 in 20 by means of a cable, parallel to the incline, wrapped round a drum at the top of the slope. The hoist drum has a mass of 450 kg, a radius of gyration of 750 mm and effective diameter 2 m. The mass of the cable* *may be neglected, but the friction couple at the drum bearings is 1.6 kN-m, and the resistance to motion of the wagon is 1.4 kN. Find the braking torque on the hoist drum to bring the wagon to rest from 25 km/h in 7 m. The rotational inertia of the wagon wheels may be neglected.*

SOLUTION

$$25 \text{ km/h} = 6.94 \text{ m/s}$$

$$\text{retardation, } a = \frac{v^2}{2s} = \frac{6.94^2}{2 \times 7} = 3.44 \text{ m/s}^2$$

Thus the angular retardation of the drum

$$\alpha = \frac{a}{r} = \frac{3.44}{1} = 3.44 \text{ rad/s}^2$$

The pull E in the cable together with the tractive resistance of 1.4 kN balances the resolved part of the weight down the incline ($12g/20$ kN) together with the inertia force due to the retardation of the wagon, ma, Fig. 9.10. Thus:

$$E + 1,400 = \frac{12 \times 1,000 \times 9.8}{20} + 12 \times 1,000 \times 3.44$$

and $E = 45,760$ N

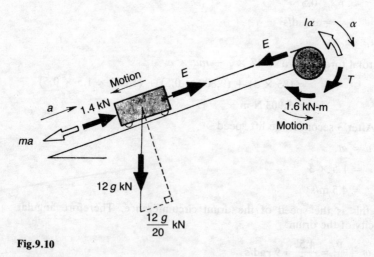

Fig.9.10

The braking torque T applied to the drum with the friction couple of 1.6 kN-m together balance the torque $E \times r$ due to the pull E in the cable and the inertia couple $I\alpha$ due to the retardation of the drum, thus:

$$T + 1,600 = E \times r + I\alpha$$
$$= 45,760 \times 1 + 450 \times 0.75^2 \times 3.44$$
$$= 46,630 \text{ N-m}$$

therefore $\quad T = $ **45.03 kN-m**

Problems

1. A load of mass 8 tonne is to be raised with a uniform acceleration of 1.1 m/s^2 by means of a light cable passing over a hoist drum of 2 m diameter. The drum has a mass of 1 tonne and a radius of gyration of 750 mm. Find the torque required at the drum if friction is neglected.

   (87.8 kN-m)

2. A mine cage of mass 4 tonne is to be raised with an acceleration of 1.5 m/s^2 using a hoist drum of 1.5 m diameter. The drum's mass is 750 kg and its radius of gyration is 600 mm. The effect of bearing friction is equivalent to a couple of 3 kN-m at the hoist drum. What is the torque required on the drum? If the driving torque ceases when the load is moving upwards at 6 m/s, find the deceleration of the load and how far it travels before coming to rest?

   (37.44 kN-m; 9.64 m/s^2 1.87 m)

3. A hoist drum has a mass of 360 kg and a radius of gyration of 600 mm. The drum diameter is 750 mm. A mass of 1 tonne hangs from a light cable wrapped round the drum and is allowed to fall freely. If the friction couple at the bearings is 2.7 kN-m calculate the runaway speed of the load after falling for 2 seconds from rest.

   (2.69 m/s)

4. The maximum allowable pull in a hoist cable is 200 kN. Calculate the maximum load in tonnes which can be brought to rest with a retardation of 5 m/s^2. The hoist drum has a moment of inertia of 8400 kg-m^2 and a diameter of 2.4 m. What is the corresponding braking torque on the drum?

   (13.52 tonne; 275 kN-m)

5. In an experiment, a flywheel is mounted on a shaft 50 mm diameter supported in bearings. Around the shaft is wrapped a light cord from which is hung a mass of 2 kg. When allowed to fall and rotate the flywheel the load falls 2 m from rest in 3 seconds. The friction couple is 0.35 N-m. Find the moment of inertia of the flywheel.

   (0.00662 kg-m^2)

6. A load of mass 250 kg is lifted by means of a rope which is wound several times round a 1 m diameter drum and which then supports a balance mass of 150 kg. As the load rises the balance mass falls. Neglecting friction and the inertia of the drum, find the torque required on the drum to give the load an upward acceleration of 1.4 m/s^2. If the drum has a mass of 80 kg and a radius of gyration of 600 mm, find the torque to accelerate the drum only.

   (771 N-m; 81 N-m)

Appendix to Chapter 9

Gravitation: satellites

The pull of gravity obeys *Newton's Law of Gravitation* which states that a body of mass m at a distance R from the centre of the earth is under a gravitational force F given by

$$F = \frac{G.m.m_e}{R^2}$$

where m_e is the mass of the earth and G a universal gravitational constant. This law of gravitation means that the force of gravity is proportional to the mass m of the body and inversely proportional to the square of the distance of the body from the earth's centre. The force of gravity is experienced as the *weight* of the body. Thus the weight of the body is inversely proportional to R^2.

If W_o is the weight of a body at the earth's surface and the radius of the earth is R_o then the weight W of the body at any other radius R from the centre of the earth is found from the proportion

$$\frac{W}{W_o} = \frac{R_o{}^2}{R^2}$$

Also, since the mass $m = W/g$ is constant, g is proportional to W, or

$$\frac{g}{g_o} = \frac{R_o{}^2}{R^2}$$

where $g_o = 9.8$ m/s^2, is the acceleration due to gravity at the surface of the earth and g is the acceleration due to gravity at radius R.

As an example, consider the motion of a satellite of mass $m = 200$ kg, travelling in an orbit round the Equator whereby it circles the earth at a height of 36,000 km about every twenty-four hours. The satellite is in a *circular orbit** determined by the correct launching speed. The speed of the satellite, its distance above the earth's surface, and its period of revolution are related, since the centripetal force required for motion in a circle is equal to the gravitational pull, i.e. the weight of the satellite relative to earth. Since the radius of the earth's surface R_o is 6,370 km, the radius of rotation of the satellite is

---

* A satellite orbiting the earth obeys *Kepler's Laws* of *Planetary Motion*, one of which states that the satellite will move in an elliptical orbit determined by the speed of the body. When a satellite is launched into orbit, depending on its speed and direction, it takes up a circular or elliptical orbit or travels out into space in a special trajectory. A circular orbit is a particular limiting case of an ellipse. If a satellite is moving in a circular orbit and receives a forward rocket thrust increasing its speed, it moves into an elliptical orbit; if already in an elliptical orbit it moves into a larger one. A typical early warning satellite has an elliptical 12-hour orbit at a height of 500 km × 40,000 km.

$R = 6,370 + 36,000 = 42,370$ km. At this height the acceleration due to gravity is given by

$$g = g_o \left[\frac{R_o}{R} \right]^2 = 9.8 \times \left[\frac{6370 \times 10^3}{42,370 \times 10^3} \right]^2$$

$$= 0.222 \text{ m/s}^2$$

and centripetal force = weight of satellite

i.e. $\dfrac{mv^2}{R} = mg$

i.e. $v = \sqrt{g.R}$

$= \sqrt{0.222 \times 42,370 \times 10^3}$

$= 3,064$ m/s or $11,030$ km/h

The *orbital period* i.e. the time for a complete circuit of the earth is given by

$$t = \frac{\text{distance travelled in orbital path(km)}}{\text{speed (km/h)}}$$

$$= \frac{2 \times \pi \times R}{v}$$

$$= \frac{2 \times \pi \times 42,370}{11,030}$$

$= 24$ hours, approximately.

The weight of the satellite at this point, i.e. the pull of the earth, is given by

$$mg = 200 \times 0.222 = 44 \text{ N}$$

This satellite we have shown, is in an orbit at a height of 36,000 km, travelling at 11,030 km/h, and if it is moving in the *same* direction as the earth's rotation, then since it circles the earth every twenty-four hours, it remains fixed at the same point above the ground, thereby providing a stable body for communications signals. The satellite's orbital period is synchronous with the earth's rotation and the satellite is in *synchronous* or *geostationary* orbit. For various reasons such a satellite is not absolutely stationary nor is the orbit truly circular and periodically it has to be adjusted back on-station.

Other orbits are also of interest, e.g. to circle the earth near the surface requires a satellite to have a speed of about 28,440 km/h with a period of revolution of 85 minutes and a specified time of orbit of four hours is obtainable at an altitude of 6,410 km. There are certain 'parking-orbits' where a spacecraft may be maintained temporarily prior to some other operation. Air resistance affects a satellite's motion. In the rarefied atmosphere high above the earth there is negligible drag and a satellite will orbit for hundreds of years, its useful life then being limited by the power sources and instrumentation. In lower orbits a satellite's speed is soon reduced by air drag and tidal forces depending on its mass and size, and it must either be designed assuming it will burn out in a few days, or be maintained in orbit by manoeuvre engines. A typical small satellite with a mass of 100 kg in a circular orbit at a height of 800 km has an estimated lifetime of over 450 years before destruction in the atmosphere.

A man travelling in a satellite would appear to be weightless; the gravitational pull on the man is exactly balanced by the centrifugal force due to rotation. If weighed by a spring balance on the satellite his 'weight' would be zero.

Note that the above calculations may be applied to a satellite orbiting the moon except that the acceleration due to gravity at the surface of the moon is about one-sixth that at the earth (i.e. g_o for the moon is 1.62 m/s²), and the radius of the moon's surface R_o is about 1740 km.

Example   *A navigation satellite of mass 180 kg is to have an orbital period of 100 minutes. Find the height at which it orbits above the earth's surface, and its speed and weight. The radius of the earth's surface is 6,370 km.*

SOLUTION
Let R metres be the radius at which the satellite orbits the earth.

Radius of earth' surface, $R_o = 6,370 \times 10^3$ m.

Acceleration due to gravity at the satellite is

$$g = g_o \left[\frac{R_o}{R}\right]^2$$

$$= 9.8 \times \left[\frac{6,370 \times 10^3}{R}\right]^2$$

$$= \frac{397.7 \times 10^{12}}{R^2} \text{ m/s}^2$$

Centripetal force = weight

i.e. $\dfrac{mv^2}{R} = mg$

therefore
$$v^2 = gR$$

$$= \frac{397.7 \times 10^{12}}{R^2} \times R$$

therefore
$$R = \frac{397.7 \times 10^{12}}{v^2} \text{ m}$$

where the speed v is in m/s.

Time of orbit $t = 100$ mins $= 6,000$ s

and
$$t = \frac{2\pi R}{v}$$

therefore
$$6,000 = \frac{2 \times \pi \times 397.7 \times 10^{12}}{v^3}$$

hence
$$v = 7,470 \text{ m/s or } \mathbf{27,000 \text{ km/h}}$$

and
$$R = \frac{397.7 \times 10^{12}}{7470^2} \text{ m} = 7.127 \times 10^6 \text{ m} = 7,127 \text{ km}$$

therefore altitude
$$h = R - R_0 = 7127 - 6370 = \mathbf{757 \text{ km}}$$

At the satellite
$$g = 9.8 \left[\frac{6,370}{7,127}\right]^2 = 7.83 \text{ m/s}^2$$

Hence the weight of the satellite is
$$mg = 180 \times 7.83 = \mathbf{1,410 \text{ N}}.$$

Problems

(Radius of earth's surface, 6.370 km; moon's surface, 1,740 km. At earth's surface, acceleration due to gravity, 9.8 m/s²; moon's surface, 1.62 m/s²)

1. A space vehicle orbits the earth at an altitude of 300 km. Find the ratio of the vehicle's weight in orbit to that at the earth's surface, and calculate its speed and orbital period.

   (0.912; 27,800 km/h; 90.5 min)

2. The earliest satellite Sputnik 1 orbited the earth at a speed of 27,180 km/h. Find the height at which it travelled above the earth's surface and its time of orbit.

   (606 km; 96.8 mins)

3. A weather satellite of mass 9.7 kg is in a circular orbit at a distance of 800 km above the earth's surface. Find the speed of the satellite, its period of revolution, and the centripetal force exerted by the earth on the satellite.

   (26,810 km/h; 100.8 min.75 N)

4. A communications satellite is released by a powerful spring from an Orbiter craft flying at its operational ceiling of 1110 km above the earth's surface. What is the orbital period of the satellite? If the satellite is then boosted by rockets into a higher orbit where its orbital period is two hours find its new height of orbit. Assume circular orbits.

   (107.4 mins.; 1,683 km)

5. The moon has an orbital period of 27.3 days around earth. Assuming a circular orbit show that the moon circles the earth at a radius of about 383,000 km, with an orbital speed of approximately 1 km/s. What is the acceleration of the moon towards the earth?

   (0.0027 m/s²)

6. A surveillance satellite has a mass of 12 tonne and is to have an orbital period of 110 minutes. To maintain this period an on-board motor adjusts the orbit every few days. Find the height at which the satellite orbits, and its speed and weight.

(1230 km; 26,043 km/h; 82.6 kN)

7. Show that the orbital period of an earth satellite is given by

$$t = \frac{2\pi}{R_o \sqrt{g_o}} (R_o + h)^{3/2}$$

where R_o is the radius of the earth, g_o the acceleration due to gravity at the earth's surface, and h the altitude of the satellite.

8. What is the speed and period of revolution of a command module travelling in a circular orbit around the moon at an altitude of 100 km above the moon's surface?

(5878 km/h; 118 min.)

9. What is the time of circuit and altitude for a lunar module in circular orbit around the moon if its speed is 6000 km/h? If the mass of the module is 200 kg, what is the gravity pull exerted by the moon on the module?

(25.7 km; 111 min.; 315 N)

CHAPTER 10
Work, Energy and Power

10.1 Work done by a force

Work is done when a force is applied to a body and the body moves in the direction of the force. The amount of *work done* is measured by the product:

force × distance moved by point of application of force

Thus, if a uniform force F moves a body a distance s measured in the direction of the force, then

work done by $F = F \times s$

When the force is applied *gradually** so that its magnitude varies from zero to a maximum value F, then the average force is $\frac{1}{2}F$ and therefore

work done $= \frac{1}{2}Fs$

If the force is in newtons and the distance in metres then the units of work are *newton-metres*. A unit of work equal to one newton-metre is defined in the SI system as the *joule* (**J**). The joule is defined precisely as: *the work done when the point of application of a force of one newton is displaced through a distance of one metre in the direction of the force.*

The joule is also the unit of energy and heat and since it is a rather small quantity it is more often convenient to use the following multiples:

1 *kilojoule* (**kJ**) $= 10^3$**J**

1 *megajoule* (**MJ**) $= 10^6$**J**

1 *gigajoule* (**GJ**) $= 10^9$**J**

---

* This is the only case of force varying with distance that will concern us. The work done by a force F over a distance s is given in general by

$$\int_0^s F.ds$$

It may happen that the line of action of the force is at an angle to the direction of motion of the body. For example, let a uniform force of 80 N act on a body at 30° to the horizontal as shown in Fig. 10.1, and let the body move a horizontal distance OA = 4 m. The work done by the force is determined by the distance OC = 4 cos 30°, moved by the force *along its line of action*. Thus:

$$\text{work done} = 80 \times OC$$

$$= 80 \times 4 \cos 30°$$

$$= 277 \text{ J}$$

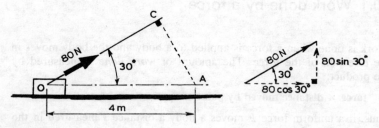

Fig. 10.1

Alternatively, the force may be resolved into components parallel and perpendicular to the direction of motion of the body, in this case, the horizontal direction OA.

$$\text{Component of force parallel to } OA = 80 \cos 30°$$

$$\text{Component of force perpendicular to } OA = 80 \sin 30°$$

The force of 80 sin 30° perpendicular to OA does not move the body in this direction and therefore does no work. The force 80 cos 30° moves the body through a distance of 4 m, therefore,

$$\text{work done} = 80 \cos 30° \times 4$$

$$= 277 \text{ J, as before}$$

10.2 Work done in particular cases

FORCE OF GRAVITY
The force required to lift a body of mass m through a height h is equal to the weight, mg. Therefore the work done in overcoming the force of gravity is $m.g.h$. For example when an aircraft climbs at an angle θ to the horizontal and travels a distance s along the line of flight, then the centre of gravity of the plane is raised through a height $s.\sin\theta$, and the

work done is $m.g. \times s \sin \theta$. Alternatively, the resolved part of the weight along the line of flight is $m.g. \sin \theta$ and the work done against this force in a distance s is $m.g. \sin \theta \times s$, as before.

RESISTING FORCE
If a body travels a distance s against a *steady* resisting force R, then

work done against resistance = resistance × distance moved

$$= Rs$$

ACCELERATING BODY
Consider a body of mass m accelerated from rest with uniform acceleration a over a distance s, when there is no resistance to motion. The accelerating force $F = ma$, and the work done by the accelerating force is

$$F \times s = ma \times s$$

This is also the work done *against* the inertia force.

BODY MOVING ON AN INCLINE
Fig. 10.2 shows the free-body diagram for the particular case of a body of mass m being accelerated up a gradient by an effort E against a constant resistance R. The work done by the effort is $E \times s$, and this is made up of the work done against the component of the weight acting down the slope, the resistance, and the inertia force. Thus

work done = $E \times s$

$$= W \sin \theta \times s + R \times s + ma \times s$$

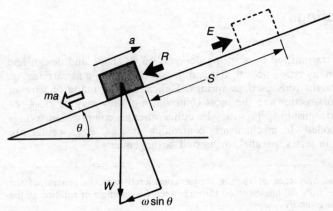

Fig. 10.2

10.3 Work done by a torque

Consider an arm OA of length r, rotating about an axis O, due to the action of a constant force F applied tangentially at A, Fig. 10.3. The applied torque about O is $Fr = T$. If the arm turns through an angle θ then the force F moves a distance $r\theta$ along the arc. Hence the work done by the force is $F \times r\theta$ or $T\theta$, i.e. *the work done by a torque is the product of the torque and the angle turned through*. Thus

work done by constant torque $T = T\theta$

If the torque is applied gradually so that it varies linearly* from zero to a maximum value T, then the average torque is $\frac{1}{2}T$ and hence

work done by a gradually applied torque $= \frac{1}{2}T\theta$

Note that 'work' and 'torque' although different in character have the same basic units, N-m. The unit for work, however, has the distinguishing name, *joule*(J).

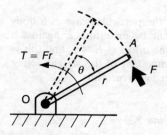

Fig. 10.3

10.4 Springs

The force transmitted by a spring depends on its shape and design and there are many types–helical, flat spiral, conical, barrel and carriage, as well as straight rods used as springs. Helical springs made of wire of circular cross-section are the most common, and they may be *close-* or *open*-coiled; tension springs may be either whereas compression springs are open-coiled. In mechanisms, controlling devices, etc., springs are often used in series, parallel, or 'nested' arrangements.

---

* This is the only case of varying torque considered here. In general, if the torque varies with the angle turned through, then for an angle of rotation θ, the work done is given by

$$\int_0^\theta T.d\theta$$

The extension of a close-coiled helical spring is directly proportional to the force when the force is applied gradually, i.e. there is a *linear* relationship, giving a straight-line graph, between load and displacement. This is not the case with all types of spring nor with a helical spring where the coils are considerably 'open'.

The *stiffness, spring constant*, or *spring rate* of a close-coiled helical spring is the load per unit extension and is approximately constant within the working range of the spring; thus if S is the stiffness, the load F required to produce an extension x is given by

$$F = Sx$$

This gives a straight-line graph of F against x, Fig. 10.4

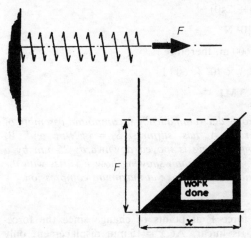

Fig. 10.4

Suppose a load gradually applied to a spring so that it varies from zero to a maximum value F and produces a maximum extension x. Then

$$\text{work done} = \text{average load} \times \text{extension}$$

$$= \tfrac{1}{2}F \times x$$

$$= \tfrac{1}{2}Sx \times x$$

$$= \tfrac{1}{2}Sx^2$$

Alternatively

$$\text{work done} = \text{area under graph (Fig. 10.4)}$$

$$= \tfrac{1}{2}Fx$$

$$= \tfrac{1}{2}Sx^2$$

Example   *An 80-tonne locomotive hauls a train of coaches of mass 240 tonne up an incline of 1 in 80 a distance of 600 m. The rolling resistance is 55 N/tonne. If the acceleration is 0.1 m/s², find the total work done.*

SOLUTION
Refer to Fig. 10.2.

Resultant force exerted along the incline

$$= \text{inertia force} + \text{component of weight} + \text{resistance}$$

$$= ma + W \sin \theta + R$$

$$= (320 \times 10^3) \times 0.1 + [(80 + 240) \times 10^3 \times 9.8] \times \frac{1}{80}$$

$$+ 55(80 + 240)$$

$$= 88.8 \times 10^3 \text{ N}$$

The distance travelled is 600 m, therefore

$$\text{work done} = 88.8 \times 10^3 \times 600 \text{ J}$$

$$= \textbf{53.3 MJ}$$

Example   *Fig. 10.5(a) shows diagrammatically a compound assembly of close-coiled springs. Spring A has stiffness $S_1 = 3$ kN/m and B, $S_2 = 7.2$ kN/m. The spring housing is moved downwards 25 mm by a gradually applied force F. Draw a graph showing how F varies with the housing movement and find the work done at maximum compression.*

SOLUTION
We need only find the force F at points of change since the force-displacement relationship is linear. At $x = 15$ mm displacement only spring A is compressed, hence the load is

$$F = S_1 x = 3000 \times 0.015$$

$$= 45 \text{ N}$$

When displacement $x = 25$ mm both springs take the load: spring A is compressed 25 mm and spring B, 10 mm. Therefore,

$$F = 3000 \times 0.025 + 7200 \times 0.01$$

$$= 147 \text{ N}$$

The F–x graph is shown in Fig. 10.5(b).

The work done by the force of 147 N is given by the area of the graph which will be found to be **1.3 J**. Alternatively, the work done on a spring of stiffness S when compressed a distance x is $\frac{1}{2}Sx^2$, hence for the two springs,

$$\text{work done} = \frac{1}{2} \times 3000 \times 0.025^2 + \frac{1}{2} \times 7200 \times 0.01^2$$

$$= \textbf{1.3 J}.$$

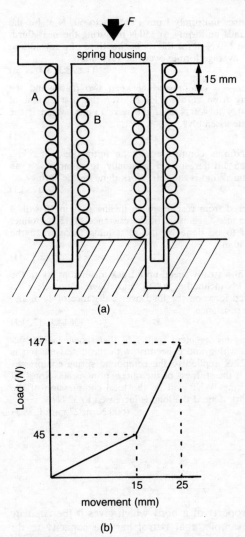

(a)

(b)

Fig. 10.5

Problems

1. A hawser wound on to a drum is used to lift a load of 300 kg through a height of 12 m. The hawser has a mass of 7 kg per metre. Find the work done in lifting the load.

(40.2 kJ)

2. A length of 40 mm diameter bar is turned at 4 revs/s in a lathe. If the cutting force is 900 N find the work done per second during the cutting operation.

(452.4 J)

3. The torque on a machine rises uniformly from 150 N-m to 600 N-m for the first third of a revolution, falls uniformly to 150 N-m during the next third, and then remains constant at 150 N-m for the remaining third of a revolution. Find the work done and the average torque over one revolution.

(1.89 kJ; 300 N-m)

4. A compound spring assembly is compressed between two plates and the applied force varies linearly from zero to 150 N over the first 20 mm of compression, then from 150 N to 450 N over the next 15 mm. What is the work done in compressing the assembly?.

(6 J)

5. A spring of stiffness 25 kN/m is compressed by an initial load of 5 kN, gradually applied, and then further loaded gradually to compress it an additional distance of 500 mm. What is the total work done on the spring?.

(6.125 kJ)

6. A 100-tonne wagon is lowered from rest down an incline of 1 in 10 with a uniform acceleration of 0.15 m/s^2, against a track resistance of 150 N/tonne. What is the work required to be done by the restraining force when the wagon travels a distance of 50 m?

(3.4 MJ)

7. A load of 2-tonne is pulled at a steady speed up a track inclined at 30° to the horizontal, by a force of 15 kN inclined at 20° to, and above, the track. Find the work done by the applied force on the load over a distance of 40 m and the work done against track resistance.

(564 kJ; 172 kJ)

8. In a compound helical spring arrangement similar to that shown in Fig.10.5 the inner spring is arranged within and concentric with the outer one but is 10 mm shorter. A load of 22 N applied to the compound spring compresses the other spring by 20 mm. If the stiffness of the outer spring is 800 N/m find the stiffness of the inner spring. What will be the total compression of the compound spring and the work done if the load is increased to 32 N?

(600 N/m; 27 mm; 0.38 J)

10.5 Energy

Energy is defined as that property of a body which gives it the *capacity to do work*. We say, for example, that petrol has the capacity to do work as a result of the chemical energy stored in it, high-pressure air in a cylinder has pressure energy which enables it to drive a piston, and high-temperature steam in a boiler has the capacity to drive a turbine by virtue of its thermal energy. Energy may take a variety of forms–chemical, nuclear, electrical, solar, sound, and so on, but we are concerned here only with *mechanical energy* which comprises **kinetic energy** possessed by a body arising from its speed, and **potential energy** as a result of its position. The term 'potential energy' usually refers to a body's energy due to its elevation above some datum, i.e. gravitational potential energy, but it may also describe other forms of stored energy due to position. A compressed spring, for example, has potential energy because when released to take up its original length it has the capacity to do work. However, because work is done initially in straining the spring the energy stored is called elastic potential energy, or simply *strain energy*. Similarly compressed air has potential energy but this is more often referred to as pressure energy. Energy and work are interchangeable quantities.

10.6 Kinetic energy: work-energy equation

The average force to accelerate a body of mass m from rest to speed v over a distance s is $F = ma$, where a is the average or uniform acceleration produced. Since $a = v^2/2s$ then the work done by the force F in moving a distance s is

$$F \times s = ma \times s$$
$$= m \times \frac{v^2}{2s} \times s$$
$$= \tfrac{1}{2}mv^2$$

The expression $\tfrac{1}{2}mv^2$ is the *kinetic energy* or *energy of motion*, of the body at speed v. Thus:

kinetic energy $= \tfrac{1}{2}mv^2$

Kinetic energy is a scalar quantity since it is not necessary to take the direction of the speed into account. Further, since F has been taken as the *average* force, we note:

1. That the kinetic energy is independent of the variation of the force during the acceleration to speed v.
2. That the work done on the body from rest is equal to the kinetic energy possessed by the body.

In general, if an effort E is applied to a body as shown in Fig. 10.6 to overcome a steady resistance R, and accelerate the body from speed u to speed v, in a distance s, then if a is the average acceleration

work done $=$ resultant force $\times$ distance moved
$$= (E - R) \times s$$
$$= F \times s$$
$$= ma \times s$$
$$= m \times \frac{(v^2 - u^2)}{2s} \times s$$
$$= \tfrac{1}{2}mv^2 - \tfrac{1}{2}mu^2$$
$$= \textbf{final kinetic energy} - \textbf{initial kinetic energy}$$

This relationship between work done by the resultant force and the change in kinetic energy produced is known as the *work-energy equation*.

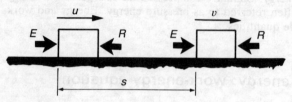

Fig. 10.6

10.7 Potential energy

The work done in lifting a load of mass m and weight $W = mg$ through a height h is Wh. This is known as the *potential energy* of the load referred to its original position and its units are those of energy, i.e. the basic unit is the joule (J). The multiples of the joule are given on page 203. Thus:

potential energy $= Wh = mgh$

Potential energy is a relative quantity since the original datum position may be chosen arbitrarily. Usually, therefore, it is the *changes* of potential energy with which we are concerned.

Evidently the work done against gravity in increasing the potential energy of a body may be recovered by allowing the body to fall back to its original position. Thus the potential energy is converted into kinetic energy by virtue of the work done on the body in falling under gravity. Fig. 10.7 shows a body falling through a vertical height h to strike the ground. The potential energy of the body measured above the ground

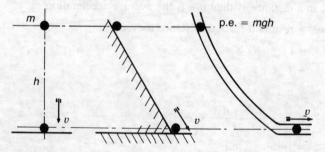

p.e. = mgh

Fig. 10.7

is *mgh* and this is converted to kinetic energy $\frac{1}{2}mv^2$, when it strikes the ground, i.e.

$$\tfrac{1}{2}mv^2 = mgh$$

i.e. **striking velocity** $v = \sqrt{2gh}$

Note: 1. Provided *there is no friction* the speed v of the body is given by the same expression for each of the cases illustrated where the mass falls down an incline or shaped tube. However, the direction of the final velocity is changed.

2. The work-energy equation must take account of any change in potential energy, i.e. work done due to gravity.

10.8 Units of energy

In the SI system the basic unit of energy is the *joule* (J) and this is the same unit as for work. Thus the units of work, kinetic energy and potential energy are the same and as has been seen the quantities are interchangeable. The joule has been defined on page 203. It is instructive to derive the unit of energy from the expressions $\frac{1}{2}mv^2$ and *mgh* for kinetic and potential energy respectively. Since m is in kilograms and v in metres per second, we see that for kinetic energy, $\frac{1}{2}mv^2$, the units are:

$$\text{kg} \times (\text{m/s})^2 = (\text{kg-m/s}^2) \times \text{m}$$
$$= \text{N-m}$$
$$= \text{J}$$

and for potential energy, *mgh*, the units are

$$\text{kg} \times (\text{m/s}^2) \times \text{m} = \text{N-m}$$
$$= \text{J}$$

A special unit of energy, used mainly in the electrical industry, is the *watt-hour* (W-h), or one of its multiples (see page 227).

Example *A drop hammer is allowed to fall from rest through a height of 6 m on to a forging. Find the downward velocity of the hammer when it strikes the forging.*

If the mass of the hammer is 500 kg what is the work done by the forging and baseplate in bringing the hammer to rest in a distance of 50 mm?

SOLUTION

Equating the kinetic energy gained to the initial potential energy for the hammer,

$\frac{1}{2}mv^2 = mg \times 6$

hence striking velocity

$$v = \sqrt{(2 \times 6 \times 9.8)}$$

$$= \textbf{10.86 m/s}$$

The work done on the hammer in bringing it to rest is made up of two parts:

1. The work done in 'destroying' the kinetic energy of the hammer.
2. The work done against the weight W in resisting the downward motion of the hammer in the final 50 mm.

Therefore

$$\text{total work done} = \tfrac{1}{2}mv^2 + W \times 0.05$$

$$= mg \times 6 + mg \times 0.05$$

$$= 500 \times 9.8 \times 6 + 500 \times 9.8 \times 0.05$$

$$= 29{,}650 \text{ J} = \textbf{29.7 kJ}$$

This work represents energy lost in permanent deformation of the forging and reappears as heat. Some of the energy loss may also be accounted for in noise and vibration.

Example *A car of mass 1,125 kg descends a hill of 1 in 5 (sine). Calculate, using an energy method, the average braking force required to bring the car to rest from 72 km/h in 50 m. The frictional resistance to motion is 250 N.*

SOLUTION
The total energy of the car when its speed is 72 km/h is the sum of its kinetic and potential energies. This total energy is destroyed by the braking force and frictional resistance acting through a distance of 50 m.

Vertical height corresponding to 50 m on the slope is $\dfrac{50}{5} = 10$ m

$$v = \frac{72}{3.6} = 20 \text{ m/s}$$

Potential energy of car $= mgh = 1{,}125 \times 9.8 \times 10 = 110{,}250$ J

Kinetic energy of car $= \tfrac{1}{2}mv^2$

$$= \frac{1{,}125 \times 20^2}{2}$$

$$= 225{,}000 \text{ J}$$

total energy $= 110{,}250 + 225{,}000 = 335{,}250$ J

work done by braking force $= F \times 50$ J

work done by frictional resistance $= 250 \times 50$ J

Equating total work done to total energy destroyed:

$$F \times 50 + 250 \times 50 = 335{,}250$$

hence $\qquad\qquad F = 6{,}455$ N $= \mathbf{6.46}$ **kN**

Problems

1. A hammer of mass 30 kg is held in a horizontal position and then released so that it swings in a vertical circle of radius 1 m. What is the kinetic energy and the speed of the hammer at its lowest point?

    If at the lowest point it strikes and breaks a metal specimen and moves on through an angle of 50° beyond the vertical before instantaneously coming to rest, how much energy has been absorbed by the specimen?

    (294 J; 4.42 m/s; 189.5 J)

2. A car descends a hill of 1 in 6 (sine). Its mass is 1,000 kg and the frictional resistance to motion is 200 N. Calculate, using an energy method, the average braking effort to bring the car to rest from 48 km/h in 30 m.

    (4.4 kN)

3. A train is moving down an incline of 1 in 120 at a speed of 40 km/h. The wheels are locked by application of the brakes on all vehicles. If the coefficient of sliding friction between wheels and track is 0.12 and the average wind resistance to motion is 90 N/tonne how far will the train move before coming to rest?

    (52.2 m)

4. A locomotive of mass 80 tonne hauls a train of twelve coaches up an incline of 1 in 80. The rolling resistance to motion is 55 N/t. Each coach is of mass 20 tonne. If the speed is increased from 24 to 48 km/h in 600 m and the time taken is 30 seconds, find: (a) the change in kinetic energy of the train; (b) the work done by the engine during this period.

    (21.3 MJ; 55.4 MJ)

5. A 10-tonne hammer is driven downwards under the influence of its own weight and by an additional steam force of 140 kN. The hammer falls through a distance of 1.8 m before striking a forging. What is the velocity of striking?

    (9.26 m/s)

6. A piston of a reciprocating engine moves with approximately simple harmonic motion. The crank speed is 1,440 rev/min and the crank arm is 150 mm long. If the mass of the piston is 5 kg find its maximum kinetic energy and the average force required to bring it to rest at inner and outer dead centres. (See Chapter 8).

    (1.28 kJ; 8.53 kN)

10.9 Strain energy

The work done in compressing or stretching a spring is stored as *strain* energy in the spring provided that there is no permanent deformation (overstretching). This strain energy is a form of 'potential energy'.

Suppose a load gradually applied to a spring so that it varies from zero to a maximum value F and produces a maximum extension x. Then the strain energy stored is

$$U = \text{work done}$$

$$= \text{average load} \times \text{extension}$$

$$= \tfrac{1}{2}F \times x$$

$$= \tfrac{1}{2}Sx^2 \text{ since } F = Sx$$

Alternatively, the strain energy is given by the area of the $F - x$ graph (*see* para 10.4). The units of strain energy are the same as those of work, i.e. joules.

Example *A wagon of mass 12 tonne travelling at 16 km/h strikes a pair of parallel spring-loaded stops. If the stiffness of each spring is 600 kN/ m, calculate the maximum compression in bringing the wagon to rest.*

Solution

kinetic energy of wagon $= \tfrac{1}{2}mv^2$

$$= \tfrac{1}{2} \times 12 \times 1,000 \times \left(\frac{16}{3.6}\right)^2$$

$$= 118,500 \text{ J}$$

This kinetic energy may be assumed to be absorbed equally by the two springs. Strain energy stored per spring is

$$\tfrac{1}{2} \times 118,500 = 59,250 \text{ J}$$

At maximum compression, the wagon is instantaneously at rest, the final kinetic energy is zero and the initial kinetic energy has been converted entirely into strain energy of the springs. Thus if x is the maximum compression of the springs,

$$\tfrac{1}{2}Sx^2 = 59,250$$

or $\quad \tfrac{1}{2} \times 600 \times 1,000 \, x^2 = 59,250$

$$x = 0.446 \text{ m} = \textbf{446 mm}$$

Example *A spring of stiffness 18 kN/m is installed between plates so that it has an initial compression of 23 mm. A body of mass 4.5 kg is dropped 150 mm from rest on to the compressed spring. Find the initial work done on the spring and the further compression neglecting loss of energy at impact.*

Solution

Let x m = further compression of the spring (Fig. 10.8)

initial spring force = stiffness × initial compression

$$= 18 \times 10^3 \times 0.023$$

$$= 414\ N$$

maximum spring force $= (414 + 18{,}000\ x)$ newton.

The load-compression graph is shown in Fig. 10.9. The line OA represents the initial compression of the spring and the area under the line gives the work done, i.e.

initial work done $=$ area OAC

$$= \tfrac{1}{2} \times 414 \times 0.023$$

$$= 4.76\ J$$

or

initial work done $=$ strain energy stored in the spring
$$= \tfrac{1}{2}Sx^2 = \tfrac{1}{2} \times 18{,}000 \times 0.023^2 = \mathbf{4.76\ J}$$

The loss of potential energy of the weight in falling is equal to the additional work done on the spring which is given by the area under the line AB. Thus:

loss of potential energy $= 4.5 \times 9.8(0.15 + x)$ joule

work done on spring $=$ area CABD

$$= 414 \times x + \tfrac{1}{2} \times 18{,}000x \times x$$

$$= (414x + 9{,}000x^2)\ \text{joule}$$

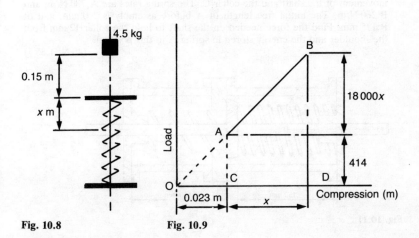

Fig. 10.8   Fig. 10.9

Equating

$$4.5 \times 9.8(0.15 + x) = 414x + 9{,}000x^2$$

hence

$$x^2 + 0.0411x - 0.000736 = 0$$

hence $x = 0.0136$ m $= 13.6$ mm (neglecting negative answer)

Problems

1. A machine is mounted on a light rigid beam AB supported by two springs as shown in Fig. 10.10. The spring at A has stiffness 20 kN/m, and that at B, 60 kN/m. As a result of the machine's action and location a vertical force $F = 4.8$ kN is applied at C. If the beam is initially horizontal what is the deflection at C when force F is applied and what is the energy stored in each spring?

(65 mm; 81 J; 75 J)

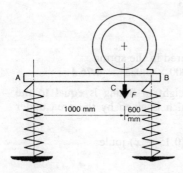

Fig. 10.10

2. Fig. 10.11 shows an assembly where two compressed springs control the movement of the shaft and the collar C. The spring rates are A,700 N/m, and B,260 N/m. The initial free length of A before assembly is 200 mm, and of B,175 mm. Find the force needed on the shaft to hold the collar 12 mm from the shoulder and the energy stored in spring A in this position.

(40 N; 1.35 J)

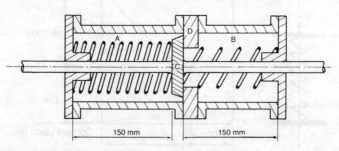

Fig. 10.11

3. A spring requires a force of 100 N, gradually applied, to compress it 20 mm. Find the amount of compression when a 1.6 kg mass falls freely from rest on to the top of the spring through a height of 80 mm. The spring is initially unloaded.

(25.8 mm)

4. A body of mass 0.9 kg falls 300 mm on to the top of a spring. The spring has

an initial compression of 30 mm and is given a further compression of 36 mm by the falling mass. Find the stiffness of the spring.

(1.715 kN/m)

5. A 5 kg mass falls from a height of 600 mm on to a plate on top of a compound spring. The spring arrangement consists of two concentric springs of which the outer spring has stiffness 2.2 kN/m and the inner one 1.8 kN/m. The outer spring is 40 mm longer than the inner one so that the falling mass does not affect the inner spring until the outer one has been compressed 40 mm. Find the total compression of the compound spring.

(152 mm)

6. A spring gun consists of a cylinder holding a spring of free length 300 mm and stiffness 200 N/m. When the gun is set, the spring is compressed to a length of 120 mm and a shot of mass 50 g is placed in the cylinder against the spring. Neglecting friction, find the speed of the shot when released.

(11.4 m/s)

7. A cylinder is inclined at 15° to the horizontal and the bottom end is rigidly capped. The cylinder houses a spring of stiffness 5 kN/m fastened to the closed end. A mass of 36 kg is released from a point in the cylinder and travels downwards a distance of 1.7 m before striking the top of the spring. Neglecting friction, what is the striking velocity of the mass and the maximum compression of the spring, assuming its free length permits it? What is the amount of energy absorbed by the spring at maximum compression?

(2.94 m/s; 268 mm; 179.7 J)

8. A train of twenty loaded wagons, each of total mass 12 tonne, is brought to rest by a pair of parallel buffer springs. The stiffness of each spring is 30 kN/m and the initial resisting force in each spring before impact is 4.5 kN. If the train speed is 2 km/h when it strikes the buffers calculate the maximum compression of the springs. *Hint*–the area under the load-compression graph is equal to the work done in compressing the spring.

(975 mm)

10.10 Conservation of energy

The *principle of conservation of energy* states that energy can be redistributed or changed in form but cannot be created or destroyed. The following examples will show how this occurs.

FALLING BODY   A falling body loses potential energy but gains a corresponding amount of kinetic energy.

MASS-SPRING   A mass vibrating at the end of a spring loses kinetic energy in stretching the spring but the spring then possesses potential or strain energy. When the motion is reversed and the spring is acting on the mass its strain energy is transferred to the mass as kinetic energy of motion.

'LOST' ENERGY
WHEN FRICTION
IS INVOLVED

A body moving along a rough surface loses kinetic energy corresponding to the work done against the friction forces. This work is generally 'lost' for mechanical purposes but reappears as heat energy in the body and surface. Thus the total energy of the system, body and surface, is conserved. The work done in overstretching a spring so that it is permanently deformed is again lost for mechanical purposes and converted into heat.

COLLISION OF
BODIES

When two perfectly elastic bodies collide, the work done in elastic deformation is recovered as they rebound. For example, when two steel balls bounce together there is compression of the steel on impact but as they move apart the steel recovers its shape and in doing so restores kinetic energy. *When perfectly elastic bodies collide the total kinetic energy before and after the collision is the same*. In a collision of inelastic bodies, as for example, when two balls of putty collide, all the kinetic energy may disappear to reappear as heat energy corresponding to the amount of work done in producing permanent deformation.

It should be noted: (i) that the principle of conservation of energy is based upon observation and experiment and not upon mathematical proof; (ii) the principle should not be used to solve problems where the *precise* measurement of energy quantities is not possible, e.g. where friction or heat energy is involved as mentioned above.

10.11 Kinetic energy of rotation

Consider the work done in accelerating a shaft of moment of inertia I from rest to a speed ω while turning through an angle θ rad. The average angular acceleration α is given by

$$\omega^2 = \omega_0^2 + 2\alpha\theta$$

$$\alpha = \frac{\omega^2}{2\theta}$$

since $\omega_0 = 0$; hence the average torque required is

$$T = I\alpha = I \times \frac{\omega^2}{2\theta}$$

The work done by the torque T in rotating the shaft through θ rad

$$= T\theta = I\frac{\omega^2}{2\theta} \times \theta = \tfrac{1}{2}I\omega^2$$

The term $\frac{1}{2}I\omega^2$ is known as the *kinetic energy of rotation*.

Similarly it may also be proved that the work done in accelerating a shaft from velocity ω_0 to velocity ω is equal to the *change* in kinetic energy of rotation, i.e.

$$T\theta = \tfrac{1}{2}I\omega^2 - \tfrac{1}{2}I\omega_0{}^2$$

When a friction couple is present, opposing motion, the total work required is

work done against friction + change in kinetic energy of rotation

As an aid to memorizing these formulae note the similarity between the expressions for kinetic energy of translation (linear motion) and kinetic energy of rotation:

kinetic energy of translation $= \tfrac{1}{2}mv^2$

kinetic energy of rotation $= \tfrac{1}{2}I\omega^2$

So far we have considered only the kinetic energy of a shaft rotating about a *fixed* axis. When the axis of rotation is in motion the shaft possesses additional energy which we shall now consider.

10.12 Total kinetic energy of a rolling wheel

If a wheel rolls then the total kinetic energy is made up of two parts:

1. The kinetic energy of translation of the centre of mass.
2. The kinetic energy of rotation about the centre of mass.

This statement requires justification but the proof is lengthy and is omitted here. A proof may be found in *Textbook of Mechanics* by J. G. Jagger. Note that it is the motion of, and rotation about, the centre of mass which must be considered; however, in many practical cases the centre of mass coincides with the axis of rotation, as in a rolling wheel.

If v is the linear velocity of the wheel and I is the moment of inertia about its axis of rotation, then

$$\text{total kinetic energy} = \tfrac{1}{2}mv^2 + \tfrac{1}{2}I\omega^2$$
$$= \tfrac{1}{2}mv^2 + \tfrac{1}{2}mk^2\omega^2$$

where m is the mass and k the radius of gyration of the wheel, and, since for rolling motion without slip $\omega = v/r$, where r is the wheel radius, then

$$\text{kinetic energy of wheel} = \tfrac{1}{2}mv^2 + \tfrac{1}{2}mk^2\frac{v^2}{r^2}$$

A particular problem, not dealt with at this stage, is the small effect of the rotational energy of a vehicle on the tractive effort or braking force required.

Example   *A shaft of moment of inertia 34 kg-m² is initially running at 600 rev/min. It is brought to rest in eighteen complete revolutions by a braking torque; reversed, and accelerated in the opposite direction by a driving torque of 675 N-m. The friction couple is 160 N-m throughout. Find the braking torque required and the revolutions turned through in attaining full speed again.*

SOLUTION

$$600 \text{ rev/min} = \frac{2\pi}{60} \times 600 = 62.83 \text{ rad/s}$$

$$\text{initial kinetic energy} = \tfrac{1}{2}I\omega^2$$

$$= \tfrac{1}{2} \times 34 \times 62.83^2$$

$$= 67{,}000 \text{ J}$$

Work done by friction torque T_f in turning through angle θ rad is

$$T_f \times \theta = 160 \times (2\pi \times 18)$$

$$= 18{,}100 \text{ J}$$

Let T be the braking torque in newton-metres, then:

$$\text{work done by braking torque} = T \times \theta$$

$$= T \times (2\pi \times 18)$$

$$= 113T \text{ joule}$$

work done by brake + work done by friction couple = kinetic energy of rotation destroyed

$$113T + 18{,}100 = 67{,}000$$

thus $T = \textbf{433 N-m}$

If ϕ rad is the angle turned through by the shaft in accelerating from rest to full speed, then

$$\text{work done against friction} = 160\phi \text{ joule}$$

$$\text{work done by accelerating torque} = 675\phi \text{ joule}$$

work done by applied accelerating torque

= work done against friction + increase of kinetic energy

$$675\phi = 160\phi + 67{,}000$$

$$\phi = 130 \text{ rad}$$

$$\simeq \textbf{20}\tfrac{1}{2} \textbf{ rev}$$

Example   *In an experiment to determine the moment of inertia of a flywheel and its shaft the wheel is allowed to roll on its shaft freely from rest down an incline formed by two parallel steel tracks (Fig. 10.12). The* *shaft rolls without slip. To allow for the work done against friction two tests are carried out, the slope of the incline being increased for the second test. The results of such an experiment were as follows: in the first test the flywheel fell through a height of 50 mm in rolling 1.5 m down the incline in 50 seconds; in the second test the flywheel fell 100 mm while rolling 1.5 m in 30 seconds. If the flywheel's mass is 20 kg and the shaft diameter is 50 mm, calculate the moment of inertia of the flywheel and shaft.*

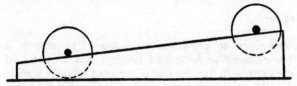

Fig. 10.12

SOLUTION

$$\text{Shaft radius } r = 0.025 \text{ m}$$

$$\text{average speed in first test} = \frac{1.5}{50} = 0.03 \text{ m/s}$$

thus

$$\text{maximum speed } v_1 = 2 \times 0.03 = 0.06 \text{ m/s}$$

and maximum angular velocity $\omega_1 = \dfrac{v_1}{r} = \dfrac{0.06}{0.025} = 2.4$ rad/s

For the second test, maximum speed $v_2 = 0.1$ m/s and maximum angular velocity $\omega_2 = 4$ rad/s. For each test:

loss of potential energy of wheel in rolling down incline

= gain of kinetic energy + work done against friction, R

Let I be the moment of inertia of the flywheel. Then

kinetic energy of flywheel = kinetic energy of rotation

+ kinetic energy of translation

$$= \tfrac{1}{2}I\omega^2 + \tfrac{1}{2} \times 20v^2$$

For the first test:

loss of potential energy = $mgh = 20 \times 9.8 \times 0.05$

thus

$$20 \times 9.8 \times 0.05 = \tfrac{1}{2}I \times 2.4^2 + \tfrac{1}{2} \times 20 \times 0.06^2 + R$$

i.e.

$$9.764 = 2.88I + R \qquad \qquad \ldots[1]$$

For the second test:

loss of potential energy = $20 \times 9.8 \times 0.1 = 19.6$ J

therefore

$$19.6 = \tfrac{1}{2}I \times 4^2 + \tfrac{1}{2} \times 20 \times 0.1^2 + R$$

i.e. $19.5 = 8I + R$ ...[2]

To eliminate R subtract equation [1] from equation [2]:

$$9.74 = 5.12I$$

therefore

$$I = 1.9 \text{ kg-m}^2$$

Example *In an experiment to determine the radius of gyration of a flywheel the shaft is mounted in horizontal bearings and a light cord wrapped around the shaft, with the free end of the cord carrying a mass of 2 kg (Fig. 10.13). When allowed to fall freely from rest the load falls 900 mm in 20 seconds before striking the floor. The flywheel eventually comes to rest owing to bearing friction. The total number of shaft revolutions from start to finish is 30. If the shaft diameter is 25 mm and the flywheel's mass is 11 kg, calculate (a) the radius of gyration of the flywheel, (b) the friction torque at the bearings, assumed constant.*

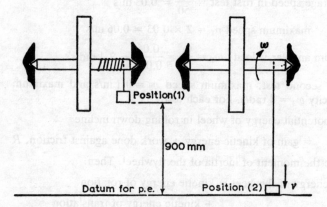

Fig. 10.13

SOLUTION

The motion of the flywheel is in two stages: (i) acceleration to a maximum angular velocity ω when the load reaches its maximum velocity v on striking the floor; (ii) deceleration to rest after the load strikes the floor due to the friction torque T_f at the bearings. Consider the first stage, Fig. 10.13:

           energy in position (1) = energy in position (2) + work done against friction between positions (1) and (2) i.e.   potential energy of load at (1) =   kinetic energy of load at (2) + kinetic energy of fly-wheel at (2) + T_f × angle turned through by shaft

i.e. $(2 \times 9.8) \times 0.9 = \frac{1}{2} \times 2v^2 + \frac{1}{2}I\omega^2 + T_f\theta_1$ ....[1]

where

I = moment of inertia of flywheel and shaft;

θ_1 = angle turned through by shaft between positions (1) and (2)

thus

$$\theta_1 = \frac{\text{length of cord unwrapped}}{\text{shaft radius}}$$

$$= \frac{0.9}{0.0125}$$

$$= 72 \text{ rad}$$

Now average velocity of falling load is

$$\frac{0.9}{20} = 0.045 \text{ m/s}$$

therefore

(maximum) $v = 2 \times 0.045 = 0.09$ m/s

and $\omega = 0.09/0.0125 = 7.2$ rad/s

Equation [1] becomes

$$17.64 = \frac{1}{2} \times 2 \times 0.09^2 + \frac{1}{2}I \times 7.2^2 + T_f \times 72$$

$$= 0.0081 + 25.9I + 72T_f \qquad ....[2]$$

Consider now the second stage. The kinetic energy of the flywheel, $\frac{1}{2}I\omega^2$, is destroyed by friction, thus

$$T_f\theta_2 = \frac{1}{2}I\omega^2$$

where θ_2 = angle turned through by the shaft in coming to rest after the load strikes the floor.

Total angle turned through by shaft = 30 rev = 188.5 rad therefore

$$\theta_2 = 188.5 - 72 = 116.5 \text{ rad}$$

$$T_f \times 116.5 = \frac{1}{2}I \times 7.2^2$$

$$T_f = 0.223I \text{ N-m}$$

Hence equation [2] becomes (neglecting the term 0.0081):

$$17.64 = 25.9I + 72 \times 0.223I$$

$$I = 0.42 \text{ kg-m}^2$$

$$mk^2 = 0.42$$

$$11 \, k^2 = 0.42$$

$$k = \textbf{0.196 m}$$

The radius of gyration of the flywheel and shaft = **196 mm**

$$T_f = 0.223I$$

$$= 0.223 \times 0.42$$

$$= \textbf{0.094 N-m}$$

Problems

1. A rotating shaft carries a load having a moment of inertia about the shaft axis of 48 kg-m^2. Calculate, using an energy method, the torque required to accelerate the shaft from rest to a speed of 10 rev/s in 12 rev. The bearing friction is equivalent to a couple of 300 N-m.

(1.56 kN-m)

2. A shaft rotating at 12 rev/s has a moment of inertia of 34 kg-m^2. It is brought to rest by a brake block acting on the rim of a 1-m diameter drum. Calculate the normal force on the brake block to bring the shaft to rest in 12 revolutions of the shaft if the coefficient of friction between block and drum is 0.45.

(5.7 kN)

3. A shaft having a moment of inertia of 16 kg-m^2 is accelerated from 1,440 rev/min to 1,500 rev/min during two revolutions of the shaft. If the friction couple is 80 N-m calculate (a) the change in kinetic energy of rotation of the shaft, (b) the average torque required to accelerate the shaft.

(15.5 kJ; 1.31 kN-m)

4. A flywheel having a mass of 25 kg is mounted on a 75 mm diameter shaft in horizontal bearings. Around the shaft is wrapped a light cord to which is attached a hanging load of mass 2 kg. If allowed to fall from rest and accelerate the flywheel the load is seen to fall 1.2 m in 24 second. Calculate by an energy method the radius of gyration of the flywheel. The effect of bearing friction may be neglected.

(513 mm)

5. A cylinder rolls freely down a slope of 1 in 50. What will be its speed after rolling 10 m from rest down the incline? What is then its total kinetic energy? The mass of the cylinder is 30 kg.

(1.62 m/s; 58.8 J)

6. In an experiment to determine the moment of inertia of a flywheel, its shaft is mounted in horizontal bearings and a mass of 10 kg is hung from a light cord attached to, and wrapped around, the shaft. It is found that when allowed to fall from rest the mass travels downwards 900 mm in 15 seconds. At the end of this period the falling mass is arrested and ceases to accelerate the flywheel, which then turns through a further sixteen complete revolutions before coming to rest owing to bearing friction. The shaft diameter is 80 mm. Find, by an energy method: (a) the moment of inertia of the flywheel; (b) the friction couple at the bearings.

(16 kg-m^2; 0.72 N-m)

7. A flywheel is mounted on a 50 mm diameter shaft. In order to determine its radius of gyration it is allowed to roll freely from rest down an incline formed by a pair of knife-edges on which the shaft may run. When one end of the track is raised 75 mm above the other the flywheel takes 30 seconds to travel 1.5 m from rest. When the raised end is 150 mm above the other end, it takes 20 seconds to travel the same distance down the incline. If the work done against frictional resistance is the same in both tests find, by an energy method, the radius of gyration of the flywheel.

(271 mm)

10.13 Power

Power is the rate of doing work. In SI units the derived unit of power is the *watt* (W) which is defined as a rate of working equal to one joule per second, i.e. to the work done by a force of one newton in moving through a distance of one metre in one second. Thus

1 watt = 1 J/s = 1 N-m/s

The watt is a small quantity and the higher multiples are more often used; these are:

1 *kilowatt* (**kW**) = 10^3W

1 *megawatt* (**MW**) = 10^6W

1 *gigawatt* (**GW**) = 10^9W

Thus if one watt of power is developed by a force of one newton moving at one metre per second then a force F newtons moving at v metres per second will develop Fv watts i.e. the *instantaneous* power developed by a force F moving at a speed v is given by

power = force × speed = Fv

Energy is a specific quantity and does not involve time. Power does involve time and the watt is a *rate* of working equal to one joule per second. If, however, this rate of working is kept up for one hour then the quantity of work done or the energy emitted or absorbed in one hour is

1 watt × 1 hour = 1 J/s × 3,600 s

= 3,600 J

i.e.

1 watt-hour = 3.6 kilojoule

The most useful unit is the kilowatt-hour, used in the electrical industry, and

1 kW-h = 1,000 W-h = 3.6 MJ

Note that 1 kW-h is a *specific quantity of energy*.

10.14 Power developed by a torque

Consider a torque T applied to rotate an axle through an angle θ in time t, then

$$\text{work done by the torque} = \text{torque} \times \text{angle turned through}$$
$$= T\theta$$

and   power or rate of working $= \dfrac{T\theta}{t}$

But $\dfrac{\theta}{t} = \omega$ the angular speed of rotation, hence

power developed by a torque $= T\omega =$ torque $\times$ angular speed

If the torque is in newton-metres and the angular speed in rad/s, then the unit of power is the watt. If the axle rotates at n rev/s then

$$\omega = 2\pi n \text{ rad/s}$$

and **power developed** $= 2\pi nT\text{ watts} = \dfrac{2\pi nT}{1000}\text{ kW}$

10.15 Efficiency

The *mechanical efficiency* of a machine or engine is defined as the ratio of the useful work done to the actual work input *in a given time*. The difference between the input and output quantities of work is because of losses due to friction, leakage, etc. Thus:

$$\text{efficiency } (\eta) = \frac{\text{work output}}{\text{work input}} \text{ in a given time}$$

and the work done in unit time is the power. Hence the efficiency may be expressed as

$$\eta = \frac{\text{power output}}{\text{power input}}$$

In the case of *plant efficiency*, such as a power station, the efficiency may be stated in terms of energy, i.e.

$$\eta = \frac{\text{energy output}}{\text{energy input}} \text{ in a given time}$$

Each part of the system, such as boilers, turbines, generators, etc, will have separate efficiencies and these may be combined to give the overall efficiency of the plant.

Example   *The cutter of a broaching machine travels at 0.1 m/s. It is operated by a square-threaded screw of 10 mm pitch, and the operating* *nut takes the axial load. The torque to be overcome due to thread load and friction is 2 N-m and the torque due to friction at the bearing surface of the operating nut is 2.25 N-m. Find the power required to rotate the operating nut.*

SOLUTION

Total resisting torque, $T = 2 + 2.25 = 4.25$ N-m

$$\text{Angular speed of nut} = \frac{\text{linear speed of screw}}{\text{pitch}}$$

$$= \frac{0.1}{0.01}$$

$$= 10 \text{ rev/s}$$

$$\omega = 2\pi 10 = 20\,\pi \text{ rad/s}$$

therefore

$$\text{power} = T\omega$$

$$= 4.25 \times 20\pi$$

$$= \mathbf{267\ W}$$

Example *A mine cage of mass 4 tonne is to be raised with an acceleration of 1.5 m/s² by a cable passing over a hoist drum of 1.5 m diameter. What is the power required when the load has reached a velocity of 6 m/s, neglecting the inertia of the drum and the weight of the cable, but allowing for a bearing friction torque of 3 kN-m at the drum? What is the power required at a uniform velocity of 6 m/s?*

SOLUTION

Let E be the tension in the cable. Fig. 10.14 shows the free-body diagram for the cage. Then

$$E = \text{weight } (W) + \text{inertia force } (ma)$$

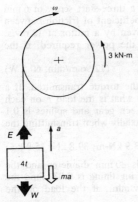

Fig. 10.14

$$= 4 \times 10^3 \times 9.8 + 4 \times 10^3 \times 1.5$$

$$= 45200 \text{ N}$$

Torque at drum = torque to overcome weight and inertia force + torque to overcome bearing friction

$$= (45200 \times 0.75) + 3000$$

$$= 36900 \text{ N-m}$$

Angular speed of drum, $\omega = \dfrac{v}{r} = \dfrac{6}{0.75} = 8$ rad/s

power $= T\omega$

$$= \frac{36900 \times 8}{1000} \text{ kW}$$

$$= \textbf{295 kW}$$

At uniform speed, there is no inertia force, hence

$$E = 4 \times 10^3 \times 9.8 = 39200 \text{ N}$$

and power $= \dfrac{(39200 \times 0.75 + 3000) \times 8}{1000} = \textbf{259 kW}$

Problems

1. A force is applied parallel to a plane inclined at 30° to the horizontal to drag a mass of 100 kg down the plane at a *steady* speed of 4 m/s against a resistance of 600 N. What is the power required?

(440 W)

2. A load of 2 tonne is pulled at a *steady* speed of 1.2 m/s up a track inclined at 30° to the horizontal by a force inclined at 20° to, and above, the track. Find the power developed by the engine. Take $\mu = 0.3$.

(17.2 kW)

3. A table carrying a machine tool is traversed by a three-start screw of 6 mm pitch. The mass of the table is 200 kg and the coefficient of friction between the table and its guides is 0.1. The screw is driven by a motor at 12 rev/s. Find the speed of operation of the tool, and the power required, if the efficiency is 70 per cent.

(12.96 m/min; 60.5 W)

4. A shaft transmits 300 kW at 2 rev/s. Calculate the torque transmitted. If a gear is splined to the shaft as shown (Fig. 10.15) what is the load F on each spline? If the coefficient of sliding friction between gear and splines is 0.1 what force would be required to move the gear axially when transmitting the above torque?

(23.9 kN-m; 39.8 kN; 15.9 kN)

5. The journal of a bearing housing a line-shaft is 80 mm diameter and the coefficient of friction between the shaft and bearing lining is 0.03. Find the power required to overcome friction at 180 rev/min., if the load on the bearing is 12 kN.

(270 W)

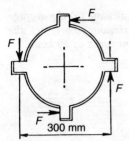

Fig. 10.15

6. A load of mass 250 kg is lifted by means of a rope which is wound several times round a 1-metre diameter drum and carries a balance mass of 150 kg. Find the tensions in the rope when the load is given a uniform acceleration of 1.4 m/s². What is the power required at the drum when the load has reached a velocity of 10 m/s from rest? If the friction torque is 50 N-m and the torque to accelerate the drum is 80 N-m what is the additional power required?
(load 2.8 kN; balance mass, 1.26 kN; 15.4 kW; 2.6 kW)

10.16 Power to drive a vehicle

A vehicle is driven forward by the engine torque being transmitted and increased through gearing to exert a torque on the *driving* axle, thereby producing a tractive effort at the road surface, (*see* para.5.11). The engine power effective at the road surface is

output power = actual work done per second

= tractive effort × road speed

= Ev

The power developed by the engine is greater than this output value because there are losses in transmission and power is absorbed in accelerating the road wheels and engine rotating parts.

The *maximum* tractive effort possible depends on the friction or adhesive force at the ground, and this in turn depends on whether power is supplied to the rear, front or all wheels (*see* para 5.13). At constant speed, the effort on the level is equal in magnitude to the total resistance. The output power is then equal to the rate of working against the resistance. If the effort is increased above that to maintain constant speed the vehicle accelerates and the power required is then equal to the rate of working against both the resistance and the inertia force. On a gradient the effort and power are affected by the component of the weight down the slope.

Example *A lorry of mass 4 tonne accelerates uniformly from 45 to 70 km/h in 11 seconds. If the tractive effort is constant during this time at*

3 kN find the average resistance to motion. Also calculate (a) the average power developed during this time, (b) the maximum power, (c) the power developed if the lorry continues at a constant speed of 70 km/h.

SOLUTION

Initial speed = 45 km/h = 12.5 m/s

Final speed = 70 km/h = 19.45 m/s

Acceleration, $a = \dfrac{19.45 - 12.5}{11} = 0.632$ m/s²

The tractive effort must balance both the tractive resistance and the inertia force. Thus:

$$E = R + ma$$

or $\qquad\qquad 3,000 = R + 4 \times 1,000 \times 0.632$

thus average resistance, $R = \textbf{472 N}$

(a) The distance travelled in 11 seconds is obtained from:

$$v^2 = u^2 + 2as$$

$$19.45^2 = 12.5^2 + 2 \times 0.632 \times s$$

hence $s = 175.5$ m

work done = $E \times s = 3,000 \times 175.5 = 526,500$ J

rate of working = $\dfrac{526,500}{11} = 47,860$ J/s

average power = $\dfrac{47,860}{1,000} = \textbf{47.9 kW}$

(b) The maximum power is developed when the speed is a maximum. Thus:

maximum power = $\dfrac{E \times v}{1,000} = \dfrac{3,000 \times 19.45}{1,000} = \textbf{58.4 kW}$

(c) At a uniform speed of 70 km/h, since there is no acceleration,

tractive effort, E = resistance, R

$$= 472 \text{ N}$$

rate of working = $\dfrac{E \times v}{1,000} = \dfrac{R \times v}{1,000}$

$$= \dfrac{472 \times 19.45}{1,000}$$

$$= \textbf{9.16 kW}$$

Example *A motor car develops 75 kW when just reaching a speed of 150 km/h on the level in still air. The total resistance to motion is shown* *by tests to be given by*

$$R = 180 + 0.6v + 0.06v^2$$

where R is in newtons and v is the speed in km/h. What is the output power and the transmission efficiency? If the car has an all-up mass of 1200 kg what is the acceleration at 60 km/h assuming the tractive effort to be constant?

SOLUTION

At 150 km/h there is no acceleration so that the tractive effort is equal to the resistance. Thus, since $v = 150$:

tractive effort $= 180 + 0.6 \times 150 + 0.06 \times 150^2$

$$= 1620 \text{ N}$$

Output power $=$ tractive effort $\times$ speed

$$= 1620 \times \frac{150}{3,6} \times \frac{1}{1000} \text{ kW}$$

$$= \mathbf{67.5 \text{ kW}}$$

Transmission efficiency $= \dfrac{\text{output power}}{\text{engine power}}$

$$= \frac{67.5}{75} \times 100$$

$$= \mathbf{90\%}$$

At 60 km/h,

resistance $= 180 + 0.6 \times 60 + 0.06 \times 60^2 = 432 \text{ N}$

The tractive effort remains at 1620 N, hence

accelerating force $= 1620 - 432 = 1188 \text{ N}$

therefore

$$1188 = ma$$
$$= 1200 \times a$$

i.e. $a = \mathbf{0.99 \text{ m/s}^2}$

Example *A car has a mass of 1200 kg and road wheels of 600 mm diameter. When accelerating up a gradient of 1 in 20 the engine to back-axle gear ratio is 9:1. The torque available at the engine output shaft is $(100 - 0.08v^2)Nm$ and the resistance to motion is $(300 + v^2)N$ where v is the speed of the car in m/s. Find the power developed and the acceleration of the car when its speed is 36 km/h, neglecting the transmission losses and the inertia effects of the road wheels and engine parts.*

SOLUTION

At 36 km/h (10 m/s),

engine torque $= 100 - 0.08 \times 10^2$

$\qquad = 92 \, \text{N m}$

axle torque $= 92 \times 9$

$\qquad = 828 \, \text{N m}$

tractive force at road $= \dfrac{\text{axle torque}}{\text{wheel radius}} = \dfrac{828}{0.3} = 2760 \, \text{N}$

Power developed = tractive force × speed

$\qquad = 2760 \times 10 = 27{,}600 \, \text{W} = \textbf{27.6 kW}$

Resistance $= 300 + 10^2$

$\qquad = 400 \, \text{N}$

Component of weight down the slope $= 1200 \times 9.8 \times \dfrac{1}{20}$

$\qquad = 588 \, \text{N}$

Accelerating force $F = 2760 - 400 - 588$

$\qquad = 1772 \, \text{N}$

and $\qquad\qquad F = ma$

i.e. $\qquad\qquad 1772 = 1200a$

i.e. $\qquad\qquad a = \textbf{1.48 m/s}^2$

Problems

1. A car of mass 1.5 tonne travels up a gradient of 1 in 10 at a maximum speed of 126 km/h against a resistance of 250 N. Find the power developed by the engine, neglecting losses.

(60.2 kW)

2. A motor car of mass 1200 kg has a road speed of 90 km/h on the level. The total resistance to motion is 1100 N. Estimate the power developed by the engine when the car has an acceleration of 1.1 m/s² under these conditions. If the transmission efficiency is 0.8 what is the engine power?

(60.5 kW; 75.6 kW)

3. A vehicle of mass 1.2 tonne climbs a gradient of 1 in 20 against a track resistance of 500 N. The wheel diameter is 600 mm. The engine torque is 110 N-m at vehicle speed 126 km/h. The gear ratio between engine and wheel is 12:1. Find the acceleration rate and the power developed by the engine at 126 km/h. Neglect losses.

(2.76 m/s²; 154 kW)

4. A 10-tonne truck is driven at a steady speed of 36 km/h on a level road. The wheels are 1 m in diameter, the tractive resistance is 200 N/tonne and the

engine speed 2100 rev/min. Calculate the power developed and the gear ratio in use, engine to axle. Neglect losses.

(20 kW; 11)

5. A car of mass 1000 kg travels at 72 km/h on the level and the engine torque at this speed is 120 N-m. The gear reduction from engine to road wheels is 13. The resistance to motion is 200 N and the road wheels are 600 mm in diameter. Find the power developed by the engine. If there is a loss of power of 15% between the engine and the road surface estimate the effective power at the road and the acceleration under these conditions.

(104 kW; 88.4 kW; 4.22 m/s^2)

6. A car is driven with the engine at full throttle, reaching a speed of 160 km/h on the level when developing 75 kW. What is the resistance due to windage and road drag at this speed?. If it is assumed that the resistance varies as the square of the road speed and the tractive effort is constant what is the acceleration achieved at 80 km/h on the level? The mass of the car is 1100 kg.

(1.69 kN; 1.15 m/s^2)

7. A typical motor car of mass 1200 kg travels at 60 km/h on the level in still air and the engine develops 30 kW. Neglecting transmission losses and assuming that the combined road and air resistance is given in newtons by the expression $(160 + 0.5v + 0.08v^2)$, where v is the road speed in km/h, find the maximum possible acceleration under these conditions.

(1.1 m/s^2)

10.17 Function of a flywheel

The function of a flywheel is to accumulate and store energy. If the moment of inertia of the wheel is large a great amount of energy can be absorbed without appreciable change in speed. For example, if the inertia of a wheel is 1,000 kg-m^2 then the energy stored at 2 rev/s is:

$$\tfrac{1}{2}I\omega^2 = \tfrac{1}{2} \times 1,000(2\pi \times 2)^2 = 80,000 \text{ J}$$

A ten per cent increase in speed (to 2.2 rev/s) would increase the energy stored to

$$80,000 \times 1.1^2 = 96,800 \text{ J}$$

That is, about 17 kJ of energy would be required to produce a ten per cent change in speed.

The flywheel serves to equalize the energy output of an engine during each cycle of operations and reduces the fluctuations of speed which would occur if it were not fitted. The cycle of operations in the engine cylinder is usually completed in one or two revolutions.

When the engine torque is greater than the resisting torque due to the load the engine speed increases; when the engine torque is less than the load torque the speed decreases. The flywheel acts as a reservoir of energy by virtue of its inertia. An excess of energy output is absorbed by the flywheel with only a small increase of speed; when the engine torque falls the absorbed energy is given up to the load.

For example, both the pressure on the piston of a reciprocating engine and the crank angle vary continuously during each revolution; the engine torque at the crankshaft therefore varies widely whereas the resistance due to the load may be constant. Similarly an electric motor delivering a constant torque may be used to drive a press or punch; the additional energy required momentarily during the punching operation is supplied by the flywheel.

The moment of inertia of a flywheel is chosen to keep the variation in speed within required limits of the operating mean speed. For example, an engine flywheel may be required to remain within ±2 per cent of mean engine speed.

The distinction between the functions of flywheel and governor should be noted: The flywheel determines the slight permissible variation of speed in one cycle of operations and acts as a store of energy. The governor controls the fuel supply to balance the average energy output with the load, while keeping the mean speed constant over a *period of time*.

Coefficient of fluctuation of speed

If a flywheel of moment of inertia I has maximum and minimum angular velocity ω_2 and ω_1 respectively, then the greatest fluctuation in energy stored is $\frac{1}{2}I(\omega_2{}^2 - \omega_1{}^2)$. If ω is the mean angular speed, then providing the speed variation is small (say 6 per cent),

$$\omega = \frac{\omega_1 + \omega_2}{2}$$

and the coefficient of fluctuation of speed, α, is defined as the ratio

$$\frac{\text{greatest fluctuation of speed per cycle}}{\text{mean speed}}$$

i.e. $\alpha = \dfrac{\omega_2 - \omega_1}{\omega}$

Thus, greatest fluctuation of energy $= \frac{1}{2}I(\omega_2{}^2 - \omega_1{}^2)$

$$= \frac{1}{2}I(\omega_2 - \omega_1)(\omega_2 + \omega_1)$$

$$= I\alpha\omega^2$$

Suppose, for example, that the variation in speed of a flywheel is ±2 per cent of the mean speed, ω, then

$$\alpha = \frac{1.02\omega - 0.98\omega}{\omega}$$

$$= 0.04$$

whereas, if the *total* variation is 2 per cent, then $\alpha = 0.02$.

Example *The greatest amount of energy which has to be stored by an engine flywheel is 2 kJ when the mean engine speed is 4 rev/s. The variation in speed is to be limited to ±1 per cent and the mass of the flywheel is 450 kg. Assuming the flywheel to be a cast iron ring having its internal diameter 0.9 of its external diameter, find the diameters and thickness of the rim. Take the density of cast iron as 7.2 Mg/m³.*

SOLUTION

$$\alpha = \frac{\omega_2 - \omega_1}{\omega} = 0.02$$

$$\omega = 2\pi \times 4 = 8\pi \text{rad/s}$$

$$I = mk^2 = 450k^2 \text{ kg-m}^2$$

Energy fluctuation $= \frac{1}{2}I(\omega_2^2 - \omega_1^2)$

$$= I\alpha\omega^2$$

therefore $2 \times 1,000 = 450 \times k^2 \times 0.02 \times (8\pi)^2$

$$k^2 = 0.352$$

therefore $\dfrac{D^2 + d^2}{8} = 0.352$

where D, d are the external and internal diameters respectively, and $d = 0.9D$. Hence

$$D = \textbf{1.25 m} \text{ and } d = \textbf{1.125 m}$$

$$\text{Mass} = 450 = \frac{\pi}{4}(D^2 - d^2) \times \text{thickness} \times \text{density}$$

$$= \frac{\pi}{4}(1.25^2 - 1.125^2) \times \text{thickness} \times 7,200$$

hence

thickness $= 0.268$ m $= \textbf{268 mm}$

Example *A 3 kW motor drives a flywheel which provides the energy for a machine press. At the start of a pressing operation the flywheel speed is 240 rev/min and each press takes 0.80 seconds and requires 5.5 kJ of energy. If the moment of inertia of the flywheel is 50 kg-m² find the reduction in speed of the flywheel after each pressing and the maximum number of pressings that can be made per minute. Assuming 80 per cent of the energy lost by the flywheel to be taken up at the press tool find the average force exerted when the stroke of the tool is 40 mm.*

SOLUTION

Energy supplied by the motor in 0.80 s $= 3 \times 1,000 \times 0.80$

$$= 2,400 \text{ J}$$

Energy required during each press $= 5.5$ kJ $= 5,500$ J

Loss of energy of flywheel = energy for pressing operation

$$- \text{ energy supplied by motor}$$

$$= 5,500 - 2,400$$

$$= 3,100 \text{ J}$$

therefore

$$\tfrac{1}{2}I(\omega_2{}^2 - \omega_1{}^2) = 3,100$$

$$\tfrac{1}{2} \times 50 \times \left[\left(\frac{2\pi \times 240}{60}\right)^2 - \left(\frac{2\pi N}{60}\right)^2\right] = 3,100$$

hence　　　　　　　　　　　　$N = 215 \text{ rev/min}$

reduction in speed = $240 - 215 = 25$ rev/min

Energy supplied by motor per minute = $3,000 \times 60 = 180,000$ J
therefore

$$\text{maximum number of pressings per minute} = \frac{180,000}{5,500}$$

$$= 32.7 \text{ say } \mathbf{32}$$

Energy actually used in pressing = $0.8 \times 5,500 = 4,400$ J

Average force × stroke = 4,400 J

$$\text{Average force} = \frac{4,400}{0.04} = 110,000 \text{ J}$$

$$= \mathbf{110 \text{ kJ}}$$

Problems

1. Find the mass of a flywheel required to keep the speed of an engine between 397 and 403 rev/min if its radius of gyration is 600 mm and the greatest fluctuation of energy is 8 kJ.

   (847 kg)

2. A machine for punching holes in sheet metal runs at 2 rev/s when not punching. The loss of energy when a hole is punched through is 5 kJ. Calculate the mass of a flywheel required, assuming the mass to be concentrated at a radius of 600 mm, if the lowest speed must not fall below 1 rev/s.

   (235 kg)

3. A flypress has two rotating spheres, each of mass 5 kg, fixed to a horizontal arm, so that they are at a radius of 700 mm from the axis of rotation. When the press is used to punch holes in a metal plate the initial speed of the arm is 5 rev/s and the speed falls to 4.5 rev/s during one operation. Find the energy used in the punching stroke and the average force exerted if the plate is 12 mm thick.

   (460 J; 38.3 kN)

4. A punching machine is driven by a motor which exerts a constant torque on the flywheel at a mean speed of 2 rev/s. If the speed variation from maximum

to minimum is not to exceed 0.2 rev/s and the greatest fluctuation of energy during the punching and idling strokes is 7 kJ find the moment of inertia of the flywheel.

(443 kg-m²)

5. An engine runs at 100 rev/min. The total fluctuation in speed is limited to 1.5 per cent of the mean speed and the maximum variation of energy supplied to the flywheel is 6.5 kJ. Find the mass of the flywheel required if the radius of gyration is 1 m.

(3.95 tonne)

6. An engine flywheel consists essentially of a ring of cast iron of density 7.2 Mg/m³. The flywheel has to deal with a fluctuation of energy of 22.5 kJ at a mean speed of 6 rev/s with a limit of ±0.5 per cent on the variation in speed. The maximum centrifugal stress in the flywheel is not to exceed 5.5 MN/m². Find the mean diameter and cross-sectional area of the rim.

(1,464 mm; 0.089 m²; for centrifugal stress *see* Chapter 13)

7. Under running conditions an engine flywheel has a mass of 1.1 tonne and a radius of gyration of 1,050 mm. The greatest fluctuation of energy during a cycle of operations is 2.15 kJ when the mean engine speed is 200 rev/min. On test, a temporary brake wheel of mass 360 kg and radius of gyration 1,150 mm is attached to the engine flywheel. Compare the speed fluctuations under running conditions and under test assuming the energy fluctuation to be the same in both cases.

(±2%; ±1.45%)

8. A riveting machine is driven by a 5 kW motor. The flywheel on the machine has a moment of inertia of 63 kg-m². The riveting operation requires 12 kJ of energy. Find the reduction in speed of the flywheel during each riveting operation if its speed at the start of each operation is 4 rev/s. What is the maximum rate at which rivets can be driven if each operation takes one second?

(0.775 rev/s; 25 per minute)

CHAPTER 11
Impulse and Momentum

11.1 Linear momentum: impulse

Consider a body of mass m acted upon by an average force F for a time t. The average acceleration a is given by:

$$a = \frac{v - u}{t}$$

where u and v are the initial and final velocities, respectively. Therefore

$$F = ma$$

$$= m \times \frac{v - u}{t}$$

We now define the *impulse* of the force F to be the product:

average force × time

Thus

$$\text{impulse} = F \times t$$

$$= m \times \frac{v - u}{t} \times t$$

$$= m(v - u)$$

i.e.

$$Ft = mv - mu$$

The product

mass × velocity

is a measure of the 'quantity of motion', called the *momentum* of the body. The term mv is the final momentum of the body at the end of time t; the second term mu is the initial momentum; the difference is the change of momentum. Thus the impulse Ft may be measured by the change in momentum it produces, i.e.

impulse = change of momentum

Again, since

$$F = \frac{mv - mu}{t}$$

then

$$\text{force} = \frac{\text{change of momentum}}{\text{time taken}}$$

or **force = rate of change of momentum**

and the change of momentum is measured in the direction of the force.*

The following points should be noted:

1. The mass m is assumed unaltered in any way, i.e. no part falls away or is added to the body.
2. Momentum is a *vector quantity*, having direction and sense corresponding to that of the velocity.
3. The force F is an *average* force and therefore the change of momentum does not depend on how the force may vary during time t.
4. When no external impulse is applied the momentum remains unchanged.
5. A change of momentum may be due to a change of speed *or* a vector change of velocity *or* to a change of mass. We shall consider only problems where there is a change of velocity and the mass is constant *during the period of application of the force*.

11.2 Units of impulse and momentum

Since

$$\text{momentum} = mv$$

---

* Note that this is simply a statement of the Second Law–'the applied force is *proportional* to the change of momentum per unit time'. Thus the law states

force α rate of change of momentum

i.e. $F \alpha \dfrac{(mv - mu)}{t}$

or $F = \text{constant} \times \dfrac{m(v - u)}{t}$

$\qquad = kma$

where k is a constant and a the acceleration produced. The SI units of force, mass, and velocity are such to make the value of the constant k unity, i.e. a force of 1 newton gives a mass of 1 kg an acceleration of 1 m/s². Therefore, since $k = 1$, the applied force is *equal* to the rate of change of momentum when using these units.

and mass m is in kilograms and velocity v in metres per second then

the units of momentum are **kg-m/s**

This is the only SI unit of momentum. However, it may be convenient in calculations to leave the units in some other form such as tonne-km/h.

Impulse is equal to the change of momentum and therefore the units of impulse are also **kg-m/s**. However, impulse Ft, is important where a large force acts for a very short time and the use of the unit form *newton-seconds* (N-s) serves on occasion to emphasize the time element in an impulse. Note that

$$1 \text{ N-s} = 1(\text{kg-m/s}^2) \times \text{s}$$

$$= 1 \text{ kg-m/s}$$

Example   *A shunting locomotive provides an impulse of 40 kN-s to set in motion a stationary 8-tonne wagon which then moves off freely at velocity u against a track resistance of 60 N/tonne and finally reaches a velocity v after 20 s. Find the values of u and v.*

SOLUTION
$$\text{Impulse} = \text{change of momentum of wagon}$$

therefore
$$40 \times 10^3 = 8 \times 10^3 \times (u - 0)$$

$$u = \textbf{5 m/s}$$

Retarding force, $F = 60 \times 8 = 480$ N

If the deceleration on the track is a, then

$$F = ma$$

i.e. $480 = 8 \times 10^3 \times a$

hence $a = 0.06$ m/s^2

and $v = u - at$

$$= 5 - 0.06 \times 20$$

$$= \textbf{3.8 m/s}$$

Problems

1. An 18-tonne truck has a maximum speed of 36 km/h on a track for which the resistance to motion is 250 N/tonne. The engine exerts a pull of 6 kN. What is the momentum at maximum speed? Using the impulse-momentum equation find the time taken to reach full speed from rest.

$$(180 \times 10^3 \text{ kg-m/s; } 120 \text{ s})$$

2. A planing machine-table has a mass of 800 kg and is to have a cutting speed of 0.4 m/s which it must reach in 0.8 s. The friction force is 40 N/tonne. Find the impulse and constant force required on the table.

$$(320 \text{ N-s; } 432 \text{ N})$$

3. An experimental passenger train consists of two power units and six carriages with a total mass of 456 tonne. The mass supported by the *driving* wheels on the power units is 140 tonne and the limiting coefficient of adhesion between wheel and track on the power units is 0.2. What is the maximum possible tractive effort?. If the resistance to motion is 120 N/tonne of total mass, find, using the impulse-momentum equation, the time taken to reach the maximum speed of 200 km/h from rest when the train climbs a gradient of 1 in 100.

(274.4 kN; 145 s)

11.3 Force varying with time

Consider a force F varying *linearly* with time t,* i.e. $F = kt$, where k is a constant, and the $F - t$ graph is a straight line, Fig. 11.1. Thus when $t = 0$, $F = 0$, and when $t = t_1$, $F = F_1 = kt_1$. The elementary impulse is $F\,dt$, the area of the elementary vertical strip. The total impulse is the sum of the elementary impulses, i.e. the total area of the graph. Thus:

impulse = area of graph

$$= \tfrac{1}{2}F_1 t_1$$

The average or constant force over the same *time period* to give the same impulse as the variable force is $\tfrac{1}{2}F_1$, then

impulse = average force × time

$$= \tfrac{1}{2}F_1 t_1$$

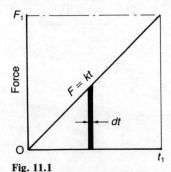

Fig. 11.1

---

* This is the only case of force varying with time that will concern us. The impulse of a force F is in general given by $\displaystyle\int_0^t F\,dt$ where F is a function of t. When

$F = kt$,

$$\text{impulse} = \int_0^{t_1} kt\,dt = \tfrac{1}{2}kt_1^2 = \tfrac{1}{2}F_1 t_1.$$

Note that this average force is different from the average force to give the same change in kinetic energy (work done) as the variable force.

Example   *A 210-kg mass initially at rest on a smooth horizontal plane is acted upon by a force F parallel to the plane. If F = 1.4t N where t is the time in seconds, find, for an interval of 30 s from rest, (a) the impulse, (b) the velocity, (c) the average force producing the same impulse.*

SOLUTION

(i) Fig. 11.2 shows the force-time graph. For an interval of 30 s from rest, the impulse is given by the area of the graph, therefore

impulse $= \frac{1}{2} \times (1.4 \times 30) \times 30 =$ **630 N-s**

(ii) change of momentum = impulse

$210(v - 0) = 630$

i.e., final velocity, $v =$ **3 m/s**

(iii) If F is the average force producing this impulse, then

average force $\times$ time = impulse

i.e. $F \times 30 = 630$

i.e. $F =$ **21 N**

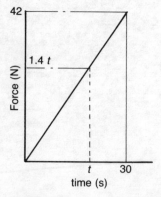

Fig.11.2

Problems

1. A 1200 kg mass has a net force applied to it which varies uniformly from 600 N to a peak of 900 N in 12s from rest and then falls uniformly to zero in the following 18s. Find the speed attained in km/h after 30s from rest.

(51.3 km/h)

2. A 4-kg mass moving at 3 m/s on a smooth horizontal plane is acted upon by a force which is applied in the same sense and direction as the motion of the mass and rises uniformly from zero to 200 N in 0.025 s before falling uniformly to zero in a further 0.025 s. Find the velocity of the mass after 0.05 s. What is the average force producing the same impulse?

(4.25 m/s; 100 N)

3. A force F acts on a mass of 400 kg initially at rest on a smooth horizontal plane. If $F = 60t$ N, where t is in seconds, and the line of action of the force is parallel to the plane, determine for a period of 20 s from rest (*a*) the impulse; (*b*) the velocity attained; (*c*) the average force to produce the impulse. If the distance travelled by the mass during the 20 s impulse period is 200 m what is the constant force to produce the same work done?

(12 kN-s; 30 m/s; 600 N; 900 N)

4. A 10-kg mass initially at rest on a smooth horizontal plane is acted upon by a force F parallel to the plane such that $F = 10 + 8t$ N where t is the time in seconds. Determine (*a*) the total impulse on the mass after the first 3 s of motion and the average force to produce this impulse; (*b*) the final kinetic energy and the average force to produce this energy if the mass travels a distance of 8 m during the period of the impulse.

(66 N-s; 22 N; 218 J; 27 N)

11.4 Conservation of linear momentum

The total momentum of a system of bodies is obtained by adding together the momentum of each individual body. When the external forces acting on a body are in balance the body is at rest with zero momentum, or moving at constant speed with constant momentum, providing the mass remains unaltered. There can be no change in the momentum of the body unless the external forces are unbalanced. If a number of bodies impinge on one another, or interact in some way, they may be treated as a system isolated from its surroundings, a 'closed' system in effect. There can be no change in the total linear momentum of the system in any given direction unless there is a resultant external force acting on the system *in that direction*. This is the *principle of conservation of linear momentum* which may be stated as

> **The total linear momentum of a body or system of bodies in any one direction remains constant unless acted upon by a resultant force in that direction.**

11.5 Impulsive forces

When a constant force F is applied to a body for a time t the impulse is Ft. When the force is large and the time interval very short, near instantaneous, it is called an *impulsive* or *impact* force. An impact force varies in magnitude from zero to a peak value and back to zero but the variations are incalculable and the force is assumed to be constant. Impulsive or impact forces arise in collisions, explosions, in the sudden tightening of a tow-rope, or the driving of a pile. The principle of conservation of linear momentum may be applied for the very short period of impact even although external forces, such as gravity and friction, may be acting, because they are normally negligible in comparison with the large impact force. Similarly where a spring is involved it may be assumed that the spring force does not come into action until after the impact is over.

11.6 Note on the use of momentum and energy equations

Both the momentum-impulse and work-energy equations can be used to solve problems in dynamics but usually one or other will be the most suitable and in some cases only one method will be possible. Key points to realise are:

1. Momentum is the easier concept to use when 'time' is the known quantity and for some situations the only concept. When 'distance' is the known quantity the work-energy equation is probably the most useful and convenient.
2. The momentum principle has the advantage in impact problems where the impact force is not known nor required and where inevitably there is a loss of energy. The law of conservation of energy cannot be used directly unless the problem is idealised by assuming perfectly elastic bodies (*see* page 252) The law applies but the various energy changes cannot be calculated.

11.7 Explosions

When a body explodes freely, fragments fly off due to the internal forces produced. Energy is added to the system from the explosion and appears as kinetic, heat and sound energy. The distribution of this additional energy is not known hence the final motion of the fragments cannot be found simply by applying the law of conservation of energy. It is necessary also to use the momentum principle and this can be applied because there are no external forces acting except for gravity and its effect can be neglected, since an explosion is instantaneous. *The total momentum, although redistributed, remains unchanged in any given direction before and immediately after an explosion.* There are several examples of explosions:

(I) BODY EXPLODING FROM REST
The total initial and final momentum of the fragments must be zero in any given direction.

(II) ROCKET STAGE SEPARATING OUT
A rocket in flight may be separated from its payload by an explosive bolt or release of a powerful spring mechanism. If the two parts continue along the line of flight then the sum of their momenta after separation must equal the initial momentum of the rocket. The impulse on each part is the same, given by *thrust × time of separation*.

(III) RIFLE AND BULLET
When a rifle fires a bullet the rifle kicks back due to the force of recoil. Equal and opposite forces are exerted on the rifle and bullet during the time that the bullet takes to traverse the barrel, hence the same impulse is applied to both rifle and bullet. The momentum of the bullet therefore equals that of the rifle. Or, the final momentum of the rifle-bullet system must be zero hence the forward momentum of the bullet must equal the backwards momentum of the rifle. For heavy guns firing shells, additional factors have to be considered since guns may be sited on inclines, travel on rails, and have specially designed arrangements of buffer springs, piston-cylinder mechanisms and retracting barrels, to reduce or prevent recoil forces completely. The momentum of the bullet equals that of the rifle but the kinetic energy given to the bullet is very much greater than that given to the rifle. The forces are equal but the work done in each case is different since the bullet travels a greater distance during the explosion. This is the difference between the 'time effect' of a force and the 'distance effect'.

Example *A projectile of mass 56 kg is moving at constant speed 8 m/s when it explodes into three pieces, A, B and C. A (20 kg) flies on at 20 m/s along the line of flight, and B (16 kg) flies at 15 m/s at 45° to the line of flight in the forward direction. Find the velocity of mass C in magnitude and direction and estimate the energy supplied in the explosion.*

SOLUTION
Fig. 11.3(a) shows the relative positions of A and B. Let C have components of *momentum x* and *y*, along and at right angles to the line of flight respectively. The initial momentum in the original direction is unchanged by the explosion, therefore equating momenta

total initial momentum = final momentum

= momentum of A

+ components of B and C
in the line of flight

i.e. $56 \times 8 = 20 \times 20 + 16 \times 15 \cos 45° + x$

i.e. $\qquad x = -121.7 \text{ N-s}$

x is negative, therefore opposite to initial direction of motion. The momentum at right angles to the line of flight is zero before and after the explosion, therefore the y component of the momentum of C must balance that of B, i.e.

$$y = 16 \times 15 \sin 45° = 169.7 \text{ N-s}$$

Fig. 11.3(b) shows the momentum vectors. The resultant R of components x and y is 209 N-s, and since the mass of C is 20 kg. Therefore

$$\text{velocity of C} = \frac{209}{20} = \textbf{10.45 m/s}$$

The direction of motion of C is given by

$$\theta = \tan^{-1} \frac{169.7}{121.7} = \textbf{54.4°}$$

Initial kinetic energy $= \frac{1}{2} \times 56 \times 8^2 = 1792 \text{ J}$

Final kinetic energy $= \frac{1}{2} \times 20 \times 20^2 + \frac{1}{2} \times 16 \times 15^2$

$$+ \tfrac{1}{2} \times 20 \times 10.45^2$$

$$= 6892 \text{ J}$$

Assuming all the energy supplied re-appears as kinetic energy then

energy of explosion = gain in kinetic energy

$$= 6892 - 1792$$

$$= \textbf{5100 J}$$

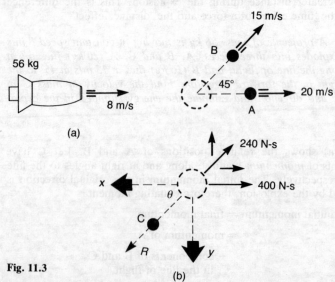

Fig. 11.3

Example *A 320 kg spacecraft is coasting at 5 km/s when an explosive charge giving a thrust of 1.5 kN separates off a 200 kg satellite which is propelled ahead of the launching craft in the line of flight. The speed of the craft falls by 15 m/s during the thrust period. Find the duration of the thrust, the final speed of the satellite, and the speed of recession.*

SOLUTION
If t s is the duration of the thrust, the separation period, then the impulse on both craft and satellite is $1500t$ N-s. The loss of speed of the craft is 15 m/s, and it's mass is 120 kg, hence for the craft, Fig.11.4.

impulse = change of momentum

i.e. $1500t = 120 \times 15$

i.e. $t = \mathbf{1.2\ s}$

The mass of the satellite is 200 kg, and if it's velocity after time t is v, then

$$1500t = 200(v - 5000)$$

and $t = 1.2$ s, hence

$$v = \mathbf{5009\ m/s}$$

Alternatively, equating the initial and final momentum of the spacecraft,

$$320 \times 5000 = 200 \times v + 120 \times (5000 - 15)$$

i.e. $v = 5009$ m/s

The satellite moves off at 9 m/s relative to the coasting speed, and the craft falls back at 15 m/s, hence the speed of recession is **24 m/s**.

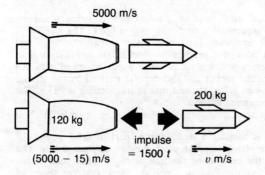

Fig. 11.4

Problems

1. A 12 kg shell is moving at a constant speed of 20 m/s when it explodes into two parts. The largest part of mass 8 kg is immediately stationary. Calculate

the energy supplied in the explosion, assuming it is all translated into kinetic energy.

(4.8 kJ)

2. A projectile of mass 40 kg is instantly at rest in mid-air when it explodes into pieces, A, B, C, D of masses 8, 14, 11 and 7 kg respectively. A flies off at 8 m/s, B at 4 m/s, and C at 3 m/s. A and B move in opposite directions and C moves at right angles to them. Find the speed and direction of mass D and the energy of the explosion.

(4.85 m/s; A, 0°, C 90°, B, 180°, D, 256.4°; 500 J)

3. A 120 kg rocket moving at constant speed 9 m/s explodes into four pieces, two of which are of equal mass, 25 kg, and continue in the same direction as the rocket, one at 8 m/s the other at 12 m/s. A third piece of mass 30 kg flies off at right angles to the original line of flight at 4 m/s. Find the speed and direction of motion of the fourth piece and estimate the energy of the explosion.

(14.8 m/s; line of flight, 0°, third piece, 90°, fourth piece, 348°; 2.36 kJ)

4. A rifle of mass 16 kg fires a 10 g bullet with a muzzle velocity of 840 m/s. What is the velocity of recoil and the energy of the explosion?

(0.525 m/s; 3.53 kJ)

5. An explosive charge giving an impulse of 4 kN-s is used to discharge a 1.3 tonne satellite from a spaceship in steady flight. The mass of vehicle remaining after separation is 1.1 tonne. What is the speed of recession of the two parts, assuming the final motions to be in the line of flight? Show that the energy of the explosion appearing as additional kinetic energy is about 13.4 kJ.

(6.7 m/s; take the speed of flight as u m/s or, more simply, zero)

6. A sounding rocket of total mass 120 kg including an 80 kg payload is coasting at 20 m/s when an explosive charge lasting 0.8 s breaks the connections mating it with its payload capsule which then moves off in the line of flight. If the energy supplied in the explosion is 3 kJ, find the velocities of the payload and rocket after 0.8 s and the thrust exerted, neglecting losses.

(25 m/s; 10 m/s; 500 N)

7. A meteorological satellite used to take photographs is connected to a launching rocket by a compression spring and tension bolts. The satellite is disconnected by a small explosion just sufficient to break the bolts and release the spring. When the rocket reaches its target height and is cruising at a constant speed complete separation takes place. Find the speed at which the satellite recedes from the rocket and the impulse exerted by the spring. The mass of the rocket alone is 210 kg, and that of the satellite is 70 kg. The strain energy stored in the spring is 420 J.

(4 m/s; 210 N-s)

8. A spacecraft of mass m is coasting a steadily in a 'parking-orbit' round the Earth when an explosive charge separates off a cargo of mass $0.6\,m$. The cargo flies off ahead of the craft along the line of flight. The speed of the craft falls by 10 m/s during the explosion period. Show that the speed of recession is 16.7 m/s and the energy of explosion is $33.3\,m$J. If $m = 300$ kg, and given that the thrust from the explosion is 1.2 kN, find the time of separation.

(1 s)

11.8 Collision of two bodies

A body A, of mass m_1, moving with velocity u_1 collides with a second body B, of mass m_2, moving along the same straight line with velocity u_2, Fig. 11.5. During the collision there is an impulse Ft exerted by one body on the other. If the time t of the impact is very short, and hence the impulsive force very large, then the change of momentum due to all other forces external to the two-body system (e.g. gravity, friction) may be neglected. *Hence, the total momentum of the system remains constant during the impact and is therefore the same after the collision as before it.*

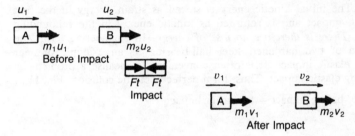

Fig. 11.5

Let v_1, v_2 be the velocities of A and B, respectively, immediately after the impact, then the *momentum equation* is

momentum before impact = momentum after impact

i.e. $m_1u_1 + m_2u_2 \qquad = m_1v_1 + m_2v_2$

We have assumed the positive direction, and both final velocities to be from left to right. Each body undergoes a change of momentum of equal magnitude but opposite direction. Thus since the impulse Ft *on* A is negative,

$$-Ft = m_1(v_1 - u_1)$$

The impulse Ft *on* B is positive, hence

$$Ft = m_2(v_2 - u_2)$$

These two equations relate the impulsive force to the time of impact and the velocities. Combining them gives the above momentum equation so that we have only one equation to find two unknown quantities–the final velocities. One other equation is required and to obtain it we must have additional information. If, say, v_1 is known then we can find v_2. If both velocities are unknown we need to know if the bodies move on together or rebound. This depends on the elasticity of the bodies or whether they are mechanically coupled on impact. In an impact the bodies first deform, then they may tend to return to their original shape and rebound due to the *force of restitution* between them. The amount of restitution depends on whether the bodies are perfectly elastic, partially elastic, or inelastic. For every case the above momentum equations apply.

11.9 Collision of perfectly elastic bodies

Perfectly elastic bodies after colliding return completely to their original shape. The initial kinetic enery is stored as strain energy in the first stage of impact and is returned as kinetic energy in the rebounding bodies. *There is therefore no loss of energy.* In engineering terms the collision of two hardened steel ball-bearings is an example of near perfect elastic impact. It is often convenient to idealise a problem by assuming elastic impact. Thus, for a perfectly elastic collision, Fig. 11.5,

k.e. before impact = k.e. after impact

hence

$$\tfrac{1}{2}m_1u_1{}^2 + \tfrac{1}{2}m_2u_2{}^2 = \tfrac{1}{2}m_1v_1{}^2 + \tfrac{1}{2}m_2v_2{}^2$$

and the momentum equation is

$$m_1u_1 + m_2u_2 = m_1v_1 + m_2v_2$$

From these two equations we arrive at the *relative velocity* equation,

$$v_1 - v_2 = -(u_1 - u_2)$$

Thus when two perfectly elastic bodies collide the *relative velocity of separation* $(v_1 - v_2)$, is equal to the *relative velocity of approach* $(u_1 - u_2)$, but in the opposite direction. For example, if a ball drops on to a surface with a velocity u_1, and the impact is elastic then for the surface $u_2 = 0$, $v_2 = 0$, and for the ball the velocity of rebound $v_1 = -u_1$ i.e. the ball rebounds reversing its velocity and rising to the same height from which it dropped.

Example *A 40-tonne rail-car travels at 4 km/h and collides with a 100-tonne wagon on the same track, moving in the opposite direction at 1.2 km/h. Find their velocities immediately after impact assuming no loss of energy. What is the impulse between them?*

SOLUTION
Refer to Fig. 11.5. The direction of motion of the rail-car (u_1) is taken as positive, then $u_1 = 4$ km/h and $u_2 = -1.2$ km/h. The final velocities v_1 and v_2 are assumed positive. Equating momenta

$$40v_1 + 100v_2 = 40 \times 4 + 100 \times (-1.2)*$$

$$v_1 + 2.5v_2 = 1$$

The relative velocity equation is

$$v_1 - v_2 = -(u_1 - u_2)$$

$$= - [4 - (-1.2)]$$

$$= - 5.2$$

From these two equations $v_1 = -3.43$ km/h and $v_2 = 1.77$ km/h i.e. the rail-car rebounds at **3.43 km/h** and the wagon rebounds at **1.77 km/h**.

The impulse on each body is equal to the change in its momentum and is the same for both. For the rail-car,

$$\text{impulse} = \text{final momentum} - \text{initial momentum}$$

$$= 40 \times 10^3 \times \left(-3.43 \times \frac{1}{3.6}\right) - 40 \times 10^3 \times \left(4 \times \frac{1}{3.6}\right)*$$

$$= - 82.6 \times 10^3 \text{ N-s}$$

$$= -82.6 \text{ kN-s (or tonne-m/s)}$$

The negative sign indicates that the impulse on the 40-tonne car is in the opposite direction to its original motion which was taken as positive.

*In the momentum equation there is no need to convert tonnes to kg or km/h to m/s since the same quantities occur on both sides of the equation but in calculating the impulse the conversion is essential.

Example *Fig. 11.6 shows a 6.5 kg sphere constrained to move along frictionless guides at a speed of 4 m/s towards a 26 kg block which is resting against a buffer spring of stiffness S = 2.2 kN/m. The spring has no initial compression and there is no loss of energy at impact. Find the velocities of the two masses immediately after impact and the maximum compression of the spring.*

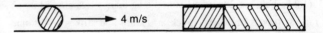

Fig. 11.6

SOLUTION
For the sphere, $u_1 = 4$ m/s, and final velocity v_1 is assumed to be from left to right (positive). *For the block*, initially at rest, $u_2 = 0$, and final velocity v_2 is from left to right (positive). Assuming the impact force to be over before the spring force acts then the momentum remains

unchanged before and after impact, therefore

$$6.5v_1 + 26v_2 = 6.5 \times 4 + 26 \times 0$$

i.e. $v_1 + 4v_2 = 4$

The relative velocity equation is

$$v_1 - v_2 = -(u_1 - u_2) = -(4 - 0) = -4$$

Solving these two equations gives

$$v_1 = -2.4 \text{ m/s}$$
$$v_2 = 1.6 \text{ m/s}$$

The sphere therefore *rebounds* with a velocity of **2.4 m/s**. The block starts to compress the spring with an initial velocity of **1.6 m/s** and its final velocity is zero when the spring has been compressed a distance x. The kinetic energy ($\frac{1}{2}mv^2$) lost by the block is equal to the strain energy stored in the spring ($\frac{1}{2}Sx^2$), hence

$$\tfrac{1}{2} \times 26 \times 1.6^2 = \tfrac{1}{2} \times 2.2 \times 10^3 \times x^2$$

therefore $x = \mathbf{0.17 \ m}$

Problems

1. A wagon of mass 20 tonne, moving along a track at 12 km/h, collides with a second wagon, of mass 10 tonne, moving at 7 km/h; both wagons are moving in the same direction. Immediately after the collision the 10-tonne wagon moves on at 11 km/h. Calculate the velocity of the 20-tonne wagon after impact and the impulse between the wagons.

   (10 km/h; 11,110 kg-m/s or 11.1 kN-s)

2. A light rod of length 2.5 m is suspended at one end and carries an iron ball of mass 3.6 kg at the other end. It is initially at rest in the vertical position. A second ball of mass 1.2 kg is carried by a similar rod suspended from the same point and held such that the angle between the two rods is 30° before being released. Assuming elastic impact find the velocity of each ball immediately after impact and the height to which each rises.

   (1.2 kg ball; −1.28 m/s, 84 mm; 3.6 kg ball; +1.28 m/s, 84 mm.)

3. A 30-tonne rail-car travels on a level track at 9 km/h and strikes a 45-tonne wagon resting against buffer springs of total stiffness 900 kN/m. The springs have no initial compression. Assuming the collision to be perfectly elastic find the maximum compression of the springs. How far will the 30-tonne car travel backwards before coming to rest if the track resistance is 75 N/tonne.

   (447 mm; 1.67 m)

4. A wagon of mass 100 tonne, moving at 4 km/h, collides with the back of a wagon, of mass 40 tonne, moving in the same direction at 1 km/h. What is the velocity of the 100-tonne wagon after the impact and the impulse between them if the 40-tonne wagon moves off at 5 km/h after the impact?

   (2.4 km/h; 44,400 N-s)

11.10 Inelastic collisions

Bodies may collide in such a way that one penetrates the other and permanently deforms it and then both move on together with a common velocity, Fig. 11.7. An example is a bullet embedding itself in a stationary target which then moves off. It is assumed that the collision time is so brief that the target does not move until the bullet is at rest relative to the target. There is no force of restitution between the bodies and the impact is said to be *inelastic*. The energy expended in deformation, resulting in heat and noise, is lost to the system. *The total momentum before and after impact remains unchanged since there is no external force acting.* Similarly when two bodies collide and move on together because they are held by a quick-coupling mechanism, a suddenly tightening tow-rope, or simply get tangled together.

If V is the common velocity immediately after impact, then the momentum equation is

$$m_1 u_1 + m_2 u_2 = (m_1 + m_2)V$$

The *loss* of kinetic energy is given by

$$(\tfrac{1}{2}m_1 u_1{}^2 + \tfrac{1}{2}m_2 u_2{}^2) - \tfrac{1}{2}(m_1 + m_2)V^2$$

In the important case where one mass, say m_2, is at rest, then $u_2 = 0$, and the momentum equation becomes

$$m_1 u_1 = (m_1 + m_2)V$$

i.e. $\qquad V = \dfrac{m_1}{m_1 + m_2} \cdot u_1$

and the loss of kinetic energy is

$$\tfrac{1}{2}m_1 u_1^2 - \tfrac{1}{2}(m_1 + m_2)V^2 = \tfrac{1}{2}m_1 u_1^2 - \tfrac{1}{2}(m_1 + m_2)\left[\frac{m_1 u_1}{m_1 + m_2}\right]^2$$

$$= \left(1 - \frac{m_1}{m_1 + m_2}\right)\tfrac{1}{2}m_1 u_1^2$$

This expression shows that there is *always* some loss of energy since the multiplying factor is always less than unity.

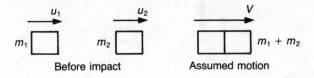

Before impact   Assumed motion

Fig. 11.7

Example *A pile-driving hammer of mass 0.5 tonne falls 2.4 m from rest on to a pile of mass 145 kg. There is no rebound and the pile is driven 150 mm into the ground. Calculate the common velocity after impact and the average resisting force of the ground in bringing the pile and driver to rest.*

SOLUTION

Equating the kinetic energy just before impact to the initial potential energy of the hammer measured above the point of impact, Fig. 11.8

$$\tfrac{1}{2}mv_1^2 = mgh$$

thus

striking velocity $v_1 = \sqrt{(2gh)}$

$$= \sqrt{(2 \times 9.8 \times 2.4)}$$

$$= 6.86 \text{ m/s}$$

Mass of pile and driver $m_1 = 145 + 0.5 \times 10^3 = 645$ kg

During impact it is assumed that the impact force between pile and driver is very much greater than the resisting force offered by the ground and the effect of gravity. The latter forces may therefore be neglected *during* impact. There is therefore no appreciable external force acting on the pile and driver. Hence

momentum of pile and driver   = momentum of driver before
    after impact   impact

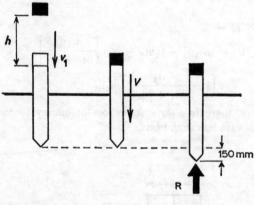

Fig. 11.8

Let v be the common velocity of pile and driver after impact, then

$$645v = 0.5 \times 10^3 \times 6.86$$

therefore

$$v = \textbf{5.32 m/s}$$

Kinetic energy of system after impact $= \frac{1}{2}m_1v^2$

$$= \frac{1}{2} \times 645 \times 5.32^2$$

$$= 9.130 \text{ J}$$

loss of potential energy in descending a *further* 150 mm:

$$= (645 \times 9.8) \times 0.15 = 948 \text{ J}$$

work done by hammer and pile = kinetic energy lost
+ loss of potential energy

$$= 9,130 + 948$$

$$= 10,078 \text{ J}$$

But the work done against the resisting force is $R \times 0.15$ J where R is in newtons, hence

$$R \times 0.15 = 10,078$$
$$R = 67,200 \text{ N} = \textbf{67 kN}$$

Example *A 25 kg package slides down a smooth chute as shown in Fig. 11.9. At the bottom it collides with a stationary trolley of mass 40 kg. The package and trolley move off together on a rough horizontal surface for which the resistance is 3N/kg. Find the common velocity after impact and the distance travelled on the horizontal before coming to rest, assuming the package and trolley remain together.*

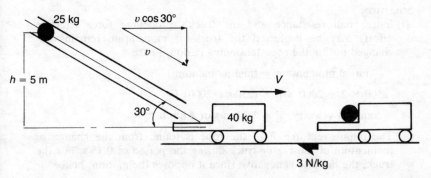

Fig. 11.9

SOLUTION
The velocity v of the package at the foot of the slope is given by

$$v = \sqrt{2gh} = \sqrt{2 \times 9.8 \times 5} = 9.9 \text{ m/s}$$

This velocity has a vertical component and the momentum due to this component is destroyed on impact. The horizontal component of velocity is $v \cos 30°$ and since the *horizontal* momentum remains unchanged, hence equating momenta

$$25 \times 9.9 \cos 30° + 40 \times 0 = 65 \, V$$

i.e. common velocity $V = \textbf{3.3 m/s}$

On the level the friction force $F = 3 \times 65 = 195$ N.

The initial velocity of the two bodies is 3.3 m/s and the deceleration a is given by

$$F = ma$$

$$195 = (25 + 40) \, a$$

$$a = 3 \text{ m/s}^2$$

If s is the distance covered before coming to rest, from the equation of motion

$$v^2 = u^2 - 2as$$

$$0 = 3.3^2 - 2 \times 3 \times \text{s}$$

i.e. $s = \textbf{1.82 m}$

Example *A 2-tonne car is towed by a 2.5 tonne truck which moves off from rest and reaches a speed of 2 m/s before the tow-rope tightens on the stationary car. The rope then stretches taking up the strain for 0.15 s. Find: (a) the common speed when the rope ceases to stretch; (b) the average impulsive force in the rope. Neglect resistance and the effect of the truck's propelling force during the stretching period.*

SOLUTION

(i) Since road resistance and the truck's propelling force (tractive effort) may be neglected the truck-car momentum remains unchanged during the rope tensioning period, hence

initial momentum = final momentum

$$2500 \times 2 + 2000 \times 0 = (2500 + 2000) \, V$$

i.e. common velocity, $V = 1.11$ m/s or **4 km/h**

(ii) The impulsive force F in the rope is found from the change of momentum of the car *or* truck during the period of 0.15 s. For the truck, the impulse is negative since it opposes the motion, hence

impulse = final momentum − initial momentum

i.e. $-F \times 0.15 = 2500(1.11 - 2)$

i.e. $F = 14,830$ N or **14.83 kN**

Problems

1. Two similar vehicles, travelling at 30 km/h in opposite directions, collide with one another. What would be their velocities after impact (a) in a perfectly

elastic collision, (b) in a completely inelastic collision? Show that in case (b) both the impulse and the total kinetic energy loss are each one-half of that which occurs when a similar vehicle collides inelastically with a wall at 60 km/h.

(30 km/h; zero)

2. A simple ballistic pendulum, a device used in the past to measure the speed of a bullet, consists of a 24 kg block of wood suspended by a light rod. The bullet of mass m is fired horizontally at close range into the block, embedding itself, and the block then swings to a height h above the lowest position. If $m = 15$ g and $h = 20$ mm find the muzzle velocity of the bullet.

(1 km/s)

3. A 5 kg package slides 6 m down a frictionless chute and lands on a stationary trolley of mass 10 kg. The chute is inclined at 30° to the horizontal and the trolley is free to roll on horizontal guides. If both package and trolley move off together from the moment of impact and the resistance of the guides is 3 N/kg find the common velocity and the distance travelled by the trolley before coming to rest.

(2.2 m/s; 8.17 mm; note that the vertial momentum of the package is destroyed on impact).

4. A rail wagon, of mass 20 tonne, starts from rest down an incline of 1 in 24 in a marshalling yard. The resistance to rolling motion is 70 N/tonne. Half-way down the incline, which is 180 m long, the wagon collides with a similar wagon at rest. Find (a) the velocity of the first wagon just before impact, (b) the velocity of the two wagons immediately after impact if they travel on coupled together, (c) their common velocity at the end of the incline.

(7.8 m/s; 3.9 m/s; 8.72 m/s)

5. A package of mass 30 kg slides 9 m down a chute of gradient 1 in 4. At the bottom it collides with a stationary package of mass 45 kg. Both parcels then travel on together on a horizontal surface. If the coefficient of friction between each package and the chute is 0.1, on *both* the gradient and the level, find (a) the common velocity immediately after impact, (b) the distance travelled on the level before they both come to rest.

(2 m/s; 2.04 m)

6. A pile is driven into the ground by a hammer of mass 400 kg, dropped from a height of 3.75 m. The pile's mass is 45 kg and the average resistance of the ground to penetration is 45 kN. Find the common velocity of pile and hammer after impact and the distance through which the pile is driven into the ground.

(7.7 m/s; 325 mm)

7. A steam hammer, of mass 8 Mg, moves vertically downwards from rest through a distance of 2 m on to a pile of mass 1 Mg. The hammer falls under the influence of its own weight and a force due to steam pressure of 120 kN. What is the velocity of striking?

If the steam pressure is cut off at impact, and there is no rebound of the hammer, find the common velocity of hammer and pile immediately after impact. What is the average resistance to penetration if the pile is driven 450 mm at each blow?

(9.95 m/s; 8.85 m/s; 870 kN)

8. A weight is dropped from a height of 1.2 m on to a stake and drives it into the ground. When the falling weight is three times as great as that of the pile to be driven it is found that it takes five impacts to drive the pile 250 mm. Calculate the number of impacts required to drive the pile the same distance

when the falling weight is five times as great as the pile, and the height through which it falls remains the same.

(three; 95 mm per impact)

9. A 600-tonne tug tows a ship of 8000 tonne from rest by means of a cable. Assuming the cable is initially slack and tightens up suddenly when the tug's speed is 2 m/s and that the cable tension reaches its maximum possible value of 80 kN find the common velocity of the two ships and the time taken neglecting resistances and the effect of the thrust of the tug's engine

(0.14 m/s; 14 s)

11.11 Collision of partially elastic bodies

When perfectly elastic bodies collide there is full restitution, i.e. no deformation, no loss of energy and the velocity of recession is equal to the velocity of approach. Where the bodies have some elasticity there will be *partial restitution* accompanied by loss of energy. The conservation of momentum equation still applies but the relative velocities of approach and recession are no longer equal. Newton's experiments on partially elastic bodies in direct collision showed that the ratio of the relative velocities was *constant for a given pair of bodies and opposite in direction*. This ratio is denoted by e. When the collision is *oblique* it is the ratio of the relative velocities in the direction of the common normal that is constant. This is *Newton's Law of Impact*, an experimental law and approximate only since so many factors affect the conditions during a collision – the line of impact, the magnitude of the velocities, and the size and shape of the bodies.

The constant 'e' is called the *coefficient of restitution*, hence, referring to Fig.11.5

$$\frac{\text{velocity of recession}}{\text{velocity of approach}} = \frac{v_1 - v_2}{u_1 - u_2} = -e$$

or

$$v_1 - v_2 = -e(u_1 - u_2)$$

The negative sign indicates that the relative velocities are opposite in direction. The value of e differs considerably for different bodies, varying from just less than unity for two hardened steel spheres to 0.2 for lead spheres. Note that the relative velocity equation applies equally to an elastic collision ($e = 1$) and to an inelastic collision ($e = 0$)

Example   *A ball-bearing is dropped on to a rigid-hard surface from a height h, Fig. 11.10. It rebounds to a heigh h_1. Find the coefficient of restitution if h = 800 mm and h_1 = 650 mm,*

SOLUTION

For the ball-bearing, the velocity with which it strikes the surface is

$$u_1 = \sqrt{2gh} = \sqrt{2 \times 9.8 \times 0.8} = 3.96 \text{ m/s}$$

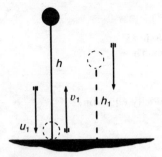

Fig. 11.10

and the velocity with which it leaves the surface to rise to a height of 650 mm is

$$v_1 = \sqrt{2gh_1} = \sqrt{2 \times 9.8 \times 0.65} = 3.6 \text{ m/s}$$

The impact is treated as a collision where the velocity of one of the masses before and after impact is zero, i.e. $u_2 = 0$ and $v_2 = 0$. Then if the direction of u_1 is taken as positive, v_1 is negative and the relative velocity equation is

$$v_1 - v_2 = -e(u_1 - u_2)$$

i.e. $-3.6 - 0 = -e(3.96 - 0)$

i.e. $e = \mathbf{0.91}$

Example *A sphere of mass 2.5 kg strikes a horizontal smooth surface obliquely with a velocity of 18 m/s, and rebounds with velocity V at an angle α to the surface, as shown in Fig. 11.11. Find the values of V and α if the coefficient of restitution is 0.8. What is the impulsive reaction of the surface on the sphere if the duration of the impulse is 0.5 s?*

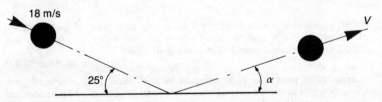

Fig. 11.11

SOLUTION

The momentum equation for the sphere in the *horizontal* direction is

$$2.5 \times 18 \cos 25° = 2.5 \times V \cos \alpha$$

i.e. $V \cos \alpha = 16.3$

Consider the velocities *normal* to the surface, using the notation of paragraph 11.11, then

for the sphere, $u_1 = + 18 \sin 25° \text{ m/s}$
and $v_1 = - V \sin \alpha$
for the surface, $u_2 = 0$
and $v_2 = 0.$

Substituting these values in the relative velocity equation,

$$v_1 - v_2 = -e(u_1 - u_2)$$

i.e. $-V \sin \alpha - 0 = -0.8(18 \sin 25° - 0)$

i.e. $V \sin \alpha = 6.09$

Solving these two equations relating V and α,

$\alpha = \textbf{20.5°}$

and $V = \textbf{17.4 m/s}$

The reaction R of the surface produces an impulse Rt which is equal to the *change* of momentum in the direction normal to the surface. Taking u_1 as positive, v_1 and the impulse are negative, therefore

$$-Rt = m(v_1 - u_1)$$

i.e. $-R \times 0.05 = 2.5(-17.4 \sin 20.5° - 18 \sin 25°)$

i.e. $R = \textbf{684 N}$

Problems

1. A metal ball is dropped on to a rigid surface from a height of 800 mm and rebounds to a height of 500 mm after the second impact. Find the coefficient of restitution.

$$(0.89; \text{ note}, \frac{h_3}{h_1} = e^4)$$

2. A ball is dropped from a height of 5 m on to a hard floor. If the coefficient of restitution is 0.4 find the height to which the ball rises after the first impact and the time it takes to reach the floor again.

(800 mm; 0.8 s)

3. A body A of mass 4 kg is constrained by frictionless guides to move at a speed of 8 m/s towards a second stationary body B of mass 12 kg constrained by the same guides. The bodies collide and the coefficient of restitution is 0.9. Find the speeds after impact and the loss of kinetic energy.

(A, −3.4 m/s; B 3.8 m/s; 18.2 J)

4. A sphere of mass 1.2 kg slides down a tube as shown in Fig. 11.12. At the bottom of the curved portion it strikes a stationary block of mass 2.4 kg. If the tube is assumed frictionless for the curved portion and to have a coefficient of friction of 0.4 for the horizontal stretch, find the distance travelled by the block before coming to rest (a) for a perfectly elastic collision (b) if $e = 0.8$ (c) when the sphere and block move on together as a single body.

(1.22 m; 0.99 m; 0.306 m)

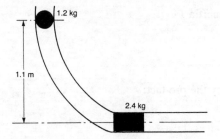

Fig. 11.12

5. A metal ball of mass 5 kg attached to a light arm of length 1.2 m falls from the horizontal position as shown in Fig. 11.13 to strike a stationary block of mass 10 kg. After impact the block is observed to move off with a velocity of 3 m/s. Find the coefficient of restitution and the velocity of the sphere immediately after impact.

(0.86; 1.15 m/s to the right)

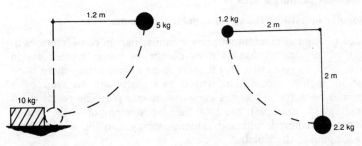

Fig. 11.13   **Fig. 11.14**

6. A mass of 1.2 kg is released from the position shown in Fig. 11.14 and swings down at the end of a string of length 2 m to strike a mass of 2.2 kg suspended similarly at rest. At impact the masses rebound; the lighter mass swings to a height h and the other mass swings to the right to a height of 900 mm. Find (a) the velocities of separation immediately after impact; (b) the coefficient of restitution; (c) the value of h.

(1.2 kg mass, 1.44 m/s to the left; 2.2 kg mass, 4.2 m/s to the right; 0.9; 105 mm)

11.12 Angular momentum and impulse

Consider a shaft rotating with initial angular velocity ω_0, and acted upon by a torque T for time t, and let the final velocity be ω. Then the average angular acceleration α is given by:

$$\alpha = \frac{\omega - \omega_0}{t}$$

and if the shaft has moment of inertia I

$$T = I\alpha$$

$$= I\frac{\omega - \omega_0}{t}$$

We define the *angular impulse* by the product:

torque × time

i.e. impulse = $T \times t$

$$= I\frac{\omega - \omega_0}{t} \times t$$

$$= I(\omega - \omega_0)$$

$$= I\omega - I\omega_0$$

The product $I\omega$ is called the *angular momentum* of the rotating body; the angular impulse is therefore measured by the change in angular momentum it produces; thus

angular impulse = change of angular momentum

The idea of angular momentum plays a similar part in considerations of changes of angular motion as does change of linear momentum in linear motion. In particular: (a) if there is no torque acting there is no change in angular momentum; (b) if two shafts are engaged by a clutch, and no external torques are brought into play (by reactions at the bearings for example), the total angular momentum of the system will remain unaltered, although kinetic energy may be lost in heat during slipping of the clutch.

The units of angular momentum and angular impulse are kg-m²/s and this is the only recommended SI form. However it is useful to note that another form of unit sometimes used for angular impulse indicates more clearly the *torque × time* meaning, i.e. N-m-s.

The units of linear momentum are kg-m/s and it is important to realize that these quantities, linear and angular momentum, *cannot be added together*.

Example *Two masses, each of mass 4 kg, rotate at a radius of 1.2 m attached at opposite ends of a light arm, Fig. 11.15. The initial speed of rotation is 400 rev/min. If the two masses are moved inwards along the arm to a radius of 0.8 m find (a) the final speed of rotation, (b) the change in kinetic energy of rotation.*

SOLUTION

(a) Since there is no external torque acting on the shaft carrying the rotating masses the angular momentum remains unchanged by the change in radius of rotation. Let the final speed of rotation be N rev/min, then

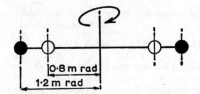

Fig. 11.15

final angular momentum = initial angular momentum

or $$I_2 \times N = I_1 \times 400$$

where I_1 and I_2 are the initial and final moments of inertia, respectively. (*Note:* It is unnecessary to convert the speeds of rotation to radians per second since the conversion factor is common to both sides of the equation and would cancel.)

Assuming each mass to be concentrated at its respective radius of rotation

$$I_1 = 2 \times 4 \times 1.2^2 \text{ kg-m}^2$$

$$I_2 = 2 \times 4 \times 0.8^2 \text{ kg-m}^2, \text{ for two masses}$$

Hence $(2 \times 4 \times 0.8^2) \times N = (2 \times 4 \times 1.2^2) \times 400$

thus $$N = \textbf{900 rev/min}$$

(b) Initial kinetic energy $= \frac{1}{2}mk^2\omega^2$

$$= \frac{1}{2} \times 8 \times 1.2^2 \times \left(\frac{2\pi}{60} \times 400\right)^2$$

$$= 10,100 \text{ J}$$

final kinetic energy $= \frac{1}{2} \times 8 \times 0.8^2 \times \left(\frac{2\pi}{60} \times 900\right)^2$

$$= 22,800 \text{ J}$$

change in kinetic energy $= 22,800 - 10,100$

$$= 12,700 \text{ J} = \textbf{12.7 kJ}$$

This is a *gain* in kinetic energy and arises from the work done in moving the masses inwards against the radial inertia force due to rotation.

Problems

1. A light bar rotates about a perpendicular axis through its centre. Attached to each arm of the rotating bar are two similar movable masses. The plane of rotation is horizontal. If, when the masses each rotate at a radius of 1 m, the speed of rotation is 6 rev/s, what will be their speed at a radius of 1.25 m when moved radially outwards during free rotation?

If each mass is 3 kg what is the loss of energy due to the change in radius?

(3.84 rev/s; 1.54 kJ)

2. A scheme proposed to conserve fuel in running a vehicle consists of a flywheel connected to the engine in such a way that energy gained when running downhill may be stored and utilized when running uphill or accelerating. If the flywheel's mass is 100 kg, its radius of gyration 400 mm and it has a maximum speed of 30,000 rev/min, calculate the corresponding angular momentum and kinetic energy of rotation.

(50.3 Mg.m^2/s; 79 MJ)

3. A rotating table has a moment of inertia about a vertical axis through its centroid of 8 kg-m^2 and turns at 6 rev/s. A servo piston of mass 1.4 kg rotates with the table at an initial radius of 1.2 m. It then moves towards the centre of rotation a distance of 600 mm in such a way that no external torque acts on the system. Assuming the piston to be a concentrated mass determine the final speed of free rotation of the table.

(7.07 rev/s)

CHAPTER 12
Aircraft and Rockets

12.1 Reaction propulsion

The principle of *reaction propulsion* whereby a jet of fluid is formed and expelled from an engine or pushed by a rotating rotor is the basis of working for propeller-driven ships and planes, jet planes, rockets, helicopters, and satellite control. Apart from the mechanical aspects and body structure, each type of vehicle differs greatly in regard to the mass of fluid dealt with and the speed and form of the jet.

For any type of jet propelled machine, the mass fluid flow, jet speed and flight speed, govern the magnitude of the propelling force or *thrust* as shown below.

12.2 Jet propulsion aircraft

The simplest form of jet propulsion is the *ramjet*, used for high-altitude, high-speed flight. The 'ram effect' is due to the forward speed of the plane or missile forcing the necessary air for combustion into the engine through a front duct and diffuser (a diverging chamber) which greatly slows it down, and then a jet of hot, extremely high-speed gas is ejected from the tail. A ramjet cannot be used on its own but must be launched at high speed or have an auxiliary power supply to bring it to its operating speed. In a *turbo-* or *straight*-jet, the incoming air has its pressure raised by a compressor followed by combustion and the issue of a jet through a nozzle in an exit jet pipe. Jet nozzles are designed according to the speed of flow required and the exit velocity usually exceeds the speed of sound (sonic velocity is about 340 m/s in air at sea-level.) Some of the hot gas is utilised to drive a turbine which in turn drives the compressor. This is called *gas-turbine* propulsion. There are various other designs of propulsion unit modified along these lines, e.g. the *turbofan* engine and there is also the *turbo-prop* plane which employs a propeller as well as a jet. The S/VTOL (Short/Vertical Take-Off and Landing) machines take off vertically or from short runways but overcome the drawback of their precursor, the helicopter, with its low forward speed. In some designs two engines are fitted, one for take-off and one for flight. The Harrier Jump Jet uses 'vectored thrust' whereby only one engine is required and the plane is propelled by several swivelling nozzles, controlling speed and direction of flight.

12.3 Thrust of a jet

The thrust of a jet depends on the rate of change of momentum given to the jet fluid. Let v be the air velocity relative to the engine at entry, Fig. 12.1 and v_e the velocity of the *gas* jet relative to the engine at exit. Then, *relative to the engine*, and neglecting the effect of the mass of fuel burned,

$$\text{initial momentum of } m \text{ kg of fluid} = mv$$
$$\text{final momentum} = mv_e$$
$$\text{change in momentum} = m(v_e - v)$$

If $\dot{m}$ is the mass of fluid passing through the engine per second*, the force exerted *on* the jet of fluid equals the change of momentum per second, i.e.

$$\text{force} = \dot{m}(v_e - v)$$

From Newton's third law, the active force exerted by the engine expelling the gas through the exit nozzle at the tail must have an equal and opposite reactive force, and this is the force due to the pressure of the combustion gases on the inside surfaces of the engine, i.e. there is a force or *thrust* T on the plane propelling it forward in reaction to the formation of the jet. When the engine is on a stationary rig the thrust is taken by the supports.
Thus

engine thrust $T = \dot{m}(v_e - v)$

There is also a small thrust due to the burnt fuel leaving in the jet. The fuel is initially *at rest* relative to the plane and its exit velocity is v_e relative to the plane, hence the additional thrust is

$$\dot{m}_f v_e$$

where $\dot{m}_f$ is the mass of fuel consumed per second.

---

* From the Law of Conservation of Mass the rate at which the mass of air enters the engine (the control volume) must equal the rate at which it leaves, if we assume the rate to be constant with time, i.e. *steady flow*. Assuming the density of the air to be constant this means that the volume per second passing along is constant at all sections. This is the *equation of continuity* (*see* Chapter 20 for further work on jets)

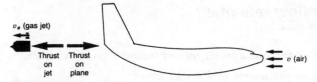

Fig. 12.1

Note that when an aircraft flies through *still* air at a forward speed v, relative to the ground, the air flows into the engine at the same speed relative to the plane. The speed of the wind affects the velocity of the air at entry, increasing it if the plane flies into a head wind, decreasing it if there is a following wind.

Specific impulse is a parameter used in jet aircraft and rocket work. It is the thrust per kg of fuel used per second, i.e.

$$\text{specific impulse} = \frac{\text{thrust}}{\text{fuel used (kg/s)}}$$

and the units are usually kN/kg/s. At just under the speed of sound a typical turbojet engine has a specific impulse of about 35 kN/kg/s.

12.4 Power developed by a turbo-jet engine

If the thrust T is known for a particular flight speed v, then if the units are newton, metre and second

power output = thrust × flight speed

$$= \frac{Tv}{1000} \text{ kW}$$

Power is only developed by a thrust when the aircraft is moving, in flight or on the runway. When stationary on the runway or on a test rig the thrust is still exerted but resisted by friction forces or restraints; since there is no movement there is no *output* power developed although some power is taken by the rotating parts of the engine. The thrust of a jet in practice is fairly constant with flight speed hence the power output varies directly with the flight speed unlike the engines of a propeller-driven plane or motor car where the power output does not depend on the speed of the vehicle itself and is called the 'shaft' or 'brake' power because work is done by a torque rotating the engine crankshaft. Shaft power may be measured when the engine is stationary by a brake resisting the crankshaft rotation. There is no shaft work for a jet engine so its power output cannot be measured and thrust is used instead to assess propulsion performance. The 'static' thrust can be measured directly on a stationary test rig but in flight the thrust must be gauged by monitoring some other engine operating quantity which is directly related to the thrust.

12.5　Mass flow rate of air

The volume of air entering a jet engine per second is given by

intake volume per second Q = intake area
× speed of the air relative to the engine

In still air or if wind speed is neglected the speed of the intake air relative to the plane is the flight speed. To find the mass of air drawn into the engine per second it may be assumed that the intake air obeys the laws of perfect gases and use the *characteristic gas equation** for a perfect gas applied to the entry conditions. This equation connects the *absolute* pressure p of the air, the volume flow per second Q, and the absolute temperature T †, with the characteristic gas constant for air, $R = 287$ J/kg K. Thus

$$pQ = \dot{m}RT$$

For example, if 80 m³ of air is drawn into an engine per second, at absolute pressure 90 kN/m² and temperature 268 K, then

$$90 \times 10^3 \times 80 = \dot{m} \times 287 \times 268$$

i.e. $\dot{m} = 94$ kg/s

12.6　Notes on lift and drag forces on an aircraft

To be airborne and fly, any type of machine must be provided with a lifting force equal to or greater than its deadweight, and this force is generated by the flow of air over the surface of wings or aerofoils when thrust forward. Alternatively it can be provided by rotating blades or downward-directed jets. When a fixed wing aircraft flies, the air flows over the wings in streamlines and a resultant force is produced on the plane because of a slight difference in air pressure at the top and bottom skins of the wings‡. This force has two components (i) a *lifting* force L, normal to the direction of airflow, counteracting the weight of the plane, (ii) a *drag* force D acting in the same direction as the flow of air, i.e. parallel to the line of flight opposing the motion of the plane. The drag is due to skin friction, turbulence, and shock effects,

---

* See authors 'Mechanical Engineering Science'.

† Symbol T is being used for both absolute temperature and thrust.

‡ A full treatment of streamline flow lift and drag forces, will be found in A. C. Kermode's 'Mechanics of Flight', Longman.

and these components vary for different aircraft depending on their shape and speed, and particularly between subsonic and supersonic types. The forces L and D are shown in Fig. 12.2(a) for a plane in level flight. The lift depends on the density of the air and on the shape and angle of attack of the wings into the air. The lift acts at the centre of pressure C (*see* page 432) and the weight W acts through the centre of gravity G. The positions of C and G may vary but are not usually far apart; the centre of gravity must lie within specified limits. The line of action of the drag force is offset, above or below the line of resultant thrust which acts along the centre line of the propeller shaft or jet, in the case of a single engine, or central longitudinal axis of a multi-engined aircraft. Fig. 12.2(a) shows (not to scale) the lift behind the line of action of the weight, and the drag force above the thrust. Fig. 12.2(b) shows an aircraft climbing; the lift is normal to the line of flight and its vertical component is equal to the deadweight. In practice, the lines of action of the resultant forces are determined largely by experiments on models, including full-scale rigs, using wind tunnels and other devices.

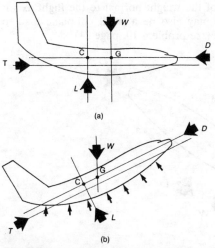

(a)

(b)

Fig. 12.2

For a rocket, lift is supplied by downward-directed jets but there are also aerodynamic forces due to airflow. Helicopters with power-driven rotors obtain their lift from the blades pushing the air downwards but with propeller-driven autogyros the lift comes from 'auto-rotation' due to the *upward* flow of air through the horizontal rotor blades. Note that the lift and drag forces on wings may be expressed in terms of the density of the air, wing area, speed of airflow, and coefficients of lift and drag (*see* page 101 for drag on a vehicle). The object of a designer is to achieve as high a lift/drag ratio as possible.

For every airborne vehicle once the values of the four main forces–thrust, lift, drag and weight, are known, the principles of dynamics can be applied to finding the motion of the vehicle.

12.7 Forces on an aircraft in flight

When an aircraft of mass m and weight W climbs with acceleration a at an angle θ to the horizontal, the forces acting along its *longitudinal* axis at any instant are the thrust T, drag force D, component of the weight, $W \sin \theta$, and inertia force ma. These forces are shown in the free-body diagram, Fig. 12.3. The accelerating force is

$$F = ma$$
$$= T - D - W \sin \theta$$

At constant speed in level flight, the thrust is equal to the drag, i.e. $T = D$.

Note that the component of the weight normal to the flight axis is supplied by the 'lift' but there may also be a small tail-plane load to balance any pitching movement (*see* problem 10, page 21)

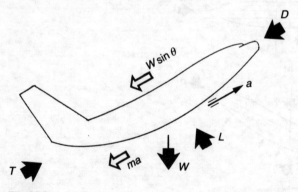

Fig. 12.3

Example   *A stationary jet engine under test is supplied with air at the rate of 80 m³/s. The air speed at entry to the engine is 50 m/s, at a pressure of 106 kN/m² (absolute) and temperature 292 K. If fuel is burned at the rate of 1.2 kg/s and the gases leave the tail at 400 m/s, find the thrust. Characteristic gas constant for air, R = 287 J/kg K.*

SOLUTION
Applying the characteristic gas equation to the air at entry,

$$\dot{m} = \frac{pQ}{RT}$$

$$= \frac{106 \times 10^3 \times 80}{287 \times 292}$$

$$= 101.2 \text{ kg/s}$$

The speed of the air changes from 50 to 400 m/s in passing through the stationary engine, hence the thrust is

$$\dot{m}(v_e - v) = 101.2(400 - 50)$$

$$= 35,420 \text{ N}$$

The fuel has its speed increased from rest to 400 m/s, hence the additional thrust due to the fuel is

$$\dot{m}_f(v_e - 0) = 1.2 \times 400$$

$$= 480 \text{ N}$$

Total thrust = 35,420 + 480 N

$$= \textbf{36 kN}$$

Note that the additional thrust due to the effect of the fuel is only 1.3% of the total thrust.

Example *A jet aeroplane of mass 4200 kg travels in level flight at a speed of 250 m/s, drawing in air to the engine at the rate of 70 m³/s. The jet speed relative to the plane is 600 m/s and the atmospheric conditions are 750 mbar and 261 K. If the lift/drag ratio is 10, estimate the acceleration of the plane and the power output. Characteristic gas constant for air, R = 287 J/kg K; 1 mbar = 100 N/m².*

If the plane climbs at 20° to the horizontal what is then its acceleration assuming the same thrust and drag as in level flight?

SOLUTION
Applying the gas equation to the air at entry to the engine,

$$\dot{m} = \frac{pQ}{RT}$$

$$= \frac{(750 \times 100) \times 70}{287 \times 261}$$

$$= 70 \text{ kg/s}$$

Thrust = $\dot{m}(v_e - v)$

$$= 70(600 - 250)$$

$$= 24,500 \text{ N}$$

Lift = Weight

$$= 4200 \times 9.8 \text{ N}$$

$$= 41,160 \text{ N}$$

$$\text{Drag force} = \frac{\text{lift}}{10} = \frac{41,160}{10} = 4,116 \text{ N}$$

Accelerating force

$$F = \text{thrust} - \text{drag}$$

$$= 24,500 - 4116$$

$$= 20,384 \text{ N}$$

And $F = ma$

where m is the mass of the aircraft, hence

$$20,384 = 4200 \times a$$

i.e. $a = \textbf{4.9 m/s}^2$

Power output at 250 m/s = thrust × speed

$$= 24,500 \times 250 \times \frac{1}{1000} \text{ kW}$$

$$= \textbf{6125 kW}$$

When climbing the weight component will be found to be 14078 N, reducing the accelerating force to 6306 N, giving the acceleration as 1.5 m/s².

Example   *A ramjet has a speed of 3312 km/h(Mach 3) at an altitude where the density of the air is 0.5 kg/m³. The jet velocity relative to the ramjet is 1.2 km/s. What is the frontal area of the intake scoop required for a thrust of 24 kN? If the air-fuel ratio is 55:1 what is the specific impulse?*

SOLUTION
(Mach 3 indicates that the jet's speed is three times that of the speed of sound at that particular altitude.)

Thrust $T = \dot{m}(v_e - v)$

i.e. $24 \times 10^3 = \dot{m}\left(1200 - \dfrac{3312}{3.6}\right)$

therefore, air consumption

$\dot{m} = 85.7 \text{ kg/s}$

If A is the area of the intake scoop and Q the intake volume in m³/s, then

$\dot{m} = \rho Q$

$= \rho \times$ area of scoop × flight speed

$= \rho A v$

therefore

$$85.7 = 0.5 \times A \times \frac{3312}{3.6}$$

therefore

$$A = \mathbf{0.19 \ m^2}$$

Fuel used per second $\dot{m}_f = \dfrac{85.7}{55} = 1.56 \ \text{kg/s}$

$$\text{Specific impulse} = \frac{T}{\dot{m}_f}$$

$$= \frac{24}{1.56}$$

$$= \mathbf{15.4 \ kN/kg/s}$$

Problems

1. An aircraft draws 45 kg of air per second into its engine and ejects it at a speed of 360 m/s relative to the engine. Find the thrust exerted by the jet, (a) when stationary, (b) at a forward speed of 800 km/h.

(16.2 kN; 6.2 kN)

2. A jet plane discharges a jet at the rate of 30 kg/s with a velocity of 1,000 m/s relative to the plane. If the forward speed of the plane is 900 km/h what is the thrust on the plane and the power developed at the jet?

(22.5 kN; 5.625 MW)

3. A jet plane of mass 4200 kg climbs at 20° to the horizontal when the propelling and drag forces are 24 kN and 4 kN respectively, along the line of flight. If the speed at the beginning of the climb is 250 m/s find the speed when the plane gains 800 m in altitude.

(263 m/s)

4. A jet aircraft has a forward speed of 200 m/s at a height where the absolute pressure is 800 mbar and the temperature 280 K. At these conditions the air consumption is 70 m³/s. The jet speed is 500 m/s relative to the plane and the fuel used is 0.65 kg/s. Neglecting the effect of fuel in the jet find the thrust and the specific impulse. What is the thrust effect of the fuel in the jet? Characteristic gas constant for air, $R = 287 \ \text{J/kg K}$ and 1 mbar = 100 N/m².

(20.9 kN; 32.1 kN/kg/s; 0.325 kN)

5. A twin-engined VTOL aircraft has a mass of 8500 kg and its swivelling nozzles give a vectored-thrust vertically downwards for lift-off. The exhaust jet speed is 550 m/s and the acceleration at lift-off is 2m/s². If the mass of fuel burned in each engine is 1.8 kg/s find the air/fuel ratio. Neglect drag but allow for the effect of the fuel in the jet.

(49.7:1)

6. A jet aircraft climbs in a straight line at an altitude of 60° to the horizontal. The jet thrust amounts to 90 kN, the mass of the aircraft is 8 tonne and the average air resistance amounts to 11 kN. Calculate the time taken to reach a height of 3 km if the speed at the start of the climb is 160 km/h, assuming the thrust to be constant. What is the output power at the start and at the highest point of the climb?

(45.5 s; 4 MW; 9.7 MW)

7. The specific fuel consumption of a four-engined jet aircraft is stated as 112 kg/kN of thrust/hour for each engine. At a speed of 1512 km/h in level flight the total thrust for all engines operating is estimated as 160 kN from static test results, and the jet speed is 1100 m/s. For each engine find the airflow in kg/s and the air to fuel ratio. What is the specific impulse for the aircraft? Neglect the effect of fuel in the jet.

(59 kg/s; 47:3; 32.2 kN/kg/s)

8. A turbojet airliner operating in level flight at a constant speed of 540 km/h sucks in air at the rate of 45 kg/s to each of its three engines and for each engine the jet speed relative to the plane is 950 m/s. What is the drag force? If the liner then climbs at 20° to the horizontal and its mass is 110 tonne find the acceleration assuming that the total thrust is boosted to 580 kN but with no change in the drag force. What is the airliner's speed after one minute?

(108 kN; 0.94 m/s²; 743 km/h)

9. The engine of a jet plane is tested on a stationary rig. Air is supplied at a speed of 80 m/s and at normal temperature and pressure conditions (N.T.P., 16° C, 101.3 kN/m²), and the volume of air entering the intake duct is measured as 150 m³/s. The speed of the exhaust jet is 600 m/s and the fuel consumption 4.4 kg/s. Find the thrust and the specific impulse.
   Characteristic gas constant for air, $R = 287$ J/kg K.
   The engine is fitted to a plane of mass 16,000 kg. Assuming the same total thrust as for the static conditions and a steady resistance of 15 kN in flight, find the angle of climb possible (*a*) at constant speed, (*b*) when accelerating at 3 m/s².

(97.9 kN; 22.3 kN/kg/s; 31.9°; 12.9°)

10. A ramjet has a forward speed of 900 m/s at an altitude where the density of the air is 0.6 kg/m³. The exhaust jet speed is 2.2 km/s relative to the ramjet. If the intake duct for the air has an area of 0.11 m², find the mass of air used per second and, neglecting the mass of fuel, the thrust exerted. If the fuel is consumed at the rate of 0.9 kg/s, what is the additional thrust?

(59.4 kg/s; 77.2 kN; 1.98 kN)

11. A 180-tonne jet transport has a take-off speed of 270 km/h against a drag force of 35 kN. Each of the four main engines consumes air at the rate of 90 kg/s and the jet speed is 845 m/s relative to the plane. Each of the two auxiliary jets consumes air at the rate of 63 kg/s and the jet speed is 600 m/s relative to the plane. Find the total thrust, the acceleration at take-off, and the minimum length of runway required.

(343.4 kN; 1.71 m/s²; 1642 m)

12. A supersonic jet aircraft in level flight cruises at 2190 km/h(Mach 2). For each of the four engines the jet speed is 900 m/s relative to the engine, the mass of air sucked into the combustion chamber is 160 kg/s, and the specific fuel consumption is 123 kg/kN of thrust/hour. Find the total thrust and the specific impulse in these conditions.

(187 kN; 29.3 kN/kg/s.)

13. A jet engine aircraft flies at a height where the air pressure (absolute) is 470 mbar and the temperature −24° C. The total intake scoop area is 0.2 m². Find the volume of air flowing through the scoop at a flight speed of 900 km/h in still air. If the aircraft is flying level find its acceleration given that the plane has a mass of 8 tonne, the exhaust speed of the jet is 950 m/s relative to the plane, and the average drag force opposing the motion is 8 kN. Characteristic gas constant for air, $R = 287$ J/kg K, and 1 mbar = 100 N/m².

(50 m³/s; 1.88 m/s²)

14. The total wing area of a supersonic aircraft is 360 m² and the wing loading is 5 kN/m². The plane flies at 2900 km/h in steady level flight in still air and the lift/drag ratio is 8. Find the power required.

(181 MW)

15. An aircraft has a mass of 11 tonne and a total wing area of 50 m². It flies in steady level flight at 1180 km/h with the engine exerting 22 kN thrust. Find the lift/drag ratio, the wing loading, and the power required.

(5; 2.16 kN/m²; 7.2 MW)

16. Each engine of a four-engined turbojet is tested on a static rig and uses 1.8 kg/s of fuel with an air/fuel ratio 45, and jet speed 950 m/s. The air is drawn into the engine at 80 m/s. If the plane takes off at 288 km/h when at its maximum allowable load of 240 tonne with the 'static' thrust boosted by 60% what is the minimum length of runway required. Neglect drag.

(1.7 km)

12.8 Propeller and rotor-driven aircraft

Propeller-drives (Fig. 12.4)

The engines supply a torque to the propeller shaft, rotating the blades, and pushing a large quantity of air backwards, giving it a moderate increase in speed. The torque is converted to a thrust due to the *reaction* on the formation of the jetstream. If the efficiency with which the propeller converts the torque to a thrust is known, then at a given speed we have

thrust × forward speed = power at propeller

= power delivered to shaft × propeller efficiency

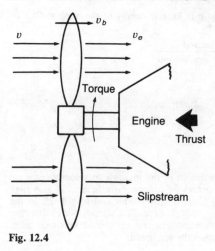

Fig. 12.4

The propulsive efficiency* may be up to 85% at moderate speeds (below 600 km/h) but at higher speeds the efficiency falls off rapidly.

The flow of air past rotating blades is complex because of the rotation and turbulence imparted to the air. However, using the simple momentum theory for a jet the thrust may be found in terms of the airflow. If v is the velocity of the air in the streamlines in front of the blades, and v_e the velocity at a point well downstream, both points being chosen where the flow is steady, then the 'propulsive velocity' is $(v_e - v)$, and the thrust is

$$T = \dot{m}(v_e - v)$$

where $\dot{m}$ is the mass of air discharged past the blades per second. It can be shown by applying Bernoulli's Theorem (Chapter 20) that the velocity v_b of the air at a section through the propeller is the average of v and v_e, i.e.

$$v_b = \tfrac{1}{2}(v_e + v)$$

If Q is the volume of air per second drawn past the blades at the propeller, and A the 'disc' area swept out by the blades, then

$$\dot{m} = \rho Q = \rho A v_b$$

where ρ is the density of the air. If d is the blade diameter then the disc area may be taken as $\pi d^2/4$.

Hence the thrust is

$$T = \rho A v_b(v_e - v)$$

i.e. $T = \tfrac{1}{2}\rho A(v_e{}^2 - v^2)$ since $v_b = \tfrac{1}{2}(v_e + v)$

---

* The work input per second is the gain in kinetic energy of the mass airflow so that the *theoretical* efficiency is given by

$$
\begin{aligned}
\eta &= \frac{\text{useful work done per second}}{\text{gain in kinetic energy of airflow/second}} \\[4pt]
&= \frac{Tv}{\tfrac{1}{2}\dot{m}(v_e{}^2 - v^2)} \\[4pt]
&= \frac{\dot{m}(v_e - v)v}{\tfrac{1}{2}\dot{m}(v_e{}^2 - v^2)} \\[4pt]
&= \frac{2}{1 + \dfrac{v_e}{v}}
\end{aligned}
$$

i.e. for a high efficiency, for a given value of v, the increase in velocity $(v_e - v)$, should be as small as possible. The actual or conversion efficiency is less than the theoretical value because of drag force on the blades and turbulence in the jet stream. The same efficiency applies to a jet-propelled aircraft but in this case the propulsion is inefficient at low speeds (e.g. take-off) and becomes more efficient as the forward speed approaches the jet speed.

Note: 1. Streamline velocities, v, v_b, v_e, are throughout *relative* to the propeller (aircraft) as previously for jets. 2. The above theory applied to propellers (and rotors below) assumes axial airflow without rotation, and is based on replacing the rotating blades by a hypothetical 'actuator' disc – a concept used to analyse problems in fluid dynamics and dealt with in specialist texts.

Helicopters (Fig. 12.5)

A helicopter may have more than one engine supplying power to one, two or three main rotors, and each rotor usually has between two and six blades. One type of rotor is the *articulated* version, where each blade is hinged and free to flap. The rotating blades of large diameter push a great quantity of air downwards for vertical movement and hovering, producing an upwards thrust in reaction to the formation of the jet. The ability to fly in all directions is through the tilting of the shaft and the deflection of the blades, but the speed of rotation of the blades is kept fairly constant. In forward flight the thrust is inclined forwards to the vertical thereby giving a horizontal component to overcome the drag force in the line of flight, and a vertical component to support the weight. A single rotor causes a torque reaction and the usual methods of counteracting this are the provision of a vertical rotor at the tail or by using a pair of contra-rotating main rotors.

The thrust in vertical flight and hovering may be found in terms of the airflow or from the shaft power as for propellers. This thrust must be greater than the total downwards force for lift-off and equal to it for hovering, allowing for the weight of the machine, any load slung below it, and down-gust or up-gust forces. When hovering the engines must develop sufficient power to provide the thrust necessary to keep the machine airborne (*see* worked example), and when close to the ground or sea there is a 'ground effect' due to the increased pressure in the cushion of air which affects the thrust and performance. Note that helicopters use the effect also of 'autorotation' when the air flows *upwards* through the blades as, for example, when the engine cuts out and the machine is descending. The examples here will be restricted to power-driven rotors in vertical flight or hovering.

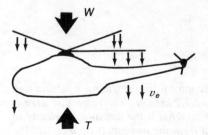

Fig. 12.5

Example   *A single-engined plane flies at 540 km/h in still air (density 1.15 kg/m³). The power supplied to the propeller shaft is 800 kW, the 'disc' area swept by the blades is 4 m², and the conversion efficiency is 85%. Find the thrust and estimate the velocity of the jet well downstream of the blades.*

SOLUTION

$$v = 540 \text{ km/h} = 150 \text{ m/s}$$

Thrust × speed = power output

i.e.

$$\text{Thrust} = \frac{0.85 \times 800 \times 10^3}{150}$$

$$= \textbf{4533 N}$$

$$\text{Thrust} = \tfrac{1}{2}\rho A(v_e^2 - 150^2)$$

i.e. $4533 = \tfrac{1}{2} \times 1.15 \times 4(v_e^2 - 150^2)$

therefore

$$v_e = \textbf{156 m/s}$$

Example   *A light commercial helicopter has two turboshaft engines each supplying 350 kW to the single main rotor, when climbing vertically at 7 m/s. The rotor diameter is 12.3 m and the rotor efficiency is 70%. Find the thrust and estimate the downwash velocity of the jetstream. Density of air = 1.2 kg/m³.*

SOLUTION

Thrust × vertical speed = total output power

i.e.   thrust × 7 = 2 × 350 × 10³ × 0.7

i.e.   thrust $T = \textbf{70} \times \textbf{10}^3$ **N**

As for a propeller (para. 12.8)

$$T = \tfrac{1}{2}\rho A(v_e^2 - v^2)$$

and assuming the disc area $A = \dfrac{\pi d^2}{4}$, then

$$70 \times 10^3 = \frac{1.2 \times \pi 12.3^2}{2 \times 4}(v_e^2 - 7^2)$$

i.e. downwash velocity $v_e = \textbf{32 m/s}$

Example   *A 22-tonne transport helicopter with a single rotor hovers well above the ground in still air (density 1.2 kg/m³). The rotor disc area is A = 255 m², and its efficiency is 80%. What is the downwash velocity of the jetstream and the power required from the turboshaft engine?*

SOLUTION

When the machine is hovering no effective work is being done but the power is supplied by the engines to produce a thrust equal to the deadweight. The approach velocity of the air, $v = 0$, and therefore the velocity of the air across a section through the rotor is $v_b = \frac{1}{2}v_e$, where v_e is the downwash velocity, Fig. 12.5. Then

$$\text{thrust } T = \dot{m}v_e$$
$$= (\rho A v_b)v_e$$
$$= \frac{1}{2}\rho A v_e^2$$

When hovering, the thrust supports the weight, i.e.,

$$\frac{1}{2}\rho A v_e^2 = W$$

i.e. $$\frac{1}{2} \times 1.2 \times 255 \times v_e^2 = 22 \times 9.8 \times 10^3$$

therefore $$v_e = \textbf{37.54 m/s}$$

Power required when hovering = gain in kinetic energy of the jet

$$= \frac{1}{2}\dot{m}v_e^2$$
$$= \frac{1}{2}(\rho A v_b)v_e^2$$
$$= \frac{1}{4}\rho A v_e^3 \quad \text{since } v_b = v_e/2$$
$$= \frac{1}{4} \times 1.2 \times 255 \times 37.54^3 \times \frac{1}{10^3}\,\text{kW}$$
$$= 4047\,\text{kW}$$

therefore $$\text{shaft power} = \frac{4047}{0.8} = \textbf{5059 kW}$$

Alternatively, when hovering the machine may be considered as moving vertically relative to the air at velocity v_b so that the power output is Tv_b at the rotor, and since $T = W$,

$$\text{shaft power} = \frac{Tv_b}{0.8} = \frac{W(v_e/2)}{0.8}$$
$$= \frac{22 \times 10^3 \times 9.8 \times 37.54}{0.8 \times 2} \times \frac{1}{10^3}\,\text{kW}$$
$$= \textbf{5059 kW}$$

Problems

1. A propeller-driven aircraft flies at 432 km/h in level flight at an altitude where the air density is 0.82 kg/m³. The engine supplies 800 kW to the propeller shaft and the conversion efficiency of the 2.8 m diameter propeller is 80%. Find the thrust and estimate the velocity of the air in the downstream well beyond the blades.

(5.33 kN; 128.5 m/s)

2. An aircraft flies at 324 km/h in still air of density 1.2 kg/m³. The propeller diameter is 2.5 m and its propulsive efficiency is 78%. The velocity of the air directly across a section through the propeller is measured as 98 m/s. Estimate (*a*) the mass of air projected per second, (*b*) the velocity of the air at a point in the downstream, (*c*) the power delivered to the propeller shaft.

(577 kg/s; 106 m/s; 1.07 MW)

3. A propeller-driven aircraft of weight 24.5 kN climbs at a steady speed of 200 km/h at an angle of 15° to the horizontal. The drag force is 6 kN. Find the thrust along the line of flight. If the propeller efficiency is 80% and the engine overall efficiency is 30% what is the energy input to the engine?

(12.34 kN; 2.86 MW)

4. An aeroplane flies in still air (density 1.18 kg/m³) at 300 km/h. The velocity at a point in the slipstream in steady flow beyond the propeller is 95 m/s and the propeller sweeps out an area of 7.06 m². Estimate the mass of air per second projected past the blades and the thrust. If the power delivered to the shaft is 950 kW find the conversion efficiency of the propeller.

(743 kg/s; 8.7 kN; 0.76)

5. A propeller-driven aircraft of mass 1800 kg flies at 180 km/h in level flight. The total wing area is 26 m², the lift/drag ratio 7, and the propulsive efficiency of the propeller, 80%. Find the thrust, the power delivered to the propeller shaft, and the wing loading.

(2.52 kN; 157.5 kW; 678 N/m²)

6. A 12-tonne helicopter has a single rotor and when airborne it climbs vertically at a steady speed of 18 km/h against a down-gust of 6 kN. What is the thrust and shaft power required, assuming a rotor efficiency of 70%?

(123.6 kN; 883 kW)

7. A single-rotor helicopter hovers well above ground level in still air (density 1.2 kg/m³) when the rotor shaft is supplied with 270 kW power. The rotor is 14.7 m diameter and its efficiency is 70%. Estimate the downwash velocity and the mass of machine being supported.

(15.5 m/s; 2.5 tonne)

8. A single-rotor helicopter is at its maximum allowable load of 25 tonne at vertical take-off. The blades are 23 m diameter and the speed of the air in the downwash below the rotor is 32 m/s. Neglecting drag, estimate the thrust at lift-off assuming the air initially to be at rest above the rotor. What is the vertical speed of the machine and the shaft power provided by the engines after 20 seconds from the start, assuming a rotor efficiency of 75%. Density of air = 1.2 kg/m³.

(255.3 kN; 29.7 km/h; 2805 kW)

12.9 Rocket propulsion: thrust

Rockets do not depend on surrounding atmospheric air for the combustion process, making them the ideal vehicles for high altitude and space travel. A rocket carries its own oxidising agent which together with the fuel is called the *propellants*. The principle of operation is simple–the propellants are burned in a combustion chamber open at one end and a continuous high-speed jet issues through the opening via an exit nozzle,

Fig. 12.6(a). A rocket firing therefore is similar to an inflated balloon suddenly released with the air exhausting from the open neck. The burning of the fuel takes a finite time and this distinguishes the combustion process from an explosion. The propellants are initially at rest *relative to the rocket* before being ignited, and if the exhaust jet speed relative to the rocket is v_e, then the thrust is given by

T = rate of change of momentum of propellants

   = mass/sec × change in speed of propellants

   = $\dot{m}v_e$

Enormous thrusts can be generated by giving large quantities of propellants a high speed relative to the rocket in the shortest possible time. The initial thrust of a Saturn V rocket system, for example, is about 34 MN for a burning time of 160 s, and the total rated thrust of a Shuttle's three main engines is about 5 MN at sea-level, boosted at lift-off by two rockets each giving 3 MN of thrust for two minutes of flight. Again, on a different scale, the thrusters used for the precise adjustment of the motion of the Orbiter craft each gives an impulsive force as low as 100 N and small gas jets with tiny thrusts are employed for minute control and orbit correction of communication satellites.

Staging increases the performance of rockets, whereby a series of rockets are placed end-on, the main booster being at the base. A stage consists of a propellant tank, motor and fuel system, and on a multi-stage vehicle as each stage burns out redundant booster rockets and equipment are discarded. The mass of fuel also decreases quickly and since the thrust remains approximately constant, a greater acceleration is achieved. An alternative to staging is *clustering*, whereby a

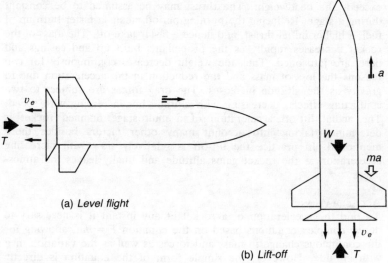

(a) *Level flight*

(b) *Lift-off*

Fig. 12.6

number of rockets are clustered around a central core 'sustainer' rocket which is the final vehicle. Without staging or clustering a space vehicle could not attain a final forward speed equal to that needed to escape from the earth's gravitational field (about 11 km/s) because of the limitation to the exhaust jet speed (about 6 km/s) of a single rocket. At the moment of launching a rocket may consist only of a casing, propellants, and a cargo of instruments carried in a satellite, but for a spaceship there will also be booster and stage rockets, a payload of passengers, several satellites and lunar modules. The propellants account for the greater part of the total mass. A Saturn V rocket, for example, had a mass at launching of over 3000 tonnes, of which nearly 80 per cent was kerosene and liquid oxygen, and this huge quantity was needed to deliver a payload of a 50-tonne Apollo spacecraft to orbit the moon. Again, the bulk of the initial mass of 2000 tonnes of a Space Shuttle consists of propellants; the re-usable Orbiter, a delta-winged aircraft, the only part to return to Earth, has a mass of 75 tonnes with its cargo of satellites, whereas the propellants for the main engines alone total 700 tonnes of liquid oxygen and hydrogen. Further large quantities of propellants are needed for 'strap-on' boosters and orbital control thrusters.

12.10 Forces on a rocket in flight

When a rocket of mass m is fired vertically from its launch pad two forces act on it immediately, the thrust T and its weight W, Fig. 12.6(b). The rocket lifts off with acceleration a when the thrust just exceeds the deadweight. The thrust may be assumed to be constant during a stage. Reducing the burning period means a faster burn-up of fuel, a higher initial thrust, and hence a faster take-off. The mass of the rocket decreases rapidly as the propellants burn up and casings and parts are jettisoned. Thus the weight decreases continuously for two reasons–the loss of mass and the reduction in the acceleration due to gravity as the altitude increases. The drag forces are subject to two conflicting effects, decreasing with altitude but increasing with speed. The actual lift-off acceleration of a multi-stage manned rocket is determined taking into account many other factors besides those mentioned. In practice the lift-off is relatively slow with increasing acceleration as the rocket gains altitude and finally leaves the atmosphere.

ACCELERATION
To find the acceleration of a rocket at any instant it is necessary to derive complex equations based on the equation $F = ma$, allowing for the continuous change in mass and forces as well as the variation in g with altitude. However, the simple form of the equation is directly

applicable at an instant when the mass and forces are known, i.e. at lift-off and at the end of a stage. Thus at lift-off the forces acting on the rocket are the thrust T, deadweight W, and the mass m of the rocket is known, hence

accelerating force $F = T - W$

$$= ma$$

which gives the acceleration a at lift-off. A typical ratio of thrust to initial weight at lift-off is 1.2.

At the end of a stage of a multi-stage rocket the thrust may be assumed to be the same as at lift-off, the mass m' is the initial mass reduced by the mass of fuel burned and the mass of rockets and equipment jettisoned, and the weight $W' = m'g$, *assuming g to be constant*. Thus, neglecting drag forces

accelerating force $F = T - W'$

$$= m'a'$$

giving the acceleration a' at the burn-out point. Note that if the rocket is fired from the moon the acceleration due to gravity at the moon's surface must be used.

Velocity and altitude achieved

As for acceleration complex equations are required when calculating the velocity and altitude achieved by a rocket but *as an exercise* some assumptions may be made to simplify the problem for a single-stage rocket or the end of a stage for a multi-stage rocket. If the time of flight t is assumed to be very short so that the loss of mass may be ignored then the value of the acceleration at lift-off as calculated above may be assumed to apply throughout the flight, hence the velocity V achieved is

$$V = at$$

and the height achieved above the earth's surface is

$$s = \frac{V^2}{2a} \text{ or } s = \tfrac{1}{2}at^2$$

The assumptions will lead to very large discrepancies where the fuel is burned at a great rate, heavy equipment is jettisoned, or the burning time lasts for several minutes. A rough approximation to the velocity reached at the end of the first stage of a multi-stage rocket can be estimated by assuming the *average* acceleration during the stage to be the average of the values at lift-off and burn-out, since the mass of the rocket and the forces acting on it are known at these two points.

Example *A 35-tonne spacecraft is launched vertically using 'strap-on' rockets which 'burn' for 8 s and expel exhaust gases at the rate of 300 kg/s with a speed of 2 km/s relative to the craft. Find the acceleration* *at lift-off and the specific impulse. If 4 tonne of equipment is jettisoned at the end of the 8 s stage, find the acceleration at the point of burn-out, assuming the thrust to be constant and resistance negligible during the stage. Estimate the velocity and altitude achieved at the end of the stage assuming the acceleration to be the average of the values found for lift-off and burn-out.*

SOLUTION

Thrust $T = \dot{m}v_e = 300 \times 2 \times 10^3 = 600 \times 10^3$ N

Weight $W = 35 \times 10^3 \times 9.8 = 343 \times 10^3$ N

Accelerating force at lift off

$$F = T - W$$
$$= (600 - 343) \times 10^3 \text{ N}$$
$$= 257 \times 10^3 \text{ N}$$

and $F = ma$

i.e. $257 \times 10^3 = 35 \times 10^3 a$

i.e. $a = \textbf{7.34 m/s}^2$

The specific impulse is the thrust per kg of propellants used per second, i.e.

$$\text{specific impulse} = \frac{T}{\dot{m}}$$
$$= \frac{600}{300}$$
$$= \textbf{2 kN/kg/s}$$

Propellants used in 8 s = $300 \times 8 = 2400$ kg

Mass of spacecraft at burn-out, m'

$$= 35 \times 10^3 - 2.4 \times 10^3 - 4 \times 10^3$$
$$= 28.6 \times 10^3 \text{ kg}$$

If a' is the acceleration at burn-out, then accelerating force

$$F' = T - m'g$$

i.e.

$$F' = 600 \times 10^3 - 28.6 \times 10^3 \times 9.8$$
$$= 320 \times 10^3 \text{ N}$$

and $\qquad F' = m'a'$

i.e. $\quad 320 \times 10^3 = 28.6 \times 10^3 a'$

i.e. $\qquad a' = \textbf{11.2 m/s}^2$

Average acceleration over 8 s period, $a_{av} = \dfrac{11.2 + 7.34}{2} = 9.27$ m/s^2

Velocity achieved $V = a_{av}t$

$$= 9.27 \times 8$$

$$= 74 \text{ m/s or } \mathbf{267 \text{ km/h}}$$

If s is the altitude achieved,

$$V^2 = 2as$$

i.e. $74^2 = 2 \times 9.27 \times s$

i.e. $s = \mathbf{295 \text{ m}}$

Example *A launch vehicle with strap-on rockets is fired at 40° to the horizontal. The total initial mass is 5 tonne, including 2.7 tonne of propellants, which are used up at a steady rate in 90 s. At the point of burn-out 900 kg of equipment is jettisoned. If the exhaust jet speed is 1.4 km/s relative to the rocket estimate the acceleration at burn-out, assuming constant thrust and a constant resistance of 8 kN parallel to the line of flight.*

SOLUTION

Fig. 12.7 shows the vehicle with the forces acting on it, along the line of flight.

$$\dot{m} = \frac{2700}{90} = 30 \text{ kg/s}$$

Thrust $T = \dot{m}v_e$

$$= 30 \times 1.4 \times 10^3 \text{ N}$$

$$= 42,000 \text{ N}$$

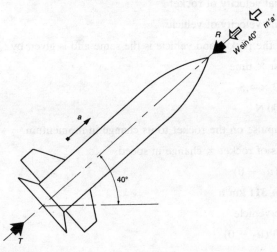

Fig. 12.7

Accelerating force

F = thrust − resistance − component of weight

$= 42,000 - 8,000 - 1400 \times 9.8 \times \sin 40°$

$= 25,180$ N

Mass of vehicle at burn-out $m' = 5000 - 2,700 - 900$

$= 1400$ kg

Hence

$F = m'a'$

i.e. $25,180 = 1,400a'$

i.e. $a' = 18$ m/s²

i.e. acceleration at point of burn-out is **18 m/s²**

Example   *A rocket of mass 250 kg is carried 'piggy-back' by a launch vehicle of mass 5 tonne. When the rocket is fired for 4.5 s in the horizontal position from the vehicle stationed on level ground, the vehicle recoils backwards a distance s m on the level against a resistance of 10 kN before coming to rest. The rocket discharges 12 kg of exhaust gases per second at a speed of 400 m/s relative to the rocket. Find the initial velocity of the rocket and the distance s, neglecting the loss of mass of fuel.*

SOLUTION

Fig. 12.8 shows the rocket and vehicle at the instant of firing.

Thrust $T = \dot{m}v_e = 12 \times 400 = 4,800$ N

Let u_1 = initial velocity of rocket

u_2 = initial velocity of vehicle

The impulse on the rocket and vehicle is the same and is given by

impulse = thrust × time

$= 4,800 \times 4.5$

$= 21,600$ N-s

Equating the impulse on the rocket to its change in momentum,

impulse = mass of rocket × change in speed

i.e. $21,600 = 250(u_1 - 0)$

i.e. $u_1 = 86.4$ m/s or **311 km/h**

Similarly for the vehicle

$21,600 = 5 \times 10^3(u_2 - 0)$

i.e. $u_2 = 4.32$ m/s

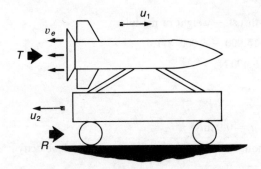

Fig. 12.8

The vehicle moves off with initial speed 4.32 m/s against a resistance of 10 kN. If it comes to rest in time t in a distance s, then

retarding impulse = resistance × time

$$= 10 \times 10^3 \times t$$

The retarding impulse is equal to the change in momentum of the vehicle,

$$-10 \times 10^3 t = 5 \times 10^3 (0 - 4.32)$$

i.e. $$t = 2.16 \text{ s}$$

and s = average speed × time

$$= \frac{(4.32 + 0)}{2} \times 2.16$$

$$= \textbf{4.7 m}$$

The student should rework this example using the equation of motion $F = ma$ throughout.

Example *A lunar module of mass 6000 kg approaches the moon at 40 m/s and has its speed reduced to 2 m/s by means of a retro-rocket. The rocket consumes propellants at the rate of 7.5 kg/s with the exhaust jet leaving at 2 km/s relative to the rocket. Find the time of combustion of the propellants. The acceleration due to gravity at the moon's surface is 1.6 m/s².*

SOLUTION

Thrust $= \dot{m} v_e$

$$= 7.5 \times 2 \times 10^3$$

$$= 15,000 \text{ N}$$

Retarding force F = thrust − weight of module

$$= 15,000 - 6,000 \times 1.6$$

$$= 5,400 \text{ N}$$

From $F = ma$

$$5,400 = 6000 \times a$$

therefore retardation $a = 0.9 \text{ m/s}^2$

Let t = time of combustion. Initial velocity $u = 40$ m/s, final velocity $v = 2$ m/s, hence from

$$v = u - at$$

$$2 = 40 - 0.9 \times t$$

$$t = \textbf{42 s}$$

Note that the mass of the module on the moon is 6000 kg as on earth, but its weight is only 9.6 kN as against 58.8 kN on earth.

Problems

1. The engine of a missile ejects 100 kg of exhaust gases per second at a speed of 800 m/s relative to the engine. Calculate the thrust of the engine at a forward speed of 200 m/s: (a) when the missile is rocket-propelled; (b) when jet propelled.

   What is the power developed when jet propelled?

   (80 kN; 60 kN; 12 MW)

2. A rocket of total mass 100 tonne is to leave its launching pad vertically with an acceleration of 2 m/s². If the velocity of the burnt gases leaving the rocket is limited to a speed of 3 km/s relative to the rocket, find the mass flow rate of gas to give the initial acceleration.

   (393 kg/s)

3. A communications satellite of mass 220 kg is orbiting at 10,500 m/s when its retro-rockets are fired for 8 s to reduce its speed to 10,475 m/s. Find the rate of discharge of exhaust gases required from the rocket to produce the necessary thrust if the speed of the exhaust jet (relative) is to be 550 m/s. Assume the only force acting on the satellite to be the thrust of the rockets.

   (1.25 kg/s)

4. A single-stage rocket has a total initial mass of 4000 kg including 2800 kg of propellants which are used up in 70 s in vertical flight. The exhaust jet velocity is 1.4 km/s relative to the rocket. Allowing for the loss of mass of fuel, estimate the acceleration at the instant of burn-out. What is the specific impulse of the rocket?

   (36.9 m/s²; 1.4 kN/kg/s)

5. A rocket has a lift-off weight of 8.8 kN and the propellants account for 60% of this weight. Vertical lift-off with negligible resistance occurs when the initial thrust exceeds the dead weight by 18%. The exhaust jet velocity relative to the rocket is limited to 1.1 km/s. Find the burning time and the acceleration at lift-off.

   (57.1 s; 1.76 m/s²)

6. A launching vehicle has a mass of 2900 tonnes including 50 tonnes of first-stage rockets which exhaust gas at a rate of 12 tonnes with a velocity of 3 km/s relative to the rocket. Find the acceleration at vertical lift-off. If the first stage burns out after 150 s and the thrust is maintained for this period, what is the acceleration at the end of the first stage, allowing for the mass of propellants exhausted and rockets jettisoned.

$$(2.6 \text{ m/s}^2; 24.5 \text{ m/s}^2)$$

7. A lunar module of mass 5 tonne is approaching the moon at 30 m/s when its two retro-rockets fire, each exhausting propellants at the rate of 3 kg/s for 20 s with a speed of 2 km/s relative to the module. Find the deceleration of the module and its speed at the end of the thrust period. For the moon's surface $g = 1.6 \text{ m/s}^2$.

$$(0.8 \text{ m/s}^2; 14 \text{ m/s})$$

8. An Ariane rocket has an all-up mass of 208 tonnes. The first-stage thrust is required to be 2.5 MN. If the exhaust jet speed is limited to 2 km/s relative to the rocket, at what rate must the propellants be ejected? At a given instant in the first stage 23 tonne of propellants have been used up and the rocket is moving vertically upwards against a resistance of 40 kN. Find the acceleration at this point.

$$(1.25 \text{ t/s}; 3.5 \text{ m/s}^2)$$

9. A Space Shuttle is fired vertically from a launch pad when mounted 'piggy-back' on a fuel tank with two 'strap-on' boosters. The total launch weight is $2000g$ kN. Lift-off is achieved by the two boosters and the three main engines all firing together for 130 s, at which point the boosters and tank are jettisoned. Assuming a constant first stage thrust of 30 MN, find the acceleration at lift-off and neglecting all loss of mass and air resistance, *estimate* the height and velocity achieved.

If each main engine uses fuel at the rate of 0.5 tonne/s each booster 2 tonnes/s and the jettisoned parts total 20 tonnes what is the acceleration at the instant of burn-out after 130 s? Taking the average of the two accelerations calculated, *estimate* the velocity and height achieved.

$$(5.2 \text{ m/s}^2, 44 \text{ km}, 2434 \text{ km/h}; 13.9 \text{ m/s}^2; 4473 \text{ km/h}; 80.8 \text{ km})$$

10. A spacecraft of mass 5.5 tonne fires its rockets to lift it vertically a distance of 10 km off the surface of the moon. If the thrust developed is 12 kN, for how long do the rockets operate? For the moon's surface, $g = 1.6 \text{ m/s}^2$

$$(185 \text{ s})$$

11. When a rocket of mass 850 kg is launched vertically from a stationary position on its pad, 9 kg/s of exhaust gases are emitted at a speed of 2 km/s relative to the rocket for a period of 5.8 s. Find the acceleration at lift-off and at the end of the 5.8 s period. Estimate the velocity achieved using the average of these two accelerations.

$$(11.4 \text{ m/s}^2; 12.75 \text{ m/s}^2; 252 \text{ km/h})$$

12. A missile of mass 2200 kg is carried in the horizontal position of a plane in level flight and released when the plane's speed is 400 m/s. The missile's rocket motor fires for 66 s and ejects 660 kg of propellants before burn-out. The exhaust jet velocity is 2.4 km/s relative to the rocket and the air resistance is 4 kN. Neglecting the decrease in mass of the rocket, find the speed of the missile at the instant of burn-out.

$$(1 \text{ km/s})$$

13. A three-stage launching vehicle and spacecraft has a total mass of 3000 tonne and its five first-stage boosters use 2040 tonne of propellants at a steady rate for 150 s. The exhaust gases eject at 2.4 km/s relative to the vehicle and 180 tonne of equipment is jettisoned at the end of the first

stage. Find the acceleration at lift-off and at the end of the first stage. Estimate the velocity of the vehicle at the end of the first stage and the height to which it is lifted, taking the average of the lift-off and end-of-stage accelerations as the constant acceleration for the stage.

$(1.08 \text{ m/s}^2, 32 \text{ m/s}^2; 8945 \text{ km/h}, 186 \text{ km})$

CHAPTER 13
Direct Stress and Strain

The strength and stiffness of materials deals with the effects of loading on machine parts and structures, particularly the nature of the internal forces and deformation produced. When a body is pulled by a tensile force or crushed by a compressive force, the loading is said to be *direct*. Such direct forces will be found to arise also when bodies are heated or cooled under constraint and in vessels under pressure.

13.1 Stress

The ability of a member to withstand load or transmit force, depends upon its dimensions. The cross-sectional area over which the load is distributed determines the *intensity of loading* or *average stress* in the member. If the intensity of loading is uniform the *direct stress*, σ, is defined as the ratio of load, F, to cross-sectional area, A, *normal to the load*, Fig. 13.1. Thus

$$stress = \frac{load}{area}$$

or $$\sigma = \frac{F}{A}$$

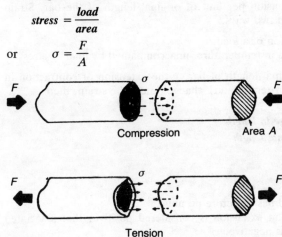

Fig. 13.1

The direct stress may be *tensile* (a pull) or *compressive* (a push). The *strength* of a member is measured by the force or stress needed to fracture it.

The SI units of force and length are the newton and the metre respectively so that the derived SI unit of stress is *newtons per square metre* (N/m^2). Other multiples of this unit used are:

1 *kilonewton per square metre* (kN/m^2) = 10^3 N/m^2

1 *meganewton per square metre* (MN/m^2) = 10^6 N/m^2

1 *giganewton per square metre* (GN/m^2) = 10^9 N/m^2

A further form now commonly used takes the area of section as (millimetre)^2. Note that

1 N/mm^2 = 10^6 N/m^2 = 1 MN/m^2

1 kN/mm^2 = 10^9 N/m^2 = 1 GN/m^2

The name *pascal* (**Pa**) is also in use for the unit N/m^2. In this text, as far as stress calculations are concerned, we shall restrict ourselves to the unit forms N/m^2, N/mm^2 and their multiples. The above units of stress are also used for pressure (see page 425).

13.2 Strain

A member under any loading experiences a change in shape or size. In the case of a bar loaded in tension the extension of the bar depends upon its total length. The bar is said to be strained and the *strain* is defined as the extension per unit of original length of the bar. Strain may be produced in two ways:

1. By application of a load.
2. By a change in temperature, unaccompanied by load or stress.

If l is the original length of bar, x the extension or contraction in length under load or temperature change, and ε the strain, then

$$strain = \frac{change\ in\ length}{original\ length}$$

or $\varepsilon = \dfrac{x}{l}$

Strain is a ratio and has therefore no units.

Strain due to an extension is considered positive, that associated with a contraction is negative.

13.3 Relation between stress and strain: Young's modulus of elasticity

If the extension or compression in a member due to a load disappears on removal of the load, then the material is said to be *elastic*. Most metals are elastic over a limited range of stress known as the *elastic range*. Elastic materials, with some exceptions, obey Hooke's law, which states that: *the strain is directly proportional to the applied stress*. Thus

$$\frac{stress}{strain} = constant, E$$

i.e. $\dfrac{\sigma}{\varepsilon} = E$ or $\varepsilon = \dfrac{\sigma}{E}$

where E is the constant of proportionality, known as the *modulus of elasticity* or *Young's modulus*.

Since strain is a ratio of two lengths, the units of E are those of stress. Values of E may be given in the basic form N/m^2, or more conveniently, since large numbers are involved, the forms GN/m^2 or kN/mm^2.

E relates to the *stiffness* or *rigidity* of a material since the higher its value the greater the load required to produce a given extension. E for steels ranges from 196 to 210 GN/m^2 but for the softer, more ductile materials such as aluminium and copper the range is lower, 70–120 GN/m^2. Many materials do not obey Hooke's Law, i.e. the load-extension diagram is not a straight line, or only a very small portion of it is straight. For those materials E is an approximation only. Thus for cast-iron E is in the range 100–125 GN/m^2, and for concrete 16–22 GN/m^2. For rubber the modulus is very low and variable, ranging from 1 MN/m^2 when soft to about 40 MN/m^2 for the harder varieties. Similarly for plastics and polymers E has a low value and can be markedly affected by the method of manufacture and type of reinforcement. For these materials the direct modulus E is not particularly applicable (*see* page 350).

For further work on the modulus of elasticity *see* paragraph 14.4 and Table 14.1.

Example *A rubber pad for a machine mounting is to carry a load of 5 kN and to compress 5 mm under this load. If the stress in the rubber is not to exceed 280 kN/m², determine the diameter and thickness of a pad of circular cross-section. Take E for rubber as 1 MN/m².*

SOLUTION

$$Stress = \frac{load}{area}$$

i.e. $\sigma = F/A$

$$280 \times 10^3 = \frac{5 \times 10^3}{\pi d^2/4}$$

hence $d^2 = 0.0227$ m^2 and $d = 0.1508$ m

i.e. diameter of pad = **151 mm**

The increase in area due to compression has been neglected. Also

$$\text{strain} = \frac{\text{reduction in length}}{\text{original length}}$$

$$\frac{\sigma}{E} = \frac{x}{l}$$

$$\frac{280 \times 10^3}{1 \times 10^6} = \frac{0.005}{l}$$

therefore thickness of pad is given by:

$$l = 0.018 \text{ m} = \textbf{18 mm}$$

Example   *Fig. 13.2 shows a steel strut with two grooves cut out along part of its length. Calculate the total compression of the strut due to a load of 240 kN. E = 200 GN/m².*

SOLUTION

Suffices 1 and 2 denote solid and grooved portions, respectively. The load at every section is the same, 240 kN.

For the solid length of 360 mm:

compression, $x_1 = \varepsilon_1 l = \varepsilon_1 \times 0.36$

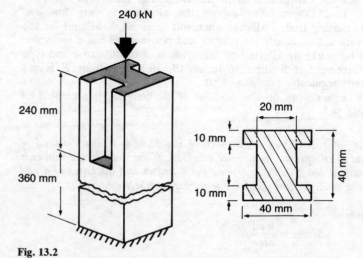

Fig. 13.2

$$\text{stress, } \sigma_1 = \frac{F}{A_1} = \frac{240 \times 10^3}{0.04 \times 0.04} = 0.15 \times 10^9 \text{ N/m}^2$$

$$\text{strain, } \varepsilon_1 = \frac{\sigma_1}{E} = \frac{0.15 \times 10^9}{200 \times 10^9} = 0.00075$$

For the grooved length of 240 mm:

$$\text{compression, } x_2 = \varepsilon_2 \times 0.24$$

$$\text{stress, } \sigma_2 = \frac{240 \times 10^3}{(0.0016 - 0.02 \times 0.02)} = 0.2 \times 10^9 \text{ N/m}^2$$

$$\text{strain, } \varepsilon_2 = \frac{\sigma_2}{E} = \frac{0.2 \times 10^9}{200 \times 10^9} = 0.001$$

The total compression of the strut is equal to the sum of the compressions of the solid and grooved portions. Therefore

$$x = x_1 + x_2$$
$$= (\varepsilon_1 \times 0.36) + (\varepsilon_2 \times 0.24)$$
$$= 0.00075 \times 0.36 + 0.001 \times 0.24$$
$$= 0.00051 \text{ m}$$
$$= \mathbf{0.51 \text{ mm}}$$

Note: It has been assumed here that the stress distribution is uniform over all sections, but at the change in cross-section the stress distribution is actually very complex. The assumption produces little error in the calculated compression.

Problems

1. A bar of 25 mm diameter is subjected to a tensile load of 50 kN. Calculate the extension on a 300 mm length. $E = 200$ GN/m^2.

   (0.153 mm)

2. A steel strut, 40 mm diameter, is turned down to 20 mm diameter for one-half its length. Calculate the ratio of the extensions in the two parts due to axial loading.

   (4 : 1)

3. When a bolt is in tension the load on the nut is transmitted through the root area of the bolt which is smaller than the shank area. A bolt 24 mm in diameter (root area = 353 mm^2) carries a tensile load. Find the percentage error in the calculated value of the stress if the shank area is used instead of the root area.

   (21.7 per cent)

4. A light alloy bar is observed to increase in length by 0.35 per cent when subjected to a tensile stress of 280 MN/m^2. Calculate Young's modulus for the material.

   (80 GN/m^2)

5. A duralumin tie, 600 mm long, 40 mm diameter, has a hole drilled out along

its length. The hole is 30 mm diameter and 100 mm long. Calculate the total extension of the tie due to a load of 180 kN. $E = 84$ GN/m^2.

(1.24 mm)

6. A steel strut of rectangular section is made up of two lengths. The first, 150 mm long, has breadth 40 mm and depth 50 mm; the second, 100 mm long, is 25 mm square. If $E = 220$ GN/m^2, calculate the compression of the strut under a load of 100 kN.

(0.107 mm)

7. A solid cylindrical bar, of 20 mm diameter and 180 mm long, is welded to a hollow tube of 20 mm internal diameter, 120 mm long, to make a bar of total length 300 mm. Determine the external diameter of the tube if, when loaded axially by a 40 kN load, the stress in the solid bar and that in the tube are to be the same. Hence calculate the total change in length of the bar. $E = 210$ kN/mm^2.

(28.3 mm; 0.184 mm)

8. A steel bar of 40 mm diameter and total length 240 mm is turned down to 30 mm diameter for a length of 60 mm and reduced to 20 mm diameter for a further length of 80 mm. Calculate the total extension of the bar when carrying a load of 70 kN in tension. $E = 210$ GN/m^2. State the stress in each portion of the bar.

(0.14 mm; 55.7, 99, 223 MN/m^2)

13.4 Compound bars

When two or more members are rigidly fixed together so that they *share* the same load and extend or compress the same amount, the two members form a compound bar. The stresses in each member are calculated using the following:

1. The total load is the sum of the loads taken by each member.
2. The load taken by each member is given by the product of its stress and its area.
3. The extension or contraction is the same for each member.

Consider a concrete column reinforced by two steel bars (Fig. 13.3) subjected to a compressive load F. Let A_s be the area of steel, A_c the area of concrete, σ_s the stress in the steel, and σ_c the stress in the concrete. Thus:

total load = load taken by steel + load taken by concrete

i.e. $F = \sigma_s A_s + \sigma_c A_c$

Since the column is a compound bar both steel and concrete compress the same amount, and since the original lengths are the same the strains are equal. Therefore

$$\varepsilon_s = \varepsilon_e$$

$$\frac{\sigma_s}{E_s} = \frac{\sigma_c}{E_c}$$

i.e. the stresses are proportional to the moduli of elasticity.

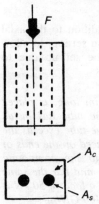

Fig. 13.3

Example *A column is made up of a steel tube, 70 mm inside diameter, filled with concrete. If the maximum stress in the concrete is not to exceed 21 N/mm² and the column is to carry a compressive load of 200 kN, calculate the minimum outside diameter of the tube. For concrete, E = 20 kN/mm². For steel E = 200 kN/mm².*

SOLUTION
Let suffices *c* and *s* denote the concrete and steel, respectively.

$$A_c = \frac{\pi \times 70^2}{4} = 3,850 \text{ mm}^2$$

$$A_s = \frac{\pi(d^2 - 70^2)}{4} \text{ mm}^2$$

where *d* mm = outside diameter of tube. Since the steel and concrete are of equal length and the compression of both is the same, the strains are equal, then working throughout in kN and mm,

$$\varepsilon_c = \varepsilon_s$$

or $\quad \dfrac{\sigma_c}{E_c} = \dfrac{\sigma_s}{E_s}$

thus

$$\sigma_s = \frac{E_s}{E_c} \times \sigma_c = \frac{200}{20} \times 21 = 210 \text{ N/mm}^2$$

$$= 0.21 \text{ kN/mm}^2$$

Total load $F = \sigma_c A_c + \sigma_s A_s$

$$200 = (21 \times 10^{-3}) \times 3,850 + 0.21 \times \frac{\pi(d^2 - 70^2)}{4}$$

hence $d^2 = 5{,}630 \text{ mm}^2$

i.e. $d = \textbf{75 mm}$

In practice a radial or lateral strain exists in addition to the axial strain due to the load. Unless the concrete shrinks on setting, to allow a small radial clearance, additional stresses may be set up due to interference between the steel and concrete.

Example   *A steel bar of 20 mm diameter and 400 mm long, is placed concentrically inside a gun-metal tube. Fig. 13.4. The tube has inside diameter 22 mm and thickness 4 mm. The length of the tube exceeds the length of the steel bar by 0.12 mm. Rigid plates are placed on the ends of the tube and an axial compressive load applied to the compound assembly. Find (a) the load which will just make tube and bar the same length; (b) the stresses in the steel and gun-metal when a load of 50 kN is applied. E for steel = 210 GN/m²; E for gun-metal = 100 GN/m².*

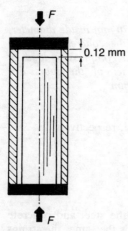

0.12 mm

Fig. 13.4

SOLUTION

$$\text{Area of gun-metal tube} = \frac{\pi}{4}(0.03^2 - 0.022^2) = 0.000327 \text{ m}^2$$

$$\text{area of steel bar} = \frac{\pi}{4} \times 0.02^2 = 0.0003142 \text{ m}^2$$

(*a*) For tube to compress 0.12 mm:

$$\text{strain} = \frac{0.12}{400} = 0.0003$$

thus $\dfrac{\sigma}{E} = 0.0003$

i.e. $\sigma = 0.0003 \times 100 = 0.03 \text{ GN/m}^2 = 30{,}000 \text{ kN/m}^2$

hence load = 30,000 × 0.000327 = **9.81 kN**

(*b*) Load available to compress bar and tube as a compound bar is given by

$$F = 50 - 9.81 = 40.19 \text{ kN}$$

Let σ_g be the *additional* stress produced in the gun-metal tube due to this load and σ_s the corresponding stress in the steel bar, then

$$F = \sigma_g A_g + \sigma_s A_s$$

i.e. $40.19 = \sigma_g \times 0.000327 + \sigma_s \times 0.0003142$

Since the lengths of tube and bar are initially the same when the load of 40.19 kN is applied, the strains are equal, then

$$\text{strain} = \frac{\sigma_g}{E_g} = \frac{\sigma_s}{E_s}$$

and $\quad \sigma_g = \dfrac{100}{210} \times \sigma_s$

From these two equations,

$$\sigma_g = 40{,}600 \text{ kN/m}^2$$
$$\sigma_s = 85{,}300 \text{ kN/m}^2$$

Therefore

final stress in the steel = **85.3 MN/m²**

final stress in gun-metal = $40{,}600 + 30{,}000 = 70{,}600$ kN/m²

$$= \textbf{70.6 MN/m}^2$$

Problems

1. A rectangular timber tie, 180 mm by 80 mm, is reinforced by a bar of aluminium of 25 mm diameter. Calculate the stresses in the timber and reinforcement when the tie carries an axial load of 300 kN. E for timber = 15 GN/m²; E for aluminium = 90 GN/m².

   (17.8 MN/m²; 106.8 MN/m²)

2. A concrete column having modulus of elasticity 20 GN/m² is reinforced by two steel bars of 25 mm diameter having a modulus of 200 GN/m². Calculate the dimensions of a square section strut if the stress in the concrete is not to exceed 7 MN/m² and the load is to be 400 kN.

   (220 mm square)

3. A cast-iron pipe is filled with concrete and used as a column to support a load W. If the outside diameter of the pipe is 200 mm and the inside diameter 150 mm, what is the maximum permissible value for W in tonnes if the compressive stress in the concrete is limited to 5 MN/m². Take E for concrete as one-tenth that of cast-iron.

   (79 tonne)

4. A concrete column is reinforced with steel bars and carries a load of 20 tonne. The overall cross-sectional area of the column is 0.1 m² and the steel

reinforcement accounts for 3 per cent of this area. Find the stress taken by the concrete. If the length of the column is 4 m how much does it shorten? Take E for steel as 200 GN/m², and concrete, 20 GN/m²

(1.54 MN/m²; 0.31 mm)

5. A cylindrical mild steel bar of 40 mm diameter and 150 mm long, is enclosed by a bronze tube of the same length having an outside diameter of 60 mm and inside diameter of 40 mm. This compound strut is subjected to an axial compressive load of 200 kN. Find: (*a*) the stress in the steel rod; (*b*) the stress in the bronze tube; (*c*) the shortening of the strut. For steel $E = 200$ kN/mm². For bronze $E = 100$ kN/mm².

(Steel 98 MN/m²; bronze 49 MN/m²; 0.0735 mm)

6. A compound assembly is formed by brazing a brass sleeve on to a solid steel bar of 50 mm diameter. The assembly is to carry a tensile axial load of 250 kN. Find the cross-sectional area of the brass sleeve so that the sleeve carries 30 per cent of the load. Find for this composite bar the stresses in the brass and steel. E for brass = 84 GN/m²; E for steel = 210 GN/m².

(2,110 mm²; steel 89.1 MN/m²; brass 35.6 MN/m²)

7. A duralumin rod of 50 mm diameter is a loose fit inside a mild steel tube of 60 mm outside diameter. If the steel tube is turned down to 55 mm diameter over one-half its length calculate the stress in the duralumin rod and the stress in each portion of the tube due to an axial load of 20 kN. Both rod and tube are the same length. E for steel = 196 GN/m²; E for duralumin = 126 GN/m². (*Hint*: strains are not equal; to solve, equate the compressions.

(7,063 kN/m²; 7,098 kN/m²; 14.9 MN/m²)

13.5 Thermal strain

Change in temperature of a material gives rise to a thermal strain. For example, a rise in temperature t in a bar of length l will cause it to extend by an amount

$$x = \alpha t l$$

where α is the coefficient of linear thermal expansion of the material. The thermal strain is given by:

$$\varepsilon = \frac{x}{l}$$

$$= \alpha t$$

There is no stress associated with this strain unless the bar is prevented from extending. In this case the load produced in the bar would be the same as the load required to compress the bar a distance equal to the free expansion of the bar. Typical values of the coefficients of linear expansion are as follows:

carbon steel	$12 \times 10^{-6}/°C$
austenitic stainless steel	$18 \times 10^{-6}/°C$

aluminium	$24 \times 10^{-6}/°C$
copper	$17 \times 10^{-6}/°C$
cast iron	$10 \times 10^{-6}/°C$
brass	$16 \times 10^{-6}/°C$

For example, carbon steel expands 0.000012 m per m length per °C rise in temperature.

The magnitude of the temperature strain is of the same order as the elastic strain in a metal due to stress and is therefore of importance. For example, the elastic strain at a tensile stress of 200 MN/m² for a carbon steel having an elastic modulus E of 200 GN/m² is:

$$\varepsilon = \frac{\sigma}{E} = \frac{200 \times 10^6}{200 \times 10^6} = 0.001$$

A temperature rise of 83° C in the same steel gives a thermal strain

$$\varepsilon = \alpha t = 12 \times 10^{-6} \times 83 \simeq 0.001$$

If the extension of a bar due to this temperature rise were completely prevented, a *compressive* stress of 200 MN/m² would be set up in the bar, since it would require this stress to return the bar to its original length.

For large temperature changes, α and E vary with temperature; however, it will be assumed here that the temperature changes are sufficiently small for α and E to be taken as constants.

Both the thermal strain αt and the elastic strain σ/E may exist together. The total strain ε is the sum of the two, thus:

$$\varepsilon = \alpha t + \frac{\sigma}{E}$$

and extension = $\varepsilon \times l$

$$= \left(\alpha t + \frac{\sigma}{E}\right)l$$

13.6 Sign convention

In tension, strain ε and stress σ are positive, in compression they are negative; and t is positive for a rise in temperature. When a stress is unknown σ is assumed positive, a negative answer therefore implies that it is compressive.

Fig. 13.5 shows a clamped bar restrained against extension or contraction. If the bar is subjected to a change in temperature the total strain is zero, i.e. the elastic strain must be equal and opposite to the thermal strain, thus:

total strain $\varepsilon = 0$

therefore

$$\alpha t + \frac{\sigma}{E} = 0$$

or $$\frac{\sigma}{E} = -\alpha t$$

The stress in the clamped bar is therefore compressive for a rise in temperature (t positive).

Fig. 13.5

13.7 Effects of thermal strain

Thermal strain may be of engineering importance because of the deformation it produces or because, if the deformation is resisted, thermal stresses result.

Thermal expansion of a metal is utilized when a collar is shrunk on to a shaft. The collar is bored out to a diameter slightly smaller than that of the shaft, it is then heated so that when expanded it may be slipped into position on the shaft. When cooled the collar grips the shaft firmly and the collar and shaft remain stressed after cooling. Again, a thermostat switch may be actuated by a rise in temperature deforming a bimetal strip. For example, when strips of copper and steel are riveted together a rise in temperature causes the copper to expand more than the steel, the strip bends and the resulting deflexion can be used to operate the switch.

The expansion of a metal must be allowed for in high temperature piping. A straight pipe will produce high loads at the pipe end connexions if not allowed to expand freely along its length. In order to overcome this a loop is inserted in the pipe-line. The flexibility of the loop permits the expansion to be taken up. Similarly special arrangements are made to allow free expansion of long exposed pipe-lines in hot climates; the pipes may be looped or the lines staggered. Gaps are often left in rail tracks to permit free expansion in hot weather without buckling of the tracks. Present-day practice, however, is to weld the joints for considerable lengths and to rely on the clamping effect of sleepers and ballast to prevent buckling.

When dissimilar metals are bonded together each tends to resist the change in length of the other and high stresses may be induced. If two parts of the same structure are at different temperatures, or if a body of non-uniform thickness is subject to a sudden change in temperature, again high stresses or excessive deformation may result.

Many examples of these effects will come to mind: cold water poured into a hot cylinder block may crack it: foundry castings of complex shapes allowed to cool too quickly may shatter; tools may be cracked by the process of quench-hardening.

Finally it may be remarked that when the change of temperature is large the properties of metals change also and this may have to be taken into account. A rise in temperature is usually accompanied by a drop in the values of the modulus of elasticity, the ultimate tensile stress and the yield stress. The ductility of the metal may increase (*see* Chapter 14). The reverse is true for a drop in temperature; in particular, at low temperatures mild steel may become relatively brittle.

Example *A steel bar 300 mm long, 24 mm diameter, is turned down to 18 mm diameter for one-third of its length. It is heated 30° C above room temperature, clamped at both ends and then allowed to cool to room temperature. If the distance between the clamps is unchanged find the maximum stress in the bar. $\alpha = 12.5 \times 10^{-6}/° C$; $E = 200 \ GN/m^2$.*

SOLUTION
If allowed to contract freely without constraint the contraction of the whole length is given by:

$$\alpha \times t \times l = 12.5 \times 10^{-6} \times 30 \times 0.3$$
$$= 112.5 \times 10^{-6} \ m$$

Contraction is prevented by a tensile force F exerted by the clamps.

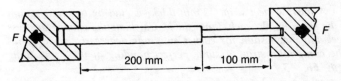

Fig. 13.6

Each portion of the bar carries the total load F but the extension of each portion is different since the lengths and section areas are different. The load is the same throughout and the maximum stress will therefore occur in the portion of smaller diameter. Since

$$\frac{\sigma}{E} = \varepsilon \text{ and } \varepsilon = \frac{x}{l}$$

then, for the longer portion, using suffix 1

$$x_1 = \varepsilon_1 l_1 = \frac{\sigma_1 l_1}{E}$$

and for the shorter portion

$$x_2 = \frac{\sigma_2 l_2}{E}$$

As the load on each portion is the same, then

$$F = \sigma_1 A_1 = \sigma_2 A_2$$

i.e. $\sigma_1 = \dfrac{A_2}{A_1}\sigma_2$

$$= \dfrac{18^2}{24^2}\sigma_2$$

$$= 0.562\sigma_2$$

As the total length remains unchanged, then

total extension due to load = contraction due to temperature drop

thus $\quad x_1 + x_2 = 112.5 \times 10^{-6}$ m

$$\dfrac{\sigma_1 l_1}{E} + \dfrac{\sigma_2 l_2}{E} = 112.5 \times 10^{-6} \text{ m}$$

therefore

$$\sigma_1 \times 0.2 + \sigma_2 \times 0.1 = 112.5 \times 10^{-6} \times 200 \times 10^9$$

$$2\sigma_1 + \sigma_2 = 225 \times 10^6$$

and $\sigma_1 = 0.562\sigma_2$ hence

$$\sigma_2 = 106 \times 10^6 \text{ N/m}^2$$

Hence

maximum stress in the bar = **106 MN/m²**

Example *A narrow steel strip, 12 mm thick, is clad by two magnesium plates of the same width, each 3 mm thick. Calculate the change in stress in the steel and magnesium for each Centigrade degree rise in temperature of the compound strip. Assume perfect bonding of the strips along their length, Fig. 13.7*

For steel, $\alpha = 12 \times 10^{-6}/°C$; $E = 200$ GN/m². For magnesium, $\alpha = 27 \times 10^{-6}/°C$; $E = 45$ GN/m².

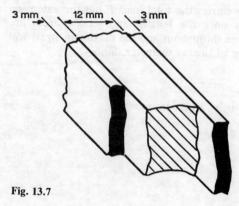

3 mm   12 mm   3 mm

Fig. 13.7

SOLUTION

The criteria for solving this problem are:

1. Since the magnesium cladding tends to extend more than the steel it will be restrained by the latter and will therefore be in compression. The steel is correspondingly stretched by the magnesium and is in tension.
2. Since there is perfect bonding the final extension of each strip is the same and, since the original lengths are equal, the total strains are equal.
3. There is no *external* load on the compound strip.

For a temperature rise of t °C, taking tensile strains as positive:

total strain in steel = total strain in magnesium

thus

$$\alpha_1 t + \frac{\sigma_1}{E_1} = \alpha_2 t + \frac{\sigma_2}{E_2}$$

or

$$\sigma_1 - \sigma_2 \times \frac{E_1}{E_2} = (\alpha_2 - \alpha_1)t \times E_1$$

i.e.

$$\sigma_1 - \sigma_2 \times \frac{200}{45} = (27 - 12) \times 10^{-6} \times 200 \times 10^9 \times t$$

therefore $\sigma_1 - 4.44\sigma_2 = 3 \times 10^6 t$

Also

total load on compound strip = 0

thus

$$\sigma_1 A_1 + \sigma_2 A_2 = 0$$

i.e.

$$\sigma_2 = -\sigma_1 \frac{A_1}{A_2}$$

$$= -\sigma_1 \times \frac{12}{2 \times 3}$$

$$= -2\sigma_1$$

From these two equations connecting σ_1 and σ_2 we find

$$\sigma_1 = 304 \times 10^3 t \text{ N/m}^2$$

$$= \textbf{304 kN/m}^2 \textbf{ per °C (tension)}$$

and $\sigma_2 = -\textbf{608 kN/m}^2 \textbf{ per °C (compression)}$

Note: These results are not accurate at the end of the strip. Further, they apply strictly only to narrow strips where restriction of expansion across the width may be neglected. The various temperature changes to which a clad strip such as this may be subjected during manufacture will often leave it in a state of stress at room temperature. Such a state of stress is termed *residual stress*. In this case it can be removed only by over-stretching beyond the yield point and cannot be removed by annealing.

Problems

1. A brittle steel rod is heated to 150 °C and then suddenly clamped at both ends. It is then allowed to cool and breaks at a temperature of 90 °C. Calculate the breaking stress of the steel. $E = 210$ GN/m²; $\alpha = 12 \times 10^{-6}/°C$.

   (151 MN/m²)

2. A steel bar of 100 mm diameter is rigidly clamped at both ends so that all axial extension is prevented. A hole of 40 mm diameter is drilled out for one-third of the length. If the bar is raised in temperature by 30 °C above that of the clamps calculate the maximum axial stress in the bar. $E = 210$ GN/m²; $\alpha = 0.000012/°C$.

   (84.7 MN/m²)

3. A metal sleeve is to be a shrink fit on a shaft of 250 mm diameter. The sleeve is bored to a diameter of 249.5 mm at 16 °C and is then heated until the bore exceeds the shaft diameter by 0.625 mm, to allow it to pass over the shaft. It is then placed on the shaft and allowed to cool. Calculate the temperature to which the sleeve must be raised. Take $\alpha = 12 \times 10^{-6}/°C$.

   (391 °C)

4. A tie-bar connects two supports in a machine assembly. The supports may be considered rigid and are 400 mm apart. A brass alloy tube is used as a spacer and sleeved over the tie-bar so that there is 4 mm clearance between the ends of the spacer and the supports at 16 °C. The spacer is 30 mm outside diameter, and 20 mm inside diameter. Find the compressive force in the spacer at the working temperature of 600 °C. For the alloy take $E = 85$ GN/m², and $\alpha = 18 \times 10^{-6}/°C$.

   (17.1 kN)

5. A phosphor-bronze spacer is a close fit in a 300 mm gap between two faces of a steel machine frame when assembled at 16 °C. Find the maximum permissible working temperature if the maximum permissible stress in the spacer is 25 MN/m² and the increase in the gap must not exceed 0.15 mm. For the bronze take $\alpha = 16.5 \times 10^{-6}/°C$, and $E = 85$ GN/m².

   (64 °C)

6. Two steel bars are connected together so as to form a rod of total length 625 mm. One, of mild steel, is 225 mm long and 25 mm diameter; the other, of stainless steel, is 400 mm long and 12 mm diameter. If the bar is heated to 50 °C above room temperature, clamped rigidly at both ends and then allowed to cool to room temperature calculate the stress in each part of the rod. For stainless steel, $E = 175$ GN/m²; $\alpha = 18 \times 10^{-6}/°C$. For mild steel, $E = 200$ GN/m²; $\alpha = 12 \times 10^{-6}/°C$.

   (Mild steel, 44.7 MN/m²; stainless steel, 195 MN/m²)

7. A steel bar of 50 mm diameter is placed between two stops, with an end clearance of 0.05 mm. The temperature of the bar is raised 60 °C and the stops are found to have been forced apart a distance of 0.05 mm. Calculate the maximum stress in the bar if its total length is 250 mm and there is a hole of 25 mm diameter drilled along its length for a distance of 100 mm. $E = 200$ kN/mm²; $\alpha = 12 \times 10^{-6}/°C$.

   (75.3 MN/m²)

8. A compound tube is formed by a stainless steel outer tube of 50 mm outside diameter and 47 mm inside diameter, together with a concentric mild steel inner tube of wall thickness 6 mm. The radial clearance between inner and outer tubes is 2 mm. The two tubes are welded together at their ends, the compound tube being free to expand when heated. Calculate the stress in each tube due to a temperature rise of 50° C. For stainless steel,

$E = 175$ GN/m²; $\alpha = 18 \times 10^{-6}/°$ C. for mild steel, $E = 200$ GN/m²; $\alpha = 12 \times 10^{-6}/°$ C.

(Mild steel, 13.4 MN/m² tensile; stainless steel, 41 MN/m² compressive)

9. A stainless steel rod of 25 mm diameter is placed inside, and concentric with, a mild steel tube of 30 mm inside diameter and 50 mm outside diameter. The tube and bar are welded together at the ends but are otherwise free to expand. Calculate the stress in each part of the compound bar so formed due to a temperature rise of 25° C. If the length of the compound bar is 250 mm what is the extension? For stainless steel $E = 170$ GN/m²; $\alpha = 18 \times 10^{-6}/°$ C. For mild steel $E = 196$ GN/m²; $\alpha = 12 \times 10^{-6}/°$ C.

(rod, -19.05 MN/m²; tube, 7.42 MN/m²; 0.0845 mm)

10. A compound bar is made up of steel plate 50 mm wide, 10 mm thick clad on both sides by copper plates each 50 mm wide, 5 mm thick. The plates are bonded together along their length and at room temperature the original length is 1.2 m. Find the load taken by *each* plate and the increase in length when the temperature rises 100° C. For steel, $E = 200$ kN/mm², $\alpha = 12 \times 10^{-6}/°$ C. For copper, $E = 110$ kN/mm²; $\alpha = 18 \times 10^{-6}/°$ C.

(Steel, $+21.28$ kN; copper, -10.64 kN; 1.41 mm)

13.8 Poisson's ratio: lateral strain

When a bar is loaded axially in tension by a force F it extends in length but at the same time the lateral dimensions contract, Fig 13.8. The extension in the direction of the force is given by Hooke's law within the limit of proportionality. The ratio of the strain in the lateral direction to that in the longitudinal direction is found to be constant for a particular material and is called *Poisson's ratio*, denoted by v. Thus:

$$v = \frac{\text{lateral strain}}{\text{longitudinal strain}}$$

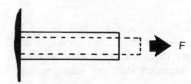

Fig. 13.8

If the longitudinal strain is ε then the lateral strain is $-v\varepsilon$.

In materials with high elasticity (elastomers) such as rubber, silicone and certain plastics, a large longitudinal extension is accompanied by appreciable reduction in cross-sectional area. In soft metals, the reduction within the elastic region is also significant but in stiff metals the lateral changes are very small.

Poisson's Ratio applies in the same way to a bar loaded in compression.

CHANGE IN VOLUME: VOLUMETRIC STRAIN

When a bar is pulled its length increases and its transverse dimensions decrease, and there is an increase in volume. Let the length of the bar be L and the cross-section to be square of side B, then if σ is the stress produced

$$\text{longitudinal strain } \varepsilon = \frac{\sigma}{E}$$

The lateral strain in the other two directions is $-ve$. If V_1 and V_2 are the initial and final volumes respectively, then

$$V_1 = LB^2$$

and V_2 = (final length) × (final thickness)^2

$$= (L + \varepsilon L)(B - v\varepsilon B)^2$$

$$= (1 - 2v\varepsilon + \varepsilon + \ldots)LB^2$$

$$= (1 - 2v\varepsilon + \varepsilon)V_1$$

neglecting the terms involving ε^2.

Therefore

$$\text{Change in volume} = V_2 - V_1$$

$$= (1 - 2v\varepsilon + \varepsilon)V_1 - V_1$$

$$= \varepsilon(1 - 2v)V_1$$

The *volumetric strain* is the change in volume per unit volume, i.e.

$$\text{volumetric strain} = \frac{V_2 - V_1}{V_1}$$

$$= \frac{\varepsilon(1 - 2v)V_1}{V_1}$$

$$= \varepsilon(1 - 2v)$$

Since a bar in tension has an increase in volume, the change in volume must be positive, therefore $(1 - 2v)$ must be positive. Thus Poisson's ratio for any material is less than 0.5. For some rubbers $v = 0.5$, and for most metals v lies between that for tungsten and chromium, 0.21, and that for gold, 0.44.

Note: The formula in terms of linear strain and Poisson's Ratio should always be used to find the *difference* of the two volumes. Calculation of V_2 directly is impracticable because of the minute changes in length involved.

Example   *A bar of titanium alloy of length 120 mm and square cross-section, 7.5 mm × 7.5 mm, is pulled axially by a force of 15 kN. Find the percentage decrease in thickness if* E = 106 GN/m², *and* v = 0.33.

SOLUTION

$$\text{Stress } \sigma = \frac{15,000}{7.5 \times 7.5 \times 10^{-6}} = 267 \times 10^6 \text{ N/m}^2$$

$$\text{Longitudinal strain } \varepsilon = \frac{\sigma}{E} = \frac{267 \times 10^6}{106 \times 10^9} = 0.0025$$

$$\text{Lateral strain} = v\varepsilon = 0.33 \times 0.0025 = 0.001$$

i.e. $\dfrac{\text{change in thickness}}{\text{original thickness}} = 0.001$

i.e. % change in thickness $= 0.001 \times 100 = \mathbf{0.1}$

Example *A bar of aluminium alloy of rectangular section 75 mm × 20 mm and 800 mm long is stretched by an axial force of 150 kN. Find the volumetric strain, the actual change in volume, and the percentage reduction in cross-sectional area of the bar. Take E = 70 GN/m², and v = 0.34.*

SOLUTION

$$\text{Stress } \sigma = \frac{150 \times 10^3}{75 \times 20 \times 10^{-6}} = 100 \times 10^6 \text{ N/m}^2$$

$$\text{Longitudinal strain } \varepsilon = \frac{\sigma}{E} = \frac{100 \times 10^6}{70 \times 10^9} = 0.00143$$

Change in volume = volumetric strain × original volume

$$= \varepsilon(1 - 2v) \times (75 \times 20 \times 800)$$

$$= 0.00143(1 - 2 \times 0.34) \times 12 \times 10^5$$

$$= \mathbf{550 \text{ mm}^3}$$

If B and b are the initial lengths of the sides of the section then the original area is Bb. The new lengths are $(B - v\varepsilon B)$ and $(b - v\varepsilon b)$, hence

$$\text{reduction in area} = Bb - (B - v\varepsilon B)(b - v\varepsilon b)$$

$$= - 2v\varepsilon Bb, \text{ neglecting small quantities}$$

$$= - 2v\varepsilon \text{ per unit area}$$

therefore percentage reduction in area $= 2v\varepsilon \times 100$

$$= 2 \times 0.34 \times 0.00143 \times 100$$

$$= \mathbf{0.1}$$

Problems

1. An axial load of 14 kN is applied to a bar of cold-drawn copper and produces an extension of 0.25 mm on a gauge length of 250 mm. If the bar is of square

section 10 mm side and the decrease in thickness is measured as 0.0034 mm find Young's modulus and Poisson's ratio for the copper.

(140 GN/m^2; 0.34)

2. A piece of gold wire of cross-sectional area 0.125 mm^2 is stretched by a force of 100 N. Find the volumetric strain and the percentage reduction in area. Take $E = 80 \text{ GN/m}^2$ and $v = 0.44$.

(0.0012; 0.88 per cent)

3. In an experiment a brass bar of 30 mm diameter and 800 mm long is subject to an axial tensile load of 60.4 kN. By optical means the contraction in diameter is measured as 0.008 mm and the extension as 0.76 mm. Find Poisson's ratio, Young's modulus, and the change in volume of the bar.

(0.28; 90 GN/m^2; 236 mm^3)

4. A bar of steel for which Poisson's ratio is 0.26 is strained in simple tension. The linear strain is 0.0015. Find the percentage change in volume of the bar.

(0.072 per cent)

5. A bar of aluminium 400 mm long is of rectangular section 50 mm by 30 mm and extends 0.8 mm when loaded in tension. Find the final volume of the bar. Take $v = 0.3$.

(600, 480 mm^3)

6. A steel bar 30 mm diameter carries a tensile load of 65 kN. Calculate the reduction in diameter of the bar. Take $E = 200 \text{ GN/m}^2$, and $v = 0.25$.

(0.0035 mm)

13.9 Strain energy: resilience

Work is done in stretching or compressing a bar of material. If the bar is elastic this work is stored as *strain energy* or energy of deformation, and is recoverable on removal of the load. The bar behaves exactly like a spring.* The energy stored per unit volume in a strained bar is also called the *resilience*.

If a force F stretches a bar a small distance dx, then the work done is Fdx. When the force F varies with the extension x the total work done in stretching the bar a distance x is

$$\int_0^x Fdx$$

and this is the area enclosed by the load-extension graph and the x-axis. It is also the total strain energy U in the elastic bar. If the bar obeys Hooke's law the load-extension graph is a straight line, Fig. 13.9, hence

U = work done

= area OAB

---

* The strain energy of springs is dealt with on p. 215.

therefore

$$U = \tfrac{1}{2}Fx$$

where F is the maximum force. The factor $\tfrac{1}{2}$ represents the fact that the force increases uniformly from zero to a maximum value F during the extension of the bar. Alternatively

U = work done

= average force × extension

= $\tfrac{1}{2}Fx$

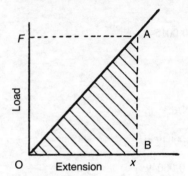

Fig. 13.9

It is convenient to write the strain energy in terms of the maximum stress σ produced by the force F. If A is the cross-sectional area of the bar and l its length, then

$$F = \sigma A$$

and $x = \sigma l/E$

thus

$$U = \tfrac{1}{2}Fx$$

$$= \tfrac{1}{2} \times (\sigma A) \times \frac{\sigma l}{E}$$

$$= \frac{\sigma^2}{2E} \times Al$$

i.e. $U = \dfrac{\sigma^2}{2E} \times \textit{volume of bar}$

The units of strain energy are those of work, i.e. *joules* (**J**).

Example *A steel strut is of square section, 100 mm by 100 mm over the middle portion, which is 150 mm long, and 70 mm by 70 mm over the remainder of its length. If the total length is 250 mm and the load 800 kN, calculate the total strain energy in the bar. E = 200 GN/m².*

SOLUTION
Since the cross-sectional areas of the two portions of the bar are different, the stresses produced in these portions differ and the strain energy for each portion must be calculated separately.

Middle portion:

$$\sigma = \frac{800 \times 10^3}{0.1 \times 0.1} = 8 \times 10^7 \text{ N/m}^2$$

volume $= 0.1 \times 0.1 \times 0.15 = 0.0015 \text{ m}^3$

strain energy $= \dfrac{\sigma^2}{2E} \times$ volume

$$= \frac{(8 \times 10^7)^2}{2 \times 200 \times 10^9} \times 0.0015$$

$$= \mathbf{24\ J}$$

End portions:

$$\sigma = \frac{800 \times 10^3}{0.07 \times 0.07} = 16.33 \times 10^7 \text{ N/m}^2$$

volume $= 0.07 \times 0.07 \times 0.1 = 0.00049 \text{ m}^3$

strain energy $= \dfrac{(16.33 \times 10^7)^2}{2 \times 200 \times 10^9} \times 0.00049$

$$= 32.7 \text{ J}$$

Total strain energy $= 24 + 32.7 = \mathbf{56.7\ J}$

Problems

1. A steel bar 1.5 m long is of 50 mm diameter for 900 mm of its length and 25 mm diameter for the remainder. What is the strain energy stored in the bar under a load of 45 kN? $E = 200 \text{ kN/mm}^2$.

   (8.5 J)

2. Compare the strain energy stored in a loaded bar 250 mm long and 50 mm diameter, with that stored in a similar bar which is turned down to 40 mm diameter for one-half its length. The maximum direct stress in each bar is to be the same.

   (1.9 : 1)

3. A bar of steel is 900 mm long and of 12 mm diameter. Calculate the strain energy stored in the bar when a tensile load of 27 kN is applied.

   Find also the additional strain energy that can be stored before the material exceeds its elastic limit stress of 325 MN/m². $E = 200 \text{ GN/m}^2$.

   (14.5 J, 12.4 J)

13.10 Application of strain energy to impact and suddenly applied loads

Impact loads

If a load is suddenly applied to a bar, as in an impact, the bar stretches and behaves as a spring, oscillating about a mean position. The strain energy stored in the bar is greatest when the bar and load are instantaneously at rest at the position of maximum displacement. At this point the total energy of the load has been absorbed as strain energy. For example, assume that a load of weight W falls through a height h on to a collar at the end of a vertical bar of length l, Fig. 13.10. Let x be the maximum instantaneous extension of the bar and σ the corresponding maximum stress. Then

$$x = \frac{\sigma l}{E}$$

At the point of maximum extension

    initial potential energy of load = strain energy in bar

$$W(h + x) = \frac{\sigma^2}{2E} \times Al$$

or $W\left(h + \frac{\sigma}{E}l\right) = \frac{\sigma^2}{2E} \times Al$

This gives a quadratic in σ.

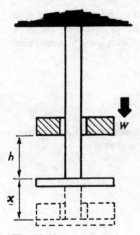

Fig. 13.10

There are two answers therefore for the stress σ; the negative answer is the compressive stress produced in the bar on rebound if the load were to lock to the collar after impact. The assumptions involved are as follows:

1. All connexions except the bar are completely rigid.
2. The limit of proportionality of stress is not exceeded.
3. There is no loss of energy at impact (mass of bar negligible).
4. The modulus of elasticity, E, is the same for impulsive loading as for steadily applied loads.

Suddenly applied loads

If the load is placed in contact with the collar without impact and suddenly let go, $h = 0$ and equating potential energy to strain energy, gives

$$Wx = \frac{\sigma^2}{2E} \times Al$$

i.e. $W\dfrac{\sigma l}{E} = \dfrac{\sigma^2}{2E} \times Al$

$$\sigma = 2\frac{W}{A}$$

Hence the maximum stress produced by the suddenly applied load is *twice* that due to the same load gradually applied. The maximum instantaneous extension will also be *twice* that for the gradually applied load.

Example   *A load of mass 50 kg falls 4 mm on to a collar at the end of a bar of 2 mm diameter and 50 mm long. The rod is made of an alloy having a modulus of elasticity of 100 kN/mm². Calculate the maximum tensile force in the rod.*

SOLUTION

$$\text{Area of section} = \frac{\pi}{4}(2)^2 = 3.142 \text{ mm}^2$$

If F kN is the maximum tensile force produced and σ the corresponding stress, then, working throughout in kN and mm,

$$\sigma = \frac{F}{3.142} = 0.318 \ F \text{ kN/mm}^2$$

and maximum extension is given by:

$$x = \varepsilon \times l$$

$$= \frac{\sigma}{E} \times l$$

$$= \frac{0.318 \ F \times 50}{100}$$

$$= 0.159 \ F \text{ mm}$$

Loss of potential energy of load equals gain of strain energy of rod. Therefore, since $W = Mg = 50 \times 9.8 = 490$ N $= 0.49$ kN,

$$W(h + x) = \tfrac{1}{2} Fx$$

$$0.49(4 + 0.159F) = \tfrac{1}{2} F \times 0.159\ F$$

thus

$$F^2 - 0.98\ F - 24.7 = 0$$

and $F = 0.49 \pm 5$

$$= 5.49 \text{ or } -4.51 \text{ kN}$$

The negative answer may be disregarded if the falling weight does not become fixed to the collar.

Maximum force in the rod = **5.49 kN**

Example *A rapidly moving free piston having a kinetic energy of 40 J suddenly seizes in the cylinder shown, Fig 13.11. Calculate the maximum tensile and compressive stresses in the cylinder due to impact. Effective length of cylinder to point of seizure, 300 mm; inside diameter, 75 mm; outside diameter, 90 mm; modulus of elasticity, 200 kN/mm².*

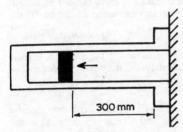

Fig. 13.11

SOLUTION

Kinetic energy lost = strain energy gained

$$= \frac{\sigma^2}{2E} \times \text{volume}$$

volume of material stressed $= \dfrac{\pi}{4}(0.09^2 - 0.075^2) \times 0.3$

$$= 5.83 \times 10^{-4} \text{ m}^3$$

kinetic energy lost = 40 J

therefore $\qquad 40 = \dfrac{\sigma^2}{2 \times 200 \times 10^9} \times 5.83 \times 10^{-4}$

and $\qquad \sigma^2 = 2.75 \times 10^{16}$

i.e. $\qquad \sigma = \pm 166 \times 10^6 \text{ N/m}^2$

$$= \pm\ \textbf{166 MN/m}^2$$

The positive answer represents the maximum tensile stress, and the negative answer the maximum compressive stress on elastic rebound, provided that the piston remains firmly fixed to the cylinder.

Problems

1. A load of mass 20 kg falls through a height of 50 mm and then starts to stretch a steel bar of 12 mm diameter and 600 mm long. If $E = 200 \times 10^9$ N/m^2, calculate the maximum stress induced.

(242 MN/m^2)

2. A load of 1 tonne is placed on a collar at the end of a vertical tie rod of 10 mm diameter. Calculate the static stress induced.

   If the load is dropped from a height of 80 mm calculate the maximum instantaneous stress induced in the rod. Length of rod = 1,150 mm, $E = 230$ GN/m^2.

   What is the maximum instantaneous stress if the load is not dropped but applied suddenly without impact?

(124.5 MN/m^2; 2.125 GN/m^2; 249 MN/m^2)

3. A collar is turned at the end of bar, 6 mm diameter, 600 mm long. The bar is hung vertically with the collar at the lower end. A load of mass 500 kg is placed just above the collar so as to be in contact but leave the bar unloaded. Calculate the maximum instantaneous stress and extension of the bar if the load is suddenly released. $E = 200$ kN/mm^2.

(347 MN/m^2; 1.04 mm)

4. A mass of 50 kg falls 150 mm on to a collar attached to the end of a vertical rod of 50 mm diameter and 2 m long. Calculate the maximum instantaneous extension of the bar. $E = 200$ GN/m^2.

(0.86 mm)

5. A mass of 200 kg falls 20 mm on to a vertical cylindrical column, thereby compressing it. The column is 800 mm long and 50 mm diameter. Find the maximum instantaneous stress produced by the impact and the total strain energy stored by the column at the instant of maximum compression. $E = 200$ GN/m^2.

(100 MN/m^2; 39.3 J)

6. A mass of 5 Mg is to be dropped a height of 50 mm on to a cast-iron column of 80 mm diameter. What is the minimum length of column if the energy of impact is to be absorbed without raising the maximum instantaneous stress above 220 MN/m^2? E for cast iron = 110 GN/m^2.

(2.42 m)

7. A mass of 9 kg falls through a height of 150 mm and then starts to stretch a steel bar of 12 mm diameter and 900 mm long. If the bar is turned down to 9 mm diameter for 300 mm of its length calculate the maximum stress induced in the bar. $E = 210$ GN/m^2.

(371 MN/m^2)

8. What is the maximum height a mass of 1,000 kg can be dropped on to a steel column of 25 mm diameter and 300 mm long, if the maximum instantaneous stress is not to exceed 0.28 kN/mm^2? $E = 210$ kN/mm^2.

(2.41 mm)

THIN-WALLED PRESSURE VESSELS

13.11 Hoop stress in a cylinder

A cylinder containing fluid under pressure is subjected to a uniform radial pressure normal to the walls, Fig. 13.12. Since the cylinder tends to expand radially there will be a tensile or *hoop stress* σ_h set up in the circumferential direction, i.e tangent to the shell wall. This stress may be found by considering the equilibrium of forces acting on one-half of the shell. Imagine the cylinder to be cut across a diameter, Fig. 13.13.

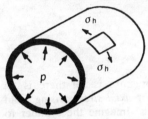

Fig. 13.12

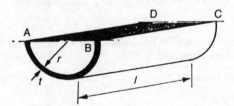

Fig. 13.13

Then there is a uniform downward pressure p acting on the diametral surface section ABCD shown; this is balanced by the upward force due to the hoop stress σ_h along the two edges.

Force due to radial pressure on area ABCD = $p \times$ area ABCD

$$= p \times \text{AB} \times \text{BC}$$

$$= p \times 2r \times l$$

where r is the cylinder internal radius and l the length. If the thickness t of the shell wall is small compared to the internal radius r (e.g. if t is less than $r/10$), then the hoop stress may be taken as uniform across the wall section. Then the upward force on the two edges due to σ_h is

$$= 2 \times \sigma_h \times \text{area of one edge}$$

$$= 2\sigma_h \times t \times l$$

Equating these two forces

$$2\sigma_h tl = 2\,prl$$

therefore

$$\sigma_h = \frac{pr}{t}$$

This is the only stress due to fluid pressure in an open ended seamless cylinder (e.g. a pipe-line) provided that the section considered is distant from an end connexion or flange. The weight of the fluid has been neglected.

13.12 Axial stress in a cylinder

In a closed pipe or cylinder, such as a pressure vessel there is, in addition to the hoop stress, a longitudinal or *axial stress* arising from the force due to pressure on the closed ends. Imagine the cylinder to be cut by a plane normal to the axis, Fig. 13.14. Then the pressure p acts on a cross-sectional area πr^2 and the corresponding axial force is:

$$p \times \pi r^2$$

This force is balanced by the force due to the axial stress σ_a acting on the area of the shell rim, which is approximately:

$$\text{circumference} \times \text{thickness} = 2\pi r \times t$$

Hence

$$\sigma_a \times 2\pi rt = p \times \pi r^2$$

$$\sigma_a = \frac{pr}{2t}$$

and $\sigma_h = \dfrac{\mathrm{pr}}{\mathrm{t}}$ hence

$$\sigma_a = \tfrac{1}{2}\sigma_h$$

i.e. the axial stress is one-half the hoop stress.

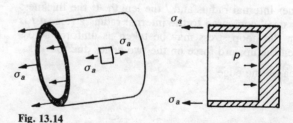

Fig. 13.14

13.13 Tangential stress in a spherical shell

If a thin spherical shell is subject to internal pressure p, a tensile stress is set up in the shell wall due to the tendency of the shell to expand under pressure. Imagine the spherical shell to be cut across a diameter and consider the forces acting on one-half of the shell, Fig. 13.15, these are:

1. The diametral force due to the pressure

$$= p \times \pi r^2$$

2. The resisting force due to the tangential stress σ_t acting on the section of the rim. If t is small compared with the internal radius r, then σ_t is nearly uniform and the area of the rim section is approximately $2\pi rt$, i.e.

   resisting force $= \sigma_t \times 2\pi rt$

Equating these two forces

$$\sigma_t \times 2\pi rt = p \times \pi r^2$$

$$\sigma_t = \frac{\mathbf{pr}}{\mathbf{2t}}$$

This applies to any diametral section of the sphere and hence at any point there is a tangential stress σ_t acting in all directions tangent to the wall.

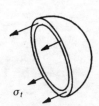

Fig. 13.15

13.14 Effect of joints on stresses in thin shells

In many cases cylindrical shells are not seamless but are jointed, the joints being along a circumferential or longitudinal seam. The distribution of stress in a riveted joint is complex and the strength of such joints cannot be calculated with any great accuracy. The design of riveted joints is largely empirical and cannot be dealt with properly here. It is possible, however, to arrive at more accurate values for the stresses by making allowances for the efficiencies of the joints. The efficiency of a joint may be defined as the ratio:

$$\frac{\text{strength of joint of given width}}{\text{strength of solid plate of same width}}$$

For example, if the efficiency of a joint is 70 per cent it means in effect that the effective area of the perforated plate is 0.7 of that of the solid plate. The average stresses calculated using the thin cylinder formulae would therefore have to be increased in the ratio 1/0.7.

A circumferential joint has to resist the axial tension whereas the longitudinal joint has to resist the hoop tension. The axial tension is one-half the hoop tension so that the longitudinal joint is potentially the weakest part of the cylinder. Circumferential joints, therefore, do not have to be of the same efficiency as the longitudinal joint and are often permitted to have a much lower efficiency.

Example *Calculate the required thickness of the shell of an experimental pressure vessel of spherical shape and 450 mm diameter, which has to withstand an internal fluid pressure of 7 MN/m² without the stress in the material of the shell exceeding 70 MN/m². If the shell is to be made by bolting together two flanged halves using sixteen bolts what should be the root area of each bolt? The tensile stress in the bolts must not exceed 150 MN/m².*

SOLUTION

$$\text{hoop stress} = \frac{pr}{2t}$$

thus
$$t = \frac{7 \times 10^6 \times 0.225}{2 \times 70 \times 10^6} = 0.01125 \text{ m}$$

$$= \textbf{11.25 mm}$$

Diametral bursting force $= p \times \pi r^2$

$$= 7 \times 10^6 \times \pi \times 0.225^2$$

$$= 1.114 \times 10^6 \text{ N}$$

force per bolt $= \dfrac{1.114 \times 10^6}{16}$

$$= 69{,}600 \text{ N}$$

therefore

$$69{,}600 = \text{stress in bolt} \times \text{root area}$$

$$= 150 \times 10^6 \times A$$

thus $A = 464 \times 10^{-6} \text{ m}^2 = \textbf{464 mm}^2$

(24 mm diameter bolts would be required.)

Example *A thin tube contains oil at a pressure of 6 MN/m². Each end is closed by a piston, the two pistons being free to move in the tube but rigidly connected by a rod as shown, Fig. 13.16. (a) Calculate the stresses in the tube if it has an inside diameter of 50 mm and a wall thickness 2.5 mm. (b) Calculate the tensile stress in the rod joining the pistons if it is of 25 mm diameter.*

Fig. 13.16

SOLUTION

(*a*) The axial force due to oil pressure is taken by the connecting rod. There is therefore no axial force or stress in the tube. The hoop stress in the tube is given by:

$$\text{hoop stress} = \frac{pr}{t}$$

$$= \frac{6 \times 10^6 \times 0.025}{0.0025}$$

$$= 60 \times 10^6 \text{ N/m}^2 = \mathbf{60 \text{ MN/m}^2}$$

(*b*) Inside area of piston $= \dfrac{\pi}{4} \times 0.05^2 - \dfrac{\pi}{4} \times 0.025^2$

$$= 1.47 \times 10^{-3} \text{ m}^2$$

axial force on piston $= 6 \times 10^6 \times 1.47 \times 10^{-3}$

$$= 8{,}820 \text{ N}$$

area of rod $= \dfrac{\pi}{4} \times 0.025^2 = 492 \times 10^{-6} \text{ m}^2$

tensile stress in rod $= \dfrac{8{,}820}{492 \times 10^{-6}}$

$$= 18 \times 10^6 \text{ N/m}^2 = \mathbf{18 \text{ MN/m}^2}$$

Example *A cylindrical boiler shell is 2 m internal diameter and is made of plate 20 mm thick. If the working pressure is 1.75 MN/m² and the efficiency of the longitudinal joint is 75 per cent find the average hoop stress in the plate at the joint.*

SOLUTION

For the riveted plate, since the joint efficiency is 0.75,

$$\text{average hoop stress} = \frac{pr}{t} \times \frac{1}{0.75}$$

$$= \frac{1.75 \times 10^6 \times 1}{0.02} \times \frac{1}{0.75}$$

$$= 116.7 \times 10^6 \text{ N/m}^2$$

$$= \textbf{117 MN/m}^2$$

Example *A cylindrical pressure vessel has an internal diameter of 1,600 mm and is subject to an internal fluid pressure of 30 bar. The plate is 15 mm thick with an ultimate tensile stress of 600 N/mm². The efficiencies of the circumferential and longitudinal joints are 50 and 80 per cent, respectively. Determine the factor of safety. 1 bar = 10⁵ N/m².*

SOLUTION

$p = 30$ bar $= 30 \times 10^5$ N/m² $= 3$ N/mm²

For the solid plate (working throughout in N and mm)

$$\text{hoop stress} = \frac{pr}{t}$$

$$= \frac{3 \times 800}{15}$$

$$= 160 \text{ N/mm}^2$$

At the longitudinal joint

$$\text{hoop stress} = \frac{160}{0.8}$$

$$= 200 \text{ N/mm}^2$$

For the solid plate

$$\text{axial stress} = \frac{160}{2} = 80 \text{ N/mm}^2$$

At the circumference joint

$$\text{axial stress} = \frac{80}{0.5}$$

$$= 160 \text{ N/mm}^2$$

Factors of safety are dealt with on page 333. In this case it is the ratio of the ultimate tensile stress to the *maximum* stress, which is the hoop stress at the longitudinal joint.

$$\text{thus factor of safety} = \frac{600}{200}$$

$$= \textbf{3}$$

Problems

1. Calculate the maximum allowable pressure in a boiler shell of 3 m diameter, 25 mm thick if the tensile stress is not to exceed 60 MN/m^2. Assume a joint efficiency of 60 per cent.

$$(600 \text{ kN/m}^2)$$

2. What is the minimum shell thickness required for a Lancashire boiler of 2,500 mm inside diameter for a pressure of 14 bar if the allowable stress is not to exceed 50 N/mm^2? Allow for a joint efficiency of 65 per cent.

$$(54 \text{ mm})$$

3. Calculate the maximum allowable diameter for a spherical pressure vessel for a nuclear reactor if it is to contain carbon dioxide at a pressure of 1 MN/m^2. The tensile stress in the shell wall is to be 75 MN/m^2 and the maximum thickness of vessel shell that can be manufactured is 75 mm.

$$(22.5 \text{ m})$$

4. A spherical copper shell is 600 mm in internal diameter and is to withstand an internal pressure of 20 bar without the stress in the copper exceeding 60 MN/m^2. Find the thickness of shell required assuming a joint efficiency of 80 per cent.

$$(6.25 \text{ mm})$$

5. A cylindrical air receiver for a compressor is 2 m in internal diameter and made of plate 15 mm thick. If the hoop stress is not to exceed 90 MN/m^2 and the axial stress is not to exceed 60 MN/m^2 find the maximum safe air pressure.

$$(1.35 \text{ MN/m}^2 \text{ or } 13.5 \text{ bar})$$

6. A bronze sleeve of 80 mm internal diameter and 6 mm thick is force fitted on to a solid steel shaft. The force fitting of the sleeve on to the shaft subjects it to an internal radial pressure. A measurement of hoop strain in the sleeve shows that the corresponding hoop stress is 96 MN/m^2. Find the radial pressure between sleeve and shaft.

$$(14.4 \text{ MN/m}^2)$$

7. A thin spherical vessel is to contain 88 m^3 of gas at a pressure of 14 bar. The stress in the material must not exceed 120 N/mm^2. Find the internal diameter of the vessel and the thickness of plate required.

$$(5.52 \text{ m}; 16.1 \text{ mm})$$

8. A metal tube of 40 mm mean diameter and 2 mm thick is tested in tension and fails at a load of 7 kN. A similar tube is used to contain fluid under pressure. Find the safe internal pressure allowing a factor of safety of 4.

$$(696 \text{ kN/m}^2 \text{ or } 6.96 \text{ bar})$$

9. A dumb-bell piston forming part of a hydraulic control valve slides freely in the cylinder shown, Fig. 13.17. The pressure in the cylinder at A is 4.2 MN/m^2 and that in B is 700 kN/m^2. Calculate the largest hoop and axial stresses in the cylinder. Internal diameter of cylinder is 50 mm, wall thickness 1.5 mm

$$(70 \text{ MN/m}^2; 5.83 \text{ MN/m}^2)$$

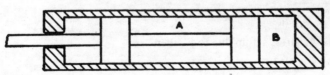

Fig. 13.17

13.15 Rotating rims

A circular thin ring rotating about an axis through its centre O with angular velocity ω rad/s is subject to inertia force acting radially outwards on every element. For any small arc AB (Fig. 13.18) subtending an angle θ rad at the centre, the inertia (centrifugal) force is

$$F = m\omega^2 r$$

where m is the mass of the element AB.

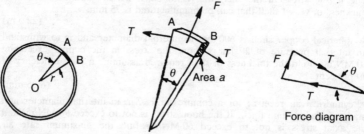

Fig. 13.18

If ρ is the density of the material and a the cross-sectional area of ring section, then

$$m = \rho \times AB \times a$$
$$= \rho \times r\theta \times a$$

hence

$$F = \rho r\theta a \times \omega^2 r$$
$$= \rho\theta a\omega^2 r^2$$

The element is maintained in equilibrium by the inertia force F and by the tangential forces T at A and B exerted by the material of the ring at these points. Thus a tensile force is set up in the ring, as in a thin cylinder under internal pressure. From the force diagram, since θ is *small*:

$$T\theta = F = \rho\theta a\omega^2 r^2$$

thus

$$T = \rho a\omega^2 r^2$$

If σ is the stress set up due to T, then

$$\sigma \times a = T$$
$$= \rho a\omega^2 r^2$$

thus $\sigma = \rho\omega^2 r^2 = \rho v^2$

where v is the linear speed of the rim and r the *mean* radius.

The tensile stress in the ring is therefore independent of the area of section of the ring. The effect of radial spokes or a disc connecting the ring to its axis of rotation has been neglected.

Example *A thin cylindrical spring steel tube of 3 mm mean radius is required to rotate at 500,000 rev/min when used in a machine for spinning nylon. Calculate the maximum tensile stress in the tube. Density of steel, 7.8 Mg/m³.*

SOLUTION

$$r = 3 \text{ mm}$$

$$\rho = 7.8 \text{ Mg/m}^3 = 7,800 \text{ kg/m}^3$$

$$\omega = \frac{2\pi \times 500,000}{60} = 52,400 \text{ rad/s}$$

$$\text{maximum tensile strength} = \rho\omega^2 r^2$$

$$= 7,800 \times 52,400^2 \times (0.003)^2$$

$$= 193 \times 10^6 \text{ N/m}^2 = \mathbf{193 \ MN/m^2}$$

Example *A flywheel may be taken as a thin ring having a rim section 50 mm by 50 mm. The flywheel is to rotate at 420 rev/min. Find the least value of the mean diameter to satisfy the following conditions: (a) the tensile stress in the material must not exceed 15 MN/m²; (b) the total mass of the flywheel must be less than 110 kg. Density of material = 7 Mg/m³.*

SOLUTION

$$\omega = \frac{2\pi \times 420}{60} = 44 \text{ rad/s}$$

If the tensile stress is limited to 15 MN/m² then

$$\rho = 7,000 \text{ kg/m}^3$$

and

$$\text{maximum tensile stress} = \rho\omega^2 r^2$$

thus

$$r^2 = \frac{15 \times 10^6}{44^2 \times 7,000}$$

$$= 1.105$$

therefore

$$r = 1.05 \text{ m}$$

and mean diameter $\leqslant 2.1$ m

If total mass is limited to 110 kg, then

$$\text{mass} = \text{area of section} \times \text{mean circumference} \times \text{density}$$

i.e. $110 = (0.05 \times 0.05) \times \pi d \times 7000$

hence, mean diameter is given by

   $d \leq 2$ m

Hence the least diameter to satisfy both conditions = **2 m**.

Problems

1. Calculate the tensile stress in a thin rim of mean diameter 900 mm rotating at 10 rev/s, if the material used has a density of 7.2 Mg/m³.

   (5.73 MN/m^2)

2. Calculate the maximum allowable speed of rotation of a cast-iron ring 1.2 m diameter if the design stress is 24 MN/m² and the density of cast iron is 7 Mg/m³.

   (933 rev/min)

3. What is the required mean diameter of a flywheel which has to rotate at a maximum speed of 2,165 rev/min with a maximum permissible stress of 75 MN/m²? Density of material = 7.5 Mg/m³.

   (883 mm)

4. A cast-steel flywheel has a rim of square cross-section 75 mm × 75 mm. The wheel has to rotate at 900 rev/min. Find the value of the least mean diameter to satisfy the following conditions: (*a*) the total mass of the wheel must be less than 225 kg; (*b*) the stress in the steel must not exceed 30 MN/m². Density of steel = 7.8 Mg/m³.

   (1.31 m)

5. A thin steel tube of 12 mm *mean* diameter and 0.5 mm thick rotates at 19,000 rev/min and carries an internal pressure of 700 kN/m². Calculate the maximum hoop stress in the tube wall if its density is 7,400 kg/m³.

   (9.105 MN/m^2)

6. A thin steel drum is required to rotate at 4,200 rev/min whilst the pressure inside the drum is 1.4 MN/m². If the drum is to be made from 6 mm plate find the maximum diameter for a limiting tensile stress of 75 MN/m². Density of steel = 7.8 Mg/m³.

   (320 mm)

CHAPTER 14

Mechanical Properties of Materials

The general properties of any material forming an engineering component depend on its chemical make-up, how it is built up from atoms and molecules into crystals, grains and solid material, and on the manufacturing processes and treatments used to produce its final form and condition. When a material is selected for a particular engineering situation a variety of these properties have to be considered including strength, machinability, corrosion resistance, electrical characteristics, thermal conductivity, melting point, etc. Often, however, these requirements have to be balanced, one against the other, and the choice of a material therefore usually involves compromise.

In this chapter the emphasis is specifically on the mechanical properties of materials and their behaviour under load. The treatment is of necessity restricted because of the proliferation of metals and plastics now in use in modern industry and the range of testing machines and techniques available. For fuller information students should refer to more specialist texts, British Standards and manufacturers' publications.

14.1 Metals and alloys

Engineering metals can be divided into two groups based on their iron content; those consisting mainly of iron are called *ferrous metals*, and all others *non-ferrous*. The 'light' metals include aluminium, magnesium, and titanium, and the 'refractory' metals with heat-resisting properties include tungsten and molybdenum. Alloys are formed by adding quantities of various elements to a basic metal, in some cases very small quantities, and the resulting materials usually have markedly different properties from those of the individual constituents. The most commonly used alloys are those of iron with a small amount of carbon to produce steel or cast-iron. The non-metallic content, less than 4 per cent *by weight*, is the primary factor in determining the nature and properties of the ferrous metal produced. Steels contain less than 1.5 per cent carbon; the 'plain' carbon steels, composed almost entirely of iron and carbon are termed low- (or mild), medium- or high-carbon steels depending on the proportion of carbon present. When a carbon steel is alloyed with other elements besides carbon it is called an 'alloy steel' and is designated according to the predominant element added, e.g. manganese steel. Each alloying element is used to produce specific effects on the properties of the steel produced or on the manufacturing process involved, e.g. to give a tough, machinable material, to resist the effects of high temperatures, or to enable a steel to be hardened. A particular example is the use of cobalt, nickel and titanium, which together with ageing processes result in high-strength, very ductile 'maraging' steels, greatly used in rocket work. Cast-irons have a higher carbon content than steels together with amounts of silicon, magnesium, sulphur and phosphorous. There is a great variety of modern cast-irons but the most common is the traditional grey iron, very brittle, easily machinable, a good conductor of heat, and useful in massive parts for damping down vibrations. Adding a small amount of magnesium in the production stages produces nodular or spheroidal iron, a strong, tough, ductile material.

Non-ferrous metals and their alloys are equally as important as the ferrous. Aluminium, a soft metal with a low melting point, is noted for its high-electrical and thermal conductivity as well as resistance to corrosion. It is the foremost metal in use after steel because of its excellent strength to weight ratio, giving light, stiff materials. Copper, alloyed with up to 40 per cent zinc and small quantities of other elements such as tin, is the basis of the various straight brasses, but when alloyed without zinc to tin, phosphorous, silicon or aluminium, it gives a range of bronzes. Phosphor-bronze, for example, is a tin bronze with added phosphorous, and like manganese bronze is particularly resistant to sea water. Further examples of non-ferrous alloys are those based on nickel, magnesium, and titanium, each of which has special properties. Nickel is noted for hardness and strength, titanium for lightness and rigidity as well as strength at high temperatures, and magnesium is the lightest of all the metals.

It is useful to define the mechanical properties of materials in general, plastics as well as metals, by considering in particular the behaviour of black mild steel when loaded in tension and compression.

14.2 Black mild steel in tension

Black mild steel is a low-carbon steel in a hot-rolled or annealed condition. A tension test on a typical specimen would give the graph of load against extension shown in Fig. 14.1.

Elastic stage

In the initial stage of the test the steel is elastic, i.e. when unloaded the test piece returns to its original unstretched length. This is represented

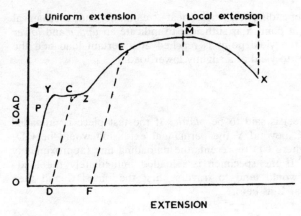

Fig. 14.1

by the line OP, Fig. 14.1. Over the major portion of this stage the material obeys Hooke's law, i.e. the extension is proportional to the load and the strain is proportional to the stress.

Limit of proportionality

The point P represents the *limit of proportionality*. Beyond P the metal no longer obeys Hooke's law.

Elastic limit

The stress at which a permanent extension occurs is the *elastic limit stress*. The metal is no longer elastic. In black mild steel the limit of proportionality and elastic limit are very close together and often cannot be distinguished. Elastic limit stress and limit of proportionality values have limited use today.

Permanent set

If the metal is loaded beyond the point P representing the elastic limit, and then unloaded, a permanent extension remains, called the *permanent set*.

Yield stress

At Y the metal stretches without further increase in load. Y is termed the *yield point* and the corresponding stress is the *yield stress*. This sharp yield is typical of mild carbon steel, wrought iron, and some plastics, but occurs with few other materials. The graph shows a very slight dip at the yield point. For a medium carbon steel and for some other metals, depending on their heat treatment and mechanical working, the dip at point Y is sufficient to indicate an *upper* and *lower* yield stress, i.e. the yield point is reached at a certain load and the material continues to yield at a slightly lower load.

Plastic stage

Beyond Y the steel is said to be *plastic*. If the test piece is unloaded from any point C beyond Y the permanent extension would be OD, approximately, where CD represents the unloading line (approximately parallel to PO). If the specimen is reloaded immediately the load–extension graph would tend to traverse first the line DC and then continue from near C as before.

Work hardening

At the point Z, further extension requires an increase in load and the steel is said to *work harden* or increase in strength. If unloaded from any point E between Z and M, the unloading graph would be approximately the line EF. If reloaded immediately, the graph would trace out approximately the same elastic line from F to E, after which it continues from E to M, as it would have done if not unloaded. The process of *cold working*, i.e. cold drawing or rolling, represents a work hardening or strengthening of this nature.

During the stage Z to M the mill scale on an unmachined specimen of black steel is seen to flake off from the stretched metal. Furthermore, the extension is now no longer small but could be measured roughly with a simple rule.

Waisting

M represents the *maximum load* which the test piece can carry. At this point the extension is no longer uniform along the length of the specimen but is localized at one portion. The test piece begins to *neck down* or *waist*, the area at the waist decreasing rapidly. Local extension continues with a decrease of load until fracture occurs at point X.

Ultimate tensile stress

The *ultimate tensile stress* (U.T.S.) is defined as:

$$\frac{\text{maximum load}}{\text{original area}}$$

Black mild steel has an U.T.S. of about $400 \, \text{MN/m}^2$. There are very few steels with a strength above $1500 \, \text{MN/m}^2$ and only a limited number with a specified U.T.S. above $1200 \, \text{MN/m}^2$. One of the strongest is the wire used in musical instruments, a very hard-drawn, high-carbon steel, the highest grade of spring wire, with a strength in the range 1800–3000 MN/m^2.

Breaking stress

The *nominal fracture* or *breaking stress* is:

$$\frac{\text{load at fracture}}{\text{original area}}$$

and this is less than the U.T.S. in a metal which necks down before fracture. This stress is seldom quoted today.

True fracture stress

The *true* or *actual fracture stress* is:

$$\frac{\text{load at fracture}}{\text{final area at fracture}}$$

and this is greater than either the nominal fracture stress or the U.T.S. in a metal which necks down, due to the reduced area at fracture. The true stress may be as much as 100 per cent higher than the U.T.S. for mild steel. Also, it may be noted that the true stress is found to be roughly constant for a given material whereas the U.T.S. varies with the treatment of the specimen before testing.

Fracture

The appearance of the fracture is shown in Fig. 14.2. It is described as a cup-and-cone fracture and is typical of a *ductile* material such as mild steel.

Fig. 14.2

Failure: factor of safety

The term 'failure' applied to a material or element in a machine can mean fracture as we have discussed here, or it can mean that the member has deformed past the elastic limit, buckled or collapsed. Fracture can also be brought about by bending or cyclic stresses as well as by direct tension or compression. In practice, engineering parts are designed with a margin of safety, e.g. by assuming a working or *allowable* stress which is a fraction of the U.T.S. or in some cases, the yield stress. A *factor of safety* based on the U.T.S. is given by

$$\text{factor of safety} = \frac{\text{U.T.S.}}{\text{allowable or working stress}}$$

Ductility

Black mild steel is a *ductile* material since it can be drawn out into a fine wire and undergo considerable plastic deformation before fracture. Ductility in a member of a structure permits it to 'give' slightly under load, which is useful where errors in workmanship or non-uniform stresses occur. Ductility is of importance in manufacture where material is to be bent or formed to shape. Ductility is measured in two ways:

1. By the *percentage reduction in area*, which is

$$\frac{\text{reduction in area}}{\text{original area}} \times 100 \text{ per cent}$$

where the reduction in area is the difference between the original area and the least area at the point of fracture.

2. By the *percentage elongation in length*. If a gauge length l is marked on the test piece before testing and the extension of this length after fracture found to be x, then, provided fracture occurred between the gauge points,

$$\text{percentage elongation} = \frac{x}{l} \times 100$$

The percentage elongation depends on the dimensions of the test piece so that, for the purposes of comparison, the dimensions have been standardized. It is found that for cylindrical test pieces, if the ratio of gauge length to diameter is kept constant, the percentage elongation is constant for a given material. By international agreement the gauge length is five diameters, i.e. $l = 5\,D$. The standard test piece for a round bar in tension (BSS. 18) is shown in Fig. 14.3. The dimensions for a 10-mm diameter test piece are shown below.

Fig. 14.3

Gauge length, l (mm)	50
Diameter, D (mm)	10
Radius, R, minimum (mm)	9
Area (mm²)	78.5
P, minimum (mm)	55

There appears to be no simple relation between percentage elongation and percentage reduction in area for steels. To estimate ductility both ratios should be found as some steels show a high percentage elongation with a low percentage reduction in area.

14.3 Stress–strain curve

So far we have considered the load-extension diagram. However, since the nominal stress is

$$\frac{\text{load}}{\text{original area}}$$

and the strain is

$$\frac{\text{extension}}{\text{original gauge length}}$$

the curve of (nominal) stress against strain will be of the same shape as the load-extension graph up to the maximum load. At the point of maximum load the test piece begins to neck down and the cross-sectional area diminishes rapidly. Beyond this point the extension to gauge length no longer measures the true strain at any point. Similarly the ratio of load to original area is an inaccurate measure of the *true stress* at the waist. Nevertheless it is convenient to sketch a *stress-strain curve* which illustrates some of the properties of the material independent of the size of the specimen.

14.4 Modulus of elasticity

The modulus of elasticity E is the ratio of stress to strain at a point on the initial straight-line portion of the load-extension diagram obtained from a tensile test on a standard test-piece.

Fig. 14.4 represents, for a metal obeying Hooke's law, the best straight line connecting load W and extension x obtained from the plotted experimental points. The modulus of elasticity E is determined

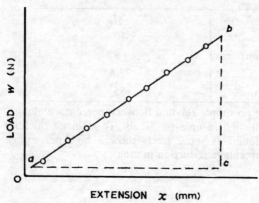

Fig. 14.4

directly from the slope of the graph as follows: If A is the cross-sectional area of the test-piece and l the gauge length, then

$$\text{strain, } \varepsilon = \frac{x}{l}$$

and

$$\text{stress, } \sigma = \frac{W}{A}$$

and

$$E = \frac{\sigma}{\varepsilon}$$

$$= \frac{W/A}{x/l}$$

$$= \frac{l}{A} \times \frac{W}{x}$$

But W/x is the slope of the load-extension graph, i.e. bc/ac. Therefore

$$E = \frac{l}{A} \times \frac{bc}{ac}$$

The load–extension graph does not usually pass through the point of zero load for two reasons:

1. The specimen is lightly loaded on first gripping in the testing machine.
2. Initial extensometer readings are slightly inaccurate at light loads.

However, since only the slope of the graph is required the zero error is unimportant when calculating the elastic modulus.

The modulus gives a very quick and accurate indication of the *stiffness* of a material. When E is large the slope of the elastic line is steep, i.e. a large load is required for a given extension. Stiffness is a measure of the amount of spring in a metal and must be distinguished from strength which is the force needed for fracture. Typical figures for E are given in Table 14.1. In practice, there is a wide spread of values around those given because of the effects of impurities, the different processes of manufacture, mechanical working and heat treatment. The stiffest of materials is the diamond, with an extremely high modulus, 1200 GN/m². The modulus for steel is in a fairly narrow range, 196–210 GN/m². For most metals, however, E lies between that of lead, 16 GN/m² and that of tungsten 360 GN/m². Where a material such as cast-iron does not obey Hooke's Law, the figures given in the table are very approximate. Plastics and rubbers have little rigidity and E for such materials may be as low as 1 GN/m², and even when reinforced with high-strength fibres, the modulus for reinforced plastic is only exceptionally more than 60 GN/m².

14.5 Specific modulus of elasticity

The denser a material the heavier it will be for a given strength and rigidity. Where the strength-to-weight ratio is critical as in aircraft and transport vehicles, it is not the absolute value of E that is important but the *specific* value which takes into account the relative density or specific gravity of the material. Thus

$$\text{specific modulus of elasticity} = \frac{E}{\text{relative density}}$$

Relative density is the density of the material relative to that of water, and since it is a ratio, the basic units of the specific modulus are the same as those of E, i.e. N/m².

Table 14.1 shows typical values of the specific modulus and it can be seen that for a surprising number of materials the specific modulus is roughly the same as that of steel, about 25 GN/m².

14.6 Black mild steel in compression: malleability

Up to the limit of proportionality, tension and compression tests on black mild steel give roughly similar stress-strain graphs, the value of the modulus of elasticity being approximately the same in compression as in tension. A well-defined yield point occurs after which the stress continues to rise with increasing strain, no maximum load or stress

Table 14.1

Material	Young's modulus (GN/m² or kN/mm²)	Relative Density	Approximate Specific Modulus (GN/m²)
Steel	196–210	7.8	25
Wrought iron	175	7.8	23
Cast-iron, grey	105–125	7.2	16
, spheroidal	180	7.2	25
Titanium alloys	110	4.5	25
Magnesium alloys	45	1.8	25
Tungsten	360	19.2	19
Aluminium alloy	70	2.7	25
Copper	80–140	8.9	9–16
Brass	84	8.4	10
Bronze	85–120	8.6	10–14
Gunmetal	80–100	8.7	9–12
Timber	7–20	0.5–0.8	14–30
Lead	16	11.3	–
Concrete	15–40	–	–
Rubber	<0.04	–	–
Unreinforced plastics	1.4	1.4	1
Glass fibre	50–85	2.5	20–34
Carbon fibre, high modulus	420	2	210
Reinforced plastic,			
glass fibre	7–60	1.9	5–30
carbon fibre	130–200	1.5	90–130

being reached before destruction. Owing to friction at the surfaces of contact between specimen and compression plattens the metal does not deform uniformly but develops a barrel shape, Fig. 14.5. To avoid buckling under load the length of a cylindrical test piece is usually less than twice the diameter.

Malleability is a very similar property to ductility and is the capacity of a metal to be forged, rolled or beaten into plates, i.e. to be shaped or deformed to a great extent when compressed. Of the common engineering metals, aluminium is the most malleable.

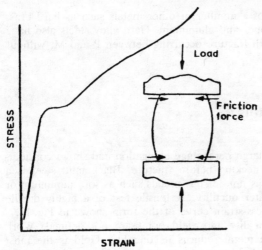

Fig. 14.5

14.7 Bright drawn mild steel

Bright drawn mild steel is again low carbon steel, but has been previously worked by cold drawing. The material is stronger but less ductile than the same steel in the form of black mild steel. A typical stress–strain curve in tension would follow the curve OPMX, Fig. 14.6(*a*). The sharp yield point has disappeared but a limit of proportionality may be determined. If sufficiently cold-worked, fracture may occur without necking. In compression the stress rises continuously and there is no fracture, Fig. 14.6(*b*). Bright drawn mild steel when annealed shows the same properties as black mild steel. The curve of

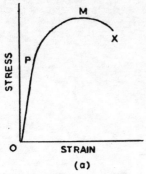
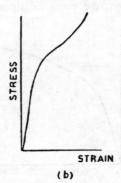

(a)   (b)

Fig. 14.6

Fig. 14.6(a) is typical of a number of other metals such as hard brass and most alloys of copper and aluminium. Hard alloy steels also have the same shape but with fracture occurring between P and M, without prior necking.

14.8 Ductile metals

A ductile metal has a large percentage elongation and shows considerable deformation and necking before fracture. Black mild steel is a ductile metal but various non-ferrous metals such as soft aluminium or copper have even greater ductility. A tensile test of a highly ductile metal would give a stress-strain curve of the form shown in Fig. 14.7. The limit of proportionality and yield point are not defined. Work hardening of a ductile metal reduces its ductility. Gold is the most ductile and malleable metal but following on the noble metals the *order* of ductility of the engineering metals is iron, copper, aluminium, zinc, tin, lead.

14.9 Proof stress

For engineering purposes it is desirable to know the stress to which a highly ductile material such as aluminium can be loaded safely before a large permanent extension takes place. This stress is known as the *proof* or *offset stress* and is defined as the stress at which a specified permanent extension has taken place in the tensile test. The extension specified may be 0.1, 0.2 or 0.5 per cent of gauge length but the 0.2 per cent figure is becoming more common.

The proof stress is found from the stress–strain curve, Fig. 14.7 as follows. From the point on the strain axis representing 0.1 per cent

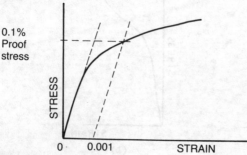

Fig. 14.7

strain draw a line parallel to the initial slope of the stress-strain diagram at O. The stress at the point where this line cuts the curve is the 0.1 per cent proof stress. The 0.2 per cent proof stress is found in a similar manner by starting from the point on the strain axis representing 0.2 per cent extension.

14.10 Brittle materials

A material which has little ductility and does not neck down before fracture is termed *brittle*. The most obviously brittle materials are the ceramic, glasses and concrete, together with some cast-irons and cold-rolled steel. Also, non-ferrous metals and alloys, when suitably worked, are brittle, as well as thermosetting plastics and some of the thermoplastics.

Fig. 14.8 shows the stress-strain curve for grey cast-iron in *tension*. The metal is elastic almost up to fracture but does not obey Hooke's law. Yielding is continuous and the total strain and elongation before fracture occurs is very small, less than 0.7 per cent elongation. Cast-iron fractures straight across the specimen as distinct from the cup-and-cone fracture of a ductile material. The modulus of elasticity for cast-iron is not a constant since there is no straight-line portion of the graph, but varies according to the point or small portion of the curve at which it is calculated.

A method of *estimating* the value of E used in rubber and plastics technology, employs the slope of the secant line OB; this gives a ratio of stress to strain at *x* per cent strain for the whole portion of the curve up to point B. This is called the *secant modulus*, and is given by BC/OC.

The stress-strain curve for cast iron in *compression* is similar to

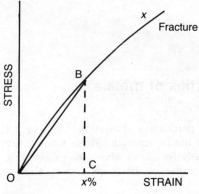

Fig. 14.8

that for a tension test. The metal fractures across planes making 55° with the axis of the specimen, except when the specimen is very short when fracture occurs across several planes.

The brittleness of a material is often best measured by the energy which it will absorb before fracture in an impact test (*see* below); the lower the energy absorbed by a standard specimen of a given material the greater the brittleness.

14.11 Resilience and toughness

When a bar is loaded within its elastic limit the work expended is stored as strain energy in the bar and is called the *resilience* of the bar. The energy is recoverable on removal of the load, i.e. the bar behaves like a spring. Resilience is a measure of the ability of the material to store energy and to withstand a blow without permanent distortion. For calculations on resilience *see* paragraph 13.9

Toughness is the converse of brittleness, and describes the ability of a material to resist the propagation of cracks and to withstand shock loads without rupturing. Both resilience and toughness are important characteristics of metals, plastics and fibres. Toughness is usually measured by the amount of energy, in joules, required to fracture a notched test piece held gripped in a vice and struck a single transverse blow by a heavy pendulum. The pendulum head strikes at a fixed height above the notch, and the 'notch-toughness' of the metal is measured by the loss of energy of the pendulum on impact and the machines are calibrated accordingly. The Izod and Charpy impact testing machines use this method. Results obtained from impact tests require care in interpretation and precise information is essential regarding the type of test, notch dimensions (which are critical), and the test conditions.

14.12 Mechanical properties of metals

Table 14.2 gives typical values of percentage elongation, yield or 0.1 per cent proof stress, and ultimate tensile strength. These values vary widely, however, not only particularly for alloys where the precise mix of elements is crucial, but also because of the many factors already discussed.

Table 14.2

	Percentage elongation (total)	Yield stress (MN/m²)	0.1% proof stress (MN/m²)	Ultimate tensile stress (MN/m²)
Copper, annealed	60	—	60	220
Copper, hard	4	—	320	400
Aluminium, soft	35	—	30	90
Aluminium, hard	5	—	140	150
Brass, soft (30% zinc)	70	—	80	320
Brass, high tensile	15	—	280	540
Phosphor bronze (cast)	10	—	150	310
Black mild steel	25–26	230–280	—	350–400
Bright mild steel	14–17	—	—	430
Structural steel	20	220–250	—	430–500
Stainless steel (cutlery)	8	1400	—	1560
Stainless steel (tool)	3	1870	—	1950
Maraging steel (high alloy)	12	1870	—	1800–3000
Cast iron, grey	—	—	120–240	280–340
Spheroidal graphite cast iron (annealed)	10–25	300–380	—	420–540
Cast-iron, malleable	20	—	—	310–500

Example *In a tensile test on a specimen of black mild steel of 12 mm diameter the following results were obtained for a gauge length of 60 mm.*

Load W(kN)	5	10	15	20	25	30	35	40
Extension $x(10^{-3}$ mm)	14	27.2	41	54	67.6	81.2	96	112

When tested to destruction, maximum load = 65 kN; load at fracture = 50 kN, diameter at fracture = 7.5 mm, total extension on gauge length = 17 mm. Find Young's modulus, specific modulus, ultimate tensile stress, breaking stress, true stress at fracture, limit of proportionality, percentage elongation, percentage reduction in area. The relative density of the steel is 7.8.

SOLUTION
The load extension graph is plotted in Fig. 14.9 and the slope of the straight line portion determined from the initial part of the best straight line drawn through the experimental points. The gradient of the straight line portion is found to be 366×10^6 N/m*. Since

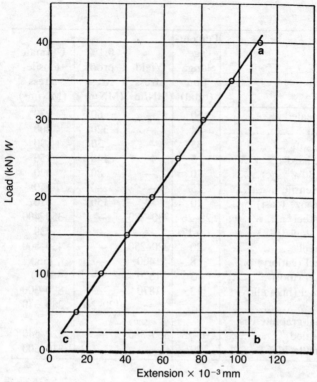

Fig. 14.9

$$E = \frac{\sigma}{\varepsilon} = \frac{W}{A} \times \frac{l}{x} = \frac{l}{A} \times \frac{W}{x}$$

and $\dfrac{W}{x} = 366 \times 10^6 \text{ N/m}$

$$l = 60 \text{ mm}$$

$$A = \frac{\pi}{4}(12 \times 10^{-3})^2 = 113 \times 10^{-6} \text{ m}^2$$

then

$$E = \frac{0.06}{113 \times 10^{-6}} \times 366 \times 10^6$$

$$= 195 \times 10^9 \text{ N/m}^2$$

$$= \textbf{195 GN/m}^2$$

Specific modulus $= \dfrac{E}{\text{relative density}} = \dfrac{195}{7.8} = \textbf{25 GN/m}^2$

Ultimate tensile stress $= \dfrac{\text{maximum load}}{\text{area}}$

$$= \frac{65}{113 \times 10^{-6}}$$

$$= 0.575 \times 10^6 \text{ kN/m}^2 = \textbf{575 MN/m}^2$$

Breaking stress $= \dfrac{50}{113 \times 10^{-6}}$

$$= 0.442 \times 10^6 \text{ kN/m}^2 = \textbf{442 MN/m}^2$$

Area at fracture $= \dfrac{\pi}{4}(7.5 \times 10^{-3})^2$

$$= 44.2 \times 10^{-6} \text{ m}^2$$

True stress at fracture $= \dfrac{\text{load at fracture}}{\text{area at fracture}} = \dfrac{50}{44.2 \times 10^{-6}}$

$$= 1.13 \times 10^6 \text{ kN/m}^2 = \textbf{1.13 GN/m}^2$$

Percentage elongation $= \dfrac{\text{extension}}{\text{gauge length}} = \dfrac{17}{60} \times 100$ per cent

$$= \textbf{28.3 per cent}$$

Percentage reduction in area

$$= \frac{113 \times 10^{-6} - 44.2 \times 10^{-6}}{113 \times 10^{-6}} \times 100 \text{ per cent}$$

$$= \textbf{61 per cent}$$

The load at the limit of proportionality = 30 kN approximately

stress at limit of proportionality $\sigma = \dfrac{30}{113 \times 10^{-6}}$

$$= 0.266 \times 10^6 \text{ kN/m}^2$$

$$= \textbf{266 MN/m}^2$$

* Gradient $\dfrac{W}{x} = \dfrac{ab}{bc} = \dfrac{38.75 - 2.5}{(106 - 7) \times 10^{-3}}$

$$= \frac{36.25}{99 \times 10^{-3}} = 366 \text{ kN/mm} = 366 \times 10^6 \text{ N/m}$$

Problems

1. The following results were recorded from a tensile test on a mild steel specimen: diameter 12 mm, gauge length 60 mm, maximum load 65 kN, diameter of neck after fracture 7.45 mm, length of broken specimen between gauge marks 77.2 mm.

Readings of load and extension were recorded as follows:

Load (kN)	5	7.5	10	12.5	15	20	25	30	32.5
Extension (10^{-3} mm)	14	20	26.9	36	40	54	68.1	82	88

Calculate the modulus of elasticity, the ultimate tensile strength, the percentage reduction in area and the percentage elongation.

(197 GN/m²; 575 MN/m²; 61.5 per cent; 28.62 per cent)

2. In a tensile test on an alloy specimen, of gauge length 70 mm, and original cross-sectional area 154 mm², the following readings of load and extension were obtained:

Load (kN)	10	20	30	40	50	60	70	80
Extension (10^{-3} mm)	28	58	88	118	148	178	210	247

   Deduce the values of Young's modulus and specific modulus for the alloy.

   In a test to destruction the maximum load recorded was 96 kN, the diameter of the neck was 10.4 mm, and the length between the gauge marks was 92.8 mm. Deduce the ultimate tensile strength, percentage elongation and percentage reduction of area. The relative density of the alloy is 6.

   (152 GN/m²; 25.3 GN/m²; 623 MN/m²; 32.5 per cent; 44.7 per cent)

3. In a tensile test on a black mild steel specimen of gauge length 60 mm and original cross-sectional area 120 mm², the load at the elastic limit was measured as 30 kN and the extension 0.075 mm. Find the stress and strain at the elastic limit, Young's Modulus, and the specific modulus for the steel. What is the work done on the specimen up to the elastic limit, stated as kJ/m³ of material between the gauge points? Relative density of the steel is 7.8.

   (250 MN/m²; 0.125 per cent; 200 GN/m²; 25.6 GN/m²; 156 kJ/m³)

14.13 Fatigue

A metal subjected to loading producing fluctuating, repeated or reversed stresses, fails at a stress level below the ultimate tensile stress. The term *fatigue failure* is applied to such a fracture. The *fatigue strength* is measured by the number of repetitions of stress before fracture occurs and depends upon the level of both the mean stress and the range of stress. Fatigue is particularly important when the stress is tensile and in the presence of impurities and stress-concentration areas such as changes in section, voids, joints, sharp corners or notches. The greatest number of fatigue failures are probably due to reversed bending stresses and this is the basis of the fatigue test most in use. A rotating test bar is held in bearings at each end and loaded at its midpoint so as to produce cyclic bending stresses.

   Fatigue failure in a metal causes a slow spreading fracture which has an appearance not unlike that of the fine granulated texture of cast iron. There is usually no sign of plastic deformation so that a fatigue failure may be mistaken for the fracture of a brittle material. On the other hand a brittle-type fracture does not necessarily mean fatigue failure has occurred. A fatigue fracture may show two areas of quite different appearance, Fig. 14.10. The first is crescent-shaped, smooth textured and may have 'beach' type markings. This represents the spreading of the fatigue crack. The remaining area is rough and jagged and represents the final tensile fracture after the metal has been greatly weakened with the spread of the fatigue crack.

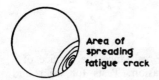

Fig. 14.10

Most ferrous metals have a 'safe' range of stress and a certain stress known as the *fatigue limit*, based on reversed bending, below which fracture does not occur even after a very large number of cycles of repetition of stress. Non-ferrous metals do not appear to have either a 'safe' range or a 'fatigue limit'. For these metals an *endurance limit* is used which is the maximum stress that can be endured for a specific frequency of loading.

There is a rough correlation between U.T.S. and the fatigue limit for many metals. For example, for a range of steels the fatigue limit is approximately one-half the U.T.S: for copper the ratio is much lower, about one-third.

Fatigue in polymers is of a similar nature to that in metals but because of the structure of the material some variables have an important effect on the cyclic stresses, e.g. the degree of crystallinity, the working temperature and the frequency of loading.

14.14 Creep

Any material when loaded constantly in tension for a long period of time will *creep*, i.e. extend slowly and steadily in a slightly plastic or viscous way. This phenomenon takes place even under the self-weight of material, and within the elastic limit. Creep can be defined as the total 'time-and-temperature' dependent strain occurring under constant load. For metals a slow steady strain takes place at any load and temperature providing the time-period is long enough, but the strain increases rapidly with increasing temperature and load; at normal temperatures it is negligible for steel and other stiff materials but significant for soft metals. Rigid plastics and polymers at normal temperatures when lightly loaded show little tendency to creep but the rate of strain increases most rapidly with increase in temperature. A creep test is similar to a tensile test except that the load and temperature are maintained constant and the test may last for years. An alternative to a long running test is a stress–rupture test which gives the stress that will produce a specified rate of strain in a relatively short period, e.g. a 1 per cent extension on the gauge length, say, in 5000 hours. A typical specification for Admiralty gunmetal shows a 0.1 per cent plastic strain at a temperature of 300° C with a low stress of 20 MN/m².

14.15 Hardness

Hardness is the term used to describe the resistance the surface of a metal offers to indentation, wear or abrasion. Tests for wear and indentation are quite different. A *scratch test* is the oldest method of determining hardness, a fairly rough method using Moh's scale of ten minerals, each of which can be scratched by the mineral above it on the scale. The softest mineral is talc (No. 1), which can be scratched by any of the others, and the hardest is diamond (No. 10) which can scratch all the others. Thus, if a tool steel can be scratched by topaz (No. 8) but not by corundum (No. 9), then its Moh number is 8.

There are several indentation testing machines which in general act by applying a load to an indenter for a given time. The tests each have their own *hardness number* and can be used for metals and plastics. They vary in the types of indenter used, the manner in which the load is applied, and the methods of measurement.

THE BRINELL HARDNESS TEST
This test is performed by pressing a hardened steel ball into the metal for 15 seconds with a load great enough to form a permanent indentation. The average diameter of the impression is measured and used to calculate the curved surface area of the impression. The *Brinell hardness number* is defined by the expression:

$$\text{Brinell hardness number} = \frac{\text{load in } \textit{kilograms}}{\text{curved surface area (mm}^2)}$$

Let P be the load (kg), D the diameter of ball (mm), d the diameter of impression (mm), h the depth of impression (mm). Then from Fig. 14.11:

$$h = \tfrac{1}{2}D - x$$

where

$$x^2 = D^2/4 - d^2/4$$

i.e. $x = \tfrac{1}{2}\sqrt{(D^2 - d^2)}$

hence

$$h = \tfrac{1}{2}D - \tfrac{1}{2}\sqrt{(D^2 - d^2)}$$

The curved surface area of a spherical cap is given by:

$$A = \pi D h$$

$$= \tfrac{1}{2}\pi D\{D - \sqrt{(D^2 - d^2)}\}$$

$$= \tfrac{1}{2}\pi D^2\{1 - \sqrt{(1 - d^2/D^2)}\}$$

Hence

$$\text{Brinell hardness number} = \frac{P}{A}$$

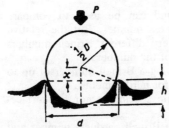

Fig. 14.11

The standard load used for steel is a mass 3,000 kg, with a 10 mm ball. For soft metals this load is usually too large and a smaller load and ball are used. In order that the Brinell hardness number for a given metal shall be independent of the load and diameter of ball used it has been shown that the load P should be varied with the diameter D according to the relation:

$$\frac{P}{D^2} = \text{constant}$$

This constant takes the following values:

Steel and cast iron	30
Copper alloys	10
Aluminium alloys	10
Copper and aluminium	5
Tin and lead	1

For example, if a 2 mm ball is to be used on steel the load P is given by

$$\frac{P}{2^2} = 30$$

i.e. $P = 120$ kg

Similarly if a load of 3,000 kg produces an indentation of 4 mm diameter with a 10 mm diameter ball, the indentation diameter d produced by a 120 kg load and a 2 mm ball is found from the principle of geometric similarity to be given by

$$\frac{d}{2} = \frac{4}{10}$$

thus

$$d = 0.8 \text{ mm}$$

The following points should be noted:
1. Ordinary steel balls are used up to a Brinell number of about 400; tungsten-carbide balls being used for harder metals up to 700.
2. The Brinell test compares the hardness of the same metal in

different conditions of cold work and can be used to compare different materials of a similar hardness number. But the relative hardness of different materials of very different Brinell numbers cannot be ascertained by comparison of the numbers.

3. The test is most useful for metals having a Brinell number of up to about 400 but is generally inadequate for harder metals.

4. The test cannot be used for very thin specimens nor for surface-hardened specimens.

5. There is a useful relation between the Brinell hardness number and the ultimate tensile strength of steel. For most carbon and alloy steels

$$\text{U.T.S.} = 0.23 \times \text{Brinell number, roughly}$$

6. When carrying out the test the following precautions should be taken.

   (*a*) The thickness of the test specimen should be at least ten times the depth of the impression.

   (*b*) The surface should be ground flat and polished.

   (*c*) The centre of the impression should be at least two and a half times the indentation diameter from the edge of the specimen.

Note: Besides the Brinell Test there are other testing machines available. Two of the most common are : (i) the *Vickers* test, which employs a diamond indenter, and is particularly useful for very thin, hard materials; (ii) the *Rockwell* test, used with a ball or diamond cone, and advantageous for rapid testing on production work because of a direct dial-reading arrangement. There is also the *Firth Hardometer* which gives the same hardness number as the Brinell, and the *Shore Sclero-scope*, a dynamic test utilising the property of resilience and measuring hardness by the rebound of a small tup from the test surface.

14.16 Polymers and plastics

A 'polymer' is a substance based mainly on the carbon atom and created by linking small molecules (monomers) to form large molecules (polymers). Depending on the method of linking, the polymer formed may be *crystalline* (hard), or *amorphous* (soft), or a mixture of both types. A diamond is a natural hard polymer; rubber and cellulose are natural soft polymers, and there are synthetic polymers with special properties. The term 'plastic' usually refers to the product manufac-tured when a polymer (also called 'resin') is mixed with various additives. Unreinforced plastics are not used for load-bearing since they have poor elastic properties, notably low rigidity, low creep resistance, and relative weakness in tension and compression. The low density of plastics, however, combined with modest strength gives them a fairly good strength-to-weight ratio, and this feature together with their advantageous properties, has led to their widespread use. The values of elastic modulus, tensile strength, and percentage elongation, for plastic materials are affected greatly by temperature conditions, and the rate and duration of loading.

Plastics fall into three main categories.

(i) *thermoplastics*, usually based on ethylene; once formed they can be softened and remoulded repeatedly by the application of heat. In general a thermoplastic is ductile, showing large elongations under load, with low tensile strength, good impact strength, but very sensitive to heat. A common engineering thermoplastic is nylon, in special grades, tough and hard, with self-lubricating properties and advantages in relation to wear, corrosion, moulding and impact strength. The critical properties for thermoplastics are the melting point and the *glass transition temperature** at which physical changes take place.

(ii) *thermosetting plastics*, formed from soft polymers by being made to set hard, through chemical change when heated to specific temperatures, depending on the polymer. They are hard and brittle, but fairly strong, and cannot be softened or moulded in any way under heat or pressure, without damage. Examples are 'bakelite' and 'epoxies' (adhesives).

(iii) *elastomers* is the name given to natural and synthetic rubbers, and to those plastics which have properties similar to rubber. When an elastomer is loaded in tension at room temperature its length increases enormously with corresponding reduction in cross-sectional area. Such materials do not obey Hooke's Law since there is a large reduction in cross sectional area as the load increases in the elastic range. Elastomers have many uses in engineering, e.g. rubber bushes, springs in anti-vibration mountings and resilient couplings.

14.17 Fibres

The original yarn and cloth organic fibres have been overtaken by inorganic fibres, mainly of glass, carbon and steel, and these modern fibres have great strength. The elastic modulus E for a steel fibre does not alter much from that of mild steel but its tensile strength, at about $1200 \, MN/m^2$, is three times greater, although still lower than the strength of the toughest alloy steels. *High-modulus* or *high-strength* fibres can be produced; for example, a high-modulus carbon fibre has an elastic modulus of $410 \, GN/m^2$ and tensile strength up to $2000 \, MN/m^2$. The difference between the properties of a material and the fibres

---

* If a soft plastic is cooled sufficiently, a temperature is reached where the plastic becomes hard and glassy. This temperature is called the *glass transition temperature*.

produced from it is shown by the most commonly used and commercially successful glass fibres. Ordinary bulk glass has a very variable tensile strength, less than $170 \, MN/m^2$, but in fibre form the strength may be raised to as much as $3000 \, MN/m^2$; after processing and being woven into strands the strength is much lower.

14.18 Fibre-reinforcement; composite materials

Fibres are used to reinforce plastics and other materials for various purposes, e.g. to increase the strength-to-weight ratio or to enable a material weak in tension to carry a tensile load. The reinforcement may be in the form of beads, long lengths of fibre made into strands or yarns, specially prepared fine fibres called 'rovings' woven into mats or cloths, but more usually as chopped up short pieces of yarn in mat form. Fibre-reinforced plastics are low-density composite materials using fibres of glass, steel or carbon, bonded into a plastic resin. The low-strength plastic protects the fibres from rubbing or chemical attack, and the tensile loads are taken by the fibres.

The mechanical properties of composite materials depend not only on those of the fibre and the host material but also on the length and weight content of the fibres, and on the nature of the bonding. The values obtained for the elastic modulus and ultimate strength for reinforced material are usually lower than those of the fibre reinforcement. A typical unreinforced thermoplastic polyester has a very low value of E, about $500 \, MN/m^2$, a low tensile strength of less than $70 \, MN/m^2$, but a high elongation, 60–110 per cent. Reinforcing with fibre can as much as triple the value of E depending on the kind of fibre used, and increase the tensile strength by an even greater factor, although there will be a loss of ductility and impact strength. Because of the ease with which plastic can be shaped, its corrosion resistance, and other unique properties, the improvement in strength when reinforced with fibre has led to the greatly increased use of these composite materials in many mechanical and structural engineering situations. Extreme examples of their use are the glass and resin nose unit of the Concorde aircraft and glass-reinforced resin coil and leaf springs. Again, in some modern aircraft composite carbon and glass-fibre reinforced plastics, together with other advanced composite materials, may account for up to one-quarter by weight of the airframe, in competition with titanium and aluminium-lithium alloys. The most advanced propeller developed through turbo-prop technology consists of six narrow ultra-light blades made of solid aluminium spar encased in glass fibre. A recent example is the construction of a two-bladed rotor for a wind energy turbine; the rotor has a span of about 60 m and the tip blades are fabricated from steel box spar with an aerofoil section consisting of a sandwich of balsa wood and plastic foam as the core, and a glass fibre reinforced plastic skin. A limitation, however, in the use of plastics, even when reinforced, is the final disposal of the material, and this is particularly so for volume products. Also the student should remember that besides strength and stiffness other mechanical properties are of great importance in engineering, particularly creep and fatigue resistance, and toughness, qualities often lacking in materials chosen for strength at high temperatures.

14.19 Non-destructive tests

The student should be aware that besides the methods of testing already mentioned which in the main 'test to destruction' there are in use a great variety of *non-destructive tests*. These tests do not destroy or impair the part undergoing test and can be employed to ascertain the soundness, quality, dimensions or tolerance of products and their coatings in all kinds of situations, e.g. coming off a production line, after heat treatment or mechanical working, where pipes or welds have to be inspected *in situ*, or for checking on the consistency of items under different operating conditions. Broadly, the tests divide into those for surface inspection and internal inspection or measurement. Each type of test tends to have its own field of application and its limitations. Some are portable, others are adaptable to being automated, designed for special situations, or restricted to ferrous materials, and so on. The range is very wide, varying from simple visual examination to the high-technology of laser beams. The following notes relate to the more common methods and indicate some of the advantages and limitations.

SURFACE INSPECTION
Simple visual examination has its obvious limitations, relying on excellent vision and illumination but is useful in the early stages of inspection; optical instruments, image-recognition, and automatic scanning systems greatly increase testing power, with some limitations caused by the wave-length of visible light. As an aid to visual examination there are various associated techniques including surface liquid dye penetrants, acid pickling and etching, and the use of magnetic particles. Eddy current methods use the principle of electromagnetic induction to measure electrical changes caused by surface cracks and voids. These methods are mainly for defects on or close to the surface.

INTERNAL INSPECTION
Apart from the more obvious tests for pressure holding and leaks there are sophisticated techniques such as radiography and ultrasonics. Radiographic methods using X-ray or gamma-rays have their associated

hazards but are essential in many cases to show up internal discontinuities or variations in thickness. Ultrasonics, a testing method useful for thick material employs sound waves with frequencies above 20,000 Hz, and is one part only of a field of tests based on the transmission or reflection of sound waves.

The division between the tests for surface and internal inspection is arbitrary as many of them can be used for both purposes. Some areas of work, such as weld testing, employ a whole range of techniques. New tests are constantly being devised, particularly in ultrasonics, acoustic and optical holography and automatic in-line systems . Most of the procedures require skill and experience in application and interpretation of results.

CHAPTER 15

Shear and Torsion

15.1 Shear stress

If two equal and opposite parallel forces F, not in the same straight line, act on parallel faces of a member (Fig. 15.1) then it is said to be loaded in *shear*. If the shaded area of cross-section parallel to the applied load is A then the *average shear stress* on the section is:

$$\tau = \frac{F}{A}$$

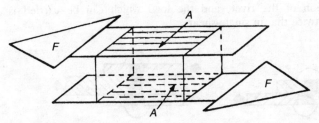

Fig. 15.1

The shear stress, or intensity of shear force, is tangential to the area over which it acts. For example in cutting plate by a guillotine (Fig. 15.2) F is the total force exerted by the blade and is balanced by an equal and opposite force provided at the edge of the table. The area resisting shear is measured by the plate thickness multiplied by the length of the blade. In a punching operation (Fig. 15.3) the area resisting shear would be the plate thickness multiplied by the perimeter of the hole punched. The ultimate strength in shear of a metal is measured in practice by a punching operation of this type.

The *ultimate shear stress or strength* is defined as

$$\frac{\text{maximum punch load}}{\text{area resisting shear}}$$

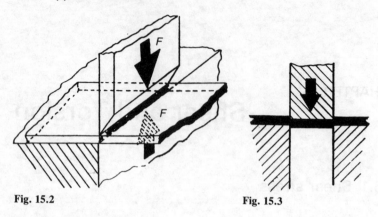

Fig. 15.2   **Fig. 15.3**

15.2 Riveted joints

A structural member commonly loaded in shear, but seldom in tension, is the rivet. Fig. 15.4 shows a riveted joint loaded in *single-shear*; Fig. 15.5 shows a joint in *double-shear*. In single-shear the area resisting shear is the cross-sectional area of the rivet, $\pi d^2/4$, where d is the diameter of the rivet. In double-shear the resisting area is twice the area of section of the rivet, and the load which can be carried is theoretically twice that in single-shear.

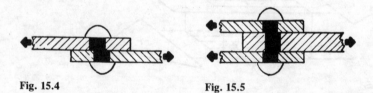

Fig. 15.4   **Fig. 15.5**

Example   *A load F of 5 kN is applied to the tensile member shown in Fig. 15.6 and is carried at the joint by a single rivet. The angle of the joint is 60° to the axis of the load. Calculate the tensile and shear stresses in a 20-mm diameter rivet.*

SOLUTION
The axis of the rivet is at 30° to the line of action of the load F.

$$\text{Area of rivet} = \frac{\pi}{4} \times 20^2 = 314.2 \text{ mm}^2$$

direct pull on rivet = component of F along axis of rivet

$$= 5 \times \cos 30°$$

$$= 5 \times 0.866$$
$$= 4.33 \text{ kN}$$

Therefore

$$\text{direct stress} = \frac{4.33}{314.2 \times 10^{-6}} = 13{,}800 \text{ kN/m}^2$$
$$= \mathbf{13.8 \text{ MN/m}^2 \text{ tensile}}$$

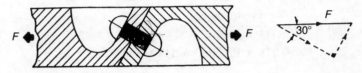

Fig. 15.6

Shear force on rivet equals component of F transverse to rivet, i.e. along joint face.

$$\text{Shear force} = 5 \times \sin 30°$$
$$= 2.5 \text{ kN}$$

thus

$$\text{shear stress on rivet} = \frac{2.5}{314.2 \times 10^{-6}} = 7{,}950 \text{ kN/m}^2$$
$$= \mathbf{7.95 \text{ MN/m}^2}$$

Note: In this case, where the stress is predominantly tensile rather than shear, the rivet would be replaced by a bolt.

Example *A solid coupling transmits 100 kW at 2 rev/s through eight equally spaced bolts (Fig. 15.7). If the bolts are 12 mm diameter and are on a pitch circle of 150 mm diameter calculate the average shear stress in each bolt.*

SOLUTION

$$\text{Power} = \frac{2\pi n T}{1{,}000} \text{ kW where } n \text{ is in rev/s and } T \text{ in N-m}$$

Thus the torque is

$$T = \frac{\text{power} \times 1{,}000}{2\pi n}$$
$$= \frac{100 \times 1{,}000}{2\pi \times 2}$$
$$= 7{,}950 \text{ N-m}$$

Total shear load at radius of 75 mm is

$$\frac{7,950}{0.075} = 106,000 \text{ N}$$

Since bolts are *ductile* it may be assumed that this load is equally distributed among the eight bolts. Therefore

$$\text{load per bolt} = \frac{106,000}{8} = 13,250 \text{ N}$$

$$\text{area of bolt} = \frac{\pi}{4} \times 12^2 = 113 \text{ mm}^2$$

thus   shear stress $= \dfrac{13,250}{113 \times 10^{-6}}$

$$= 118 \times 10^6 \text{ N/m}^2 = \textbf{118 MN/m}^2$$

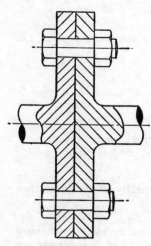

Fig. 15.7

Problems

1. Calculate the maximum thickness of plate which can be sheared on a guillotine if the ultimate shearing strength of the plate is 250 MN/m² and the maximum force the guillotine can exert is 200 kN. The width of the plate is 1 m.

    (0.8 mm)

2. A rectangular hole 50 mm by 65 mm is punched in a steel plate 6 mm thick. The ultimate shearing stress of the plate is 200 N/mm². Calculate the load on the punch.

    (276 kN)

3. Calculate the maximum diameter of hole which can be punched in 1.5 mm plate if the punching force is limited to 40 kN. The plate is aluminium having an ultimate shear strength of 90 MN/m².

    (94.5 mm)

4. A boiler is to be made of a 2-m diameter cylinder having a riveted single-lap seam. Calculate the minimum number of 14 mm diameter rivets required per metre length of longitudinal seam if the boiler pressure is 140 kN/m². The ultimate shear stress of each rivet is 320 MN/m² and a factor of safety of 7 is to be used.

(20)

5. A bar is cut at 45° to its axis and joined by two 12-mm diameter bolts, Fig. 15.8. If the pull in the bar is 80 kN calculate the direct and shear stresses in each bolt.

(250 MN/m²; 250 MN/m²)

Fig. 15.8

6. A solid coupling is to transmit 225 kW at 10 rev/s. The coupling is fastened with six bolts on a pitch circle diameter of 200 mm. If the ultimate shear stress is 300 MN/m² calculate the bolt diameter required. The factor of safety is to be 4.

(10.1 mm)

7. A shaft is to transmit 180 kW at 600 rev/min through solid coupling flanges. There are four coupling bolts, each of 15 mm diameter. If the shear stress in each bolt is to be limited to 40 MN/m² calculate the minimum diameter of the circle at which the bolts are to be placed.

(203 mm)

8. A gear wheel 25 mm wide is shrunk on to a 50 mm diameter shaft so that the radial pressure at the circle of contact is 7 MN/m². The coefficient of friction between gear wheel and shaft is 0.2 and they are also prevented from relative rotation by a key 6 mm wide, 25 mm long. If the shaft transmits 15 kW at 2 rev/s calculate the shear stress in the key.

(282 MN/m²)

9. In a flexible coupling transmitting 30 kW at 1,200 rev/min the pins transmitting the drive are set at a radius of 100 mm. If there are four pins and all pins transmit the drive equally, calculate the pin diameters. Allow a safe shear stress of 30 MN/m².

If, due to faulty machining one pin is ahead of its correct position and may be assumed to take the whole drive, what should be the pin diameter?

(5 mm; 10 mm)

10. A shaft is to be fitted with a flanged coupling having 8 bolts on a circle of diameter 150 mm. The shaft may be subject *either* to a direct tensile load of 400 kN *or* to a twisting-moment of 18 kN-m. If the maximum direct and shearing stresses permissible in the bolt material are 125 MN/m² and 55 MN/m² respectively find the minimum diameter of bolt required. Assume each bolt takes an equal share of the load or torque. Using this bolt

diameter and assuming only one bolt to carry the full torque what would then be the shearing stress in the bolt?

(26.3 mm, using torque data; 442 MN/m²)

15.3 Shear strain

Fig. 15.9 shows an element of material rigidly fixed at one face DC and subject to a shearing stress τ on the parallel face AB. The element will deform, and the deformation may be taken as similar to that which would take place if the element were made up of a number of thin independent layers, each layer slipping relative to its neighbour below. In effect, the element will deform to the rhombus DA'B'C. The *shear strain* is defined as the angle of deformation ADA' (or BCB') in radians. Since shear strain is small

$$\phi = \angle \text{ADA}'$$

$$\simeq \frac{\text{AA}'}{\text{AD}} \text{ rad}$$

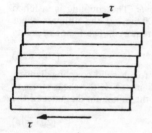
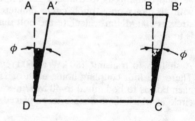

Fig. 15.9

15.4 Relation between shear stress and shear strain: modulus of rigidity

By analogy with the tensile stress-strain relation for an elastic material we write for shear:

$$\frac{\text{shear stress}}{\text{shear strain}} = \text{constant}, G$$

i.e. $$\frac{\tau}{\phi} = G$$

where the constant G is known as the modulus of rigidity of the material. The units of G are those of stress, i.e. newtons per square metre (N/m²) or one of the other forms GN/m², MN/m².

For most carbon steels the value of G is about 84×10^9 N/m² or 84 GN/m². For cast-iron and ductile materials such as copper, aluminium, bronze, G lies between 28 and 42 GN/m².

15.5 Torsion of a thin tube

Consider the thin tube shown in Fig. 15.10. The mean radius is r, and the thickness of wall, t, is very small compared with r. If a twisting moment T is applied to both ends of the tube, one end will twist *relative* to the other. A strip AB parallel to the tube axis will distort to AB′. If it is assumed that the displacement BB′ is small compared with the length of tube AB, then AB′ will be approximately straight. Then angle $\angle$BOB′ is the *angle of twist* θ of the length AB. The *shear strain* is

$$\phi \simeq \frac{BB'}{AB} = \frac{r\theta}{l} \text{ rad}$$

since BB′ $= r\theta$, and AB $=$ length of tube l.

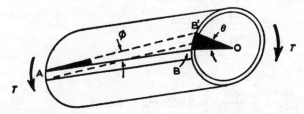

Fig. 15.10

The shear force on the cross-section of the tube is

$$F = \frac{\text{torque}}{\text{radius}} = \frac{T}{r}$$

This force acts on area $2\pi rt$, since the tube is thin. Therefore

$$\text{shear stress } \tau = \frac{\text{shear force}}{\text{area}} = \frac{F}{2\pi rt}$$

$$= \frac{T/r}{2\pi rt}$$

$$= \frac{T}{2\pi r^2 t}$$

Also since $\dfrac{\tau}{\phi} = G$, the modulus of rigidity, and $\phi = \dfrac{r\theta}{l}$ hence

$$\frac{\tau}{\phi} = \frac{\tau}{r\theta/l} = G$$

hence

$$\frac{\tau}{r} = \frac{G\theta}{l}$$

Note: (a) for a given torque the angle of twist varies directly with the length; (b) in these formulae the twist θ must be in radians.

Example　*A thin cylindrical tube, 25 mm diameter, 1.5 mm thick, 300 mm long, is subjected to a torque T. Calculate the maximum value of T if the allowable shear stress is not to exceed 35 N/mm². If the modulus of rigidity of the material is 80 kN/mm² what is the angle of twist for maximum torque?*

SOLUTION

Allowable shear force, F = area of section × shear stress

$$= 2\pi rt \times \tau = 2\pi \times 12.5 \times 1.5 \times 35$$

$$= 4,125 \text{ N}$$

Maximum torque, $T = F \times r = 4,125 \times 12.5$

$$= 51,600 \text{ N-mm}$$

$$= \mathbf{51.6 \text{ N-m}}$$

Since $\dfrac{G\theta}{l} = \dfrac{\tau}{r}$

then $\theta = \dfrac{\tau l}{rG} = \dfrac{35}{12.5} \times \dfrac{300}{80 \times 10^3} = 0.0105 \text{ rad} = \mathbf{0.6°}$

(Note that the working throughout is in N and mm.)

Problems

1. A *thin* steel tube 90 mm inside diameter is subject to a torque of 500 N-m. (a) If the shear stress is not to exceed 28 MN/m² calculate the tube thickness. (b) If the twist is not to exceed 2.5 mm of arc on a 600 mm length what would be the thickness required? $G = 84 \text{ GN/m}^2$.

(1.4 mm; 0.113 mm)

2. A *thin* tube 1.5 mm thick, 80 mm mean diameter is subjected to a torque of 350 N-m. Calculate (a) the shear stress in the tube, (b) the twist on a 1 m length. $G = 84 \text{ GN/m}^2$.

(23.2 MN/m²; 0.0069 rad or 0.39°)

15.6 Twisting of solid shafts

To derive the relation between torque, angle of twist and shear stress for a solid shaft of diameter d we make the following assumptions:

1. The shaft is composed of a succession of thin concentric tubes.
2. Each thin tube carries shear force independent of, and without interfering with, its neighbours.
3. Lines which are radial before twisting are assumed to remain radial after twisting.
4. The shaft is not stressed beyond the elastic limit.

For any elementary thin tube of thickness dr at radius r (Fig. 15.11):

area of section $= 2\pi r \times dr$

If τ is the shear stress at radius r, then

shear force on tube $= 2\pi r\, dr \times \tau$

thus

torque carried by tube $= 2\pi r \tau\, dr \times r$

$$= 2\pi r^2 \tau\, dr$$

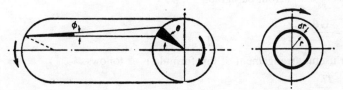

Fig. 15.11

If θ is the angle of twist of the tube in a length l, then from the results of paragraph 15.5:

$$\tau = \frac{G\theta}{l} r$$

therefore

torque carried by tube $= 2\pi r^2 \tau\, dr$

$$= 2\pi r^2 \times \frac{G\theta}{l} r \times dr$$

$$= 2\pi \frac{G\theta}{l} r^3\, dr$$

The whole torque T carried by the solid shaft is the sum of all the elementary torques, i.e.

$$T = \int_0^{d/2} 2\pi \frac{G\theta}{l} r^3 \, dr$$

Since radial lines before twisting remain radial after twisting θ is therefore the same for all the thin tubes making up the shaft. Also G and l are constant, therefore:

$$T = \frac{G\theta}{l} \int_0^{d/2} 2\pi r^3 \, dr$$

$$= \frac{G\theta}{l} J$$

where

$$J = \int_0^{d/2} 2\pi r^3 \, dr$$

$$= \frac{\pi d^4}{32}$$

which is the *polar second moment of area* of a shaft of circular section. The units of J are (metres)^4 or (millimetres)^4. Then, rearranging the above equation:

$$\frac{T}{J} = \frac{G\theta}{l}$$

and, since $\dfrac{\tau}{r} = \dfrac{G\theta}{l}$ from paragraph 15.5:

$$\frac{T}{J} = \frac{G\theta}{l} = \frac{\tau}{r}$$

Another useful arrangement of this formula is as follows:

$$\boldsymbol{\theta = \frac{Tl}{GJ}}$$

Some important points should be noted:

1. The angle of twist θ varies *directly* with length l.
2. Since $\tau = Tr/J$ for a given torque T the shear stress τ is proportional to the radius r. Thus the maximum shear stress occurs at the outside surface where $r = d/2$, and the shear stress at the centre of the shaft is zero. Fig. 15.12 shows the variation of τ across a diameter.

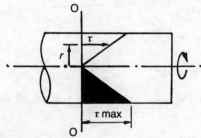

Fig. 15.12

15.7 Twisting of hollow shafts

If d_2, d_1 are the outside and inside diameters of a hollow shaft subject to a twisting moment T then the basic equation for the torque becomes

$$T = \frac{G\theta}{l} \int_{d_1/2}^{d_2/2} 2\pi r^3 \, \mathrm{d}r = \frac{G\theta}{l} J$$

as before, where now:

$$J = \int_{d_1/2}^{d_2/2} 2\pi r^3 \, \mathrm{d}r$$

$$= \frac{\pi(d_2{}^4 - d_1{}^4)}{32}$$

Since,

$$\frac{T}{J} = \frac{\tau}{r}$$

the maximum shear stress for a given torque is again at the outside fibres of the shaft where $r = d_2/2$.

Note: For a very thin tube of thickness t, radius r

J = area of section $\times r^2$

$\quad = 2\pi r t \times r^2$

$\quad = 2\pi r^3 t$

15.8 Stiffness and strength

The *stiffness* or *torsional rigidity* of a shaft is the torque to produce unit angle of twist. Thus if a torque T produces a twist θ then

$$\text{stiffness} = \frac{T}{\theta} = \frac{GJ}{l}$$

The *strength* of a shaft is measured by the torque it can transmit for a given permissible value of the maximum shear stress. For a given shear stress therefore the strengths of two shafts are in the ratio of the corresponding torques. Alternatively, for a given torque, the strengths are in the ratio of the maximum allowable shear stress produced.

15.9 Power and torque

If a shaft transmits power at n rev/s, the torque T in newton-metres (N-m) carried by the shaft is given by power = work done by torque per second

$$= \frac{\text{torque (N-m)} \times \text{speed (rad/s)}}{1,000} \text{ kW}$$

$$= \frac{2\pi n T}{1,000} \text{ kW}$$

since speed = $2\pi n$ rad/s.

Example   *Compare the torsional stiffness of a solid shaft 50 mm diameter, 300 mm long, with that of a hollow shaft of the same material having diameters 75 mm, 50 mm and length 200 mm.*

SOLUTION

Torsional stiffness $= \dfrac{T}{\theta} = \dfrac{GJ}{l}$

which is proportional to J/l since G is constant.
For the solid shaft:

$$J = \frac{\pi}{32} \times 50^4 = 613 \times 10^3 \text{ mm}^4$$

thus

$$\frac{J}{l} = \frac{613 \times 10^3}{300} = 2.043 \times 10^3 \text{ mm}^3$$

For the hollow shaft:

$$J = \frac{\pi(75^4 - 50^4)}{32} = 2.5 \times 10^6 \text{ mm}^4$$

therefore

$$\frac{J}{l} = \frac{2.5 \times 10^6}{200} = 12.5 \times 10^3 \text{ mm}^3$$

thus ratio of stiffness:

$$\frac{\text{hollow shaft}}{\text{solid shaft}} = \frac{12.5 \times 10^3}{2.043 \times 10^3}$$

$$= \textbf{6.1 : 1}$$

i.e. the hollow shaft is 6.1 times as stiff in torsion as the solid shaft.

Example   *A shaft used in an aircraft engine is of 50 mm diameter. The maximum allowable shear stress is 84 MN/m². Find the torsional strength of the shaft. If the shaft now has a hole bored in it find the percentage reductions in strength and mass, (a) if the hole is 40 mm diameter, (b) if the hole is 25 mm diameter. The hole is concentric with the shaft axis.*

SOLUTION
The strength of the shaft is the torque it can transmit for the given shear stress:

$$J = \frac{\pi}{32} \times 50^4 = 613 \times 10^3 \text{ mm}^4$$

$$\frac{T}{J} = \frac{\tau}{d/2}$$

and since $\tau = 84 \text{ MN/m}^2 = 84 \text{ N/mm}^2$

$$T = \frac{84 \times 613 \times 10^3}{25}$$

$$= 2.06 \times 10^6 \text{ N-mm} = \textbf{2.06 kN-m}$$

(*a*) When a hole 40 mm diameter is bored:

$$J = \frac{\pi(50^4 - 40^4)}{32} = 363 \times 10^3 \text{ mm}^4$$

$$T = \frac{\tau J}{\frac{1}{2}d_2}$$

$$= \frac{84 \times 363 \times 10^3}{\frac{1}{2} \times 50}$$

$$= 1.22 \times 10^6 \text{ N-mm} = 1.22 \text{ kN-m}$$

thus percentage reduction in strength is given by:

$$\frac{2.06 - 1.22}{2.06} \times 100 = \textbf{40.7 per cent}$$

For a given length the mass of shaft varies with the cross-sectional area. Therefore percentage reduction in mass equals percentage reduction in area, i.e.

$$\frac{(\pi/4) \times 50^2 - (\pi/4)(50^2 - 40^2)}{(\pi/4) \times 50^2} \times 100 = \textbf{64 per cent}$$

(*b*) When a hole of 25 mm diameter is bored:

$$J = 576 \times 10^3 \text{ mm}^4$$

$$T = 1.94 \text{ kN-m}$$

percentage reduction in strength = **5.83 per cent**

percentage reduction in mass = **25 per cent**

Example *A solid shaft is to transmit 750 kW at 200 rev/min. If the shaft is not to twist more than 1° on a length of twelve diameters, and the shear stress is not to exceed 45 MN/m², calculate the minimum shaft diameter required. G = 84 GN/m².*

SOLUTION

$$\text{Torque } T = \frac{\text{Power in kW} \times 1,000}{2\pi n}$$

$$= \frac{750 \times 1,000}{2\pi \times 200/60}$$

$$= 35,800 \text{ N-m}$$

There are two independent conditions to be satisfied in this problem. The torque is limited by *both* the twist *and* the shear stress.
For condition of twist:

$$1° = \frac{\pi}{180} \text{ rad}$$

$$l = 12 \, d$$

and $$\frac{T}{J} = \frac{G\theta}{l}$$

thus $$\frac{35,800}{\pi d^4/32} = \frac{84 \times 10^9 \times \pi/180}{12 \, d}$$

and $$d^3 = 2.98 \times 10^{-3}$$

i.e. $$d = 0.144 \text{ m} = \textbf{144 mm}$$

For condition of maximum shear stress:

$$\tau = 45 \times 10^6 \text{ N/m}^2$$

$$\frac{T}{J} = \frac{\tau}{\frac{1}{2}d}$$

therefore

$$\frac{35,800}{\pi d^4/32} = \frac{45 \times 10^6}{\frac{1}{2}d}$$

$$d^3 = 4.04 \times 10^{-3}$$

and $$d = 0.159 \text{ m} = \textbf{159 mm}$$

The *least* diameter to satisfy both conditions is therefore $d = \textbf{159 mm}$.

Problems

1. A hollow shaft, of 50 mm internal diameter and 12 mm thick, twists through an angle of 1° in a length of 2 m when subjected to a torque of 1 kN-m. Calculate the modulus of rigidity for the material.

$$(49.2 \text{ GN/m}^2)$$

2. The propeller shaft of an aircraft engine is steel tubing of 75 mm external and 60 mm internal diameter. The shaft is to transmit 150 kW at 1,650 rev/min. The failing stress in shear for this shaft is 140 MN/m². What is the factor of safety?

$$(7.87)$$

3. An aluminium alloy bar was tested in tension and torsion. The tension test on one portion of 20 mm diameter showed an extension of 0.34 mm with a load of 40 kN measured on a gauge length of 200 mm. The torsion test on a second portion of 14 mm diameter showed an angle of twist of 0.125 rad on a gauge length of 250 mm when the torque was 35 N-m. Find E and G for the material.

$$(74.7 \text{ GN/m}^2; 18.6 \text{ GN/m}^2)$$

4. A hollow shaft is to transmit 2 MW at 40 rev/s. The external diameter is to be 1.3 times the internal diameter. The maximum torque may be taken as 20 per cent greater than the average value and the maximum shear stress is limited to 150 MN/m². Find the external diameter of the shaft.

$$(79.6 \text{ mm})$$

5. For phosphor-bronze the relation between the moduli of elasticity and rigidity may be taken as

$$E = 2.6\,G$$

A tensile specimen of this material, of diameter 20 mm, extended by 0.075 mm on a gauge length of 50 mm when the load applied was 50 kN. What would be the angle of twist per metre length on a shaft of the same material (20 mm diameter) due to a torque of 15 N-m?

$$(0.0234 \text{ rad or } 1.34°)$$

6. A brass shaft of 6 mm diameter is tested in torsion. At the limit of proportionality the torque is 0.6 N-m and the angle of twist 1.13° on a gauge length of 250 mm. Calculate G for the brass.

$$(59.8 \text{ GN/m}^2)$$

7. Calculate the maximum shear stress in a 6 mm diameter bolt when tightened by a force of 50 N at the end of a 150-mm spanner. What would be the corresponding stress in a 10-mm diameter bolt?

$$(177 \text{ MN/m}^2; 38.1 \text{ MN/m}^2)$$

8. A gear wheel is keyed to a 50 mm diameter shaft by a square section key of width w mm and length 50 mm. The load on the wheel teeth amounts to 5,000 N at a radius of 150 mm. If the shear stress in the key is to be twice the maximum shear stress in the shaft calculate the width w.

$$(9.85 \text{ mm})$$

9. A solid circular shaft is connected to the drive shaft of an electric motor by a solid flanged coupling, the drive being taken through eight bolts, of 12 mm diameter, on a pitch circle diameter of 225 mm. The bolts carry the whole driving torque and are loaded in shear only. Calculate the shaft diameter if the maximum shear stress in the shaft is to be equal to the shear stress in the bolts.

$$(80.3 \text{ mm})$$

10. A length of hollow steel shaft is used to drill a hole 3 km deep in rock. The power exerted is 180 kW and the speed of rotation of the drill is 60 rev/min. If the inner and outer diameters of the shaft are 150 mm and 175 mm, respectively, calculate: (*a*) the maximum shear stress in the shaft; (*b*) the twist of one end relative to the other, in revolutions. $G = 84 \text{ kN/mm}^2$.

$$(59.3 \text{ MN/m}^2; 3.84 \text{ rev})$$

11. A length of hollow steel shaft transmits 900 kW at 165 rev/min. The maximum shear stress is not to exceed 56 MN/m² and the inside diameter is to be 0.6 times the outside diameter. Determine (*a*) the shaft diameters (*b*) the shear stress at the *inner* surface of the shaft.

$$(106 \text{ mm}; 176 \text{ mm}; 33.7 \text{ MN/m}^2)$$

12. Calculate the power which will be transmitted at 220 rev/min by a hollow shaft of 150 mm inside diameter and 50 mm thick, if the maximum shear stress is 70 MN/m². Find the percentage by which the shaft will be stronger if made solid instead of hollow and the external diameter is the same.

(4.32 MW; 14.88 per cent)

13. A hollow shaft driving a ship's screw is to carry a torque of 13 kN-m and is to be of 150 mm diameter externally. Calculate the inside diameter if the maximum shear stress is not to exceed 40 MN/m². Calculate the angle of twist in degrees on a 5.5 m length and the *minimum* shear stress. G = 84 GN/m².

(127 mm; 2°; 33.9 MN/m²)

CHAPTER 16
Shear Force and Bending Moment

16.1 Shear force

The *shear force* in a beam at any section is the force transverse to the beam tending to cause it to shear across the section. Fig. 16.1 shows a beam under a transverse load W at the free end D; the other end A is built in to the wall. Such a beam is called a *cantilever* and the load W, which is assumed to act at a point, is called a *concentrated* or *point load*.

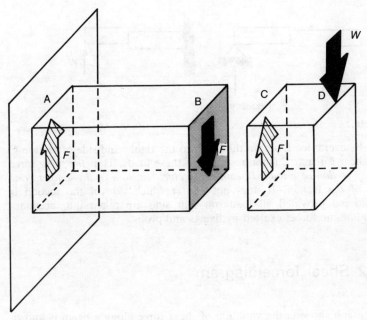

Fig. 16.1

Consider the equilibrium of any portion of beam CD. At section C for balance of forces there must be an upward force F equal and opposite to the load W at D. This force F is provided by the resistance of the beam to shear at the plane B; this plane being coincident with the plane section at C. F is the shear force at B and in this case is the same magnitude for any section in AD. Consider now the equilibrium of the portion of beam AB. There is a downward force $F = W$, exerted on plane B, so for balance there must be an upward force F at A, this latter force being exerted *on* the beam by the wall.

Sign convention

The shear force at any section is taken *positive* if the right-hand side tends to slide downwards relative to the left-hand portion, Fig. 16.2. A *negative* shear force tends to cause the right-hand portion to slide upward relative to the left.

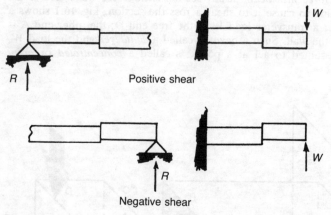

Positive shear

Negative shear

Fig. 16.2

If several loads act on the beam to the right-hand side of section C the shear force at C is the resultant of these loads. Thus *the shear force at any section of a loaded beam is the algebraic sum of the loads to one side of the section*. It does not matter which side of the section is considered provided all loads on that side are taken into account— including the forces exerted by fixings and props.

16.2 Shear force diagram

The graph showing the variation of shear force along a beam is known as the *shear force diagram*. For the beam of Fig. 16.1 the shear force was $+W$, uniform along the beam. Fig. 16.3 shows the shear force diagram for this beam, O–O being the axis of zero shear force.

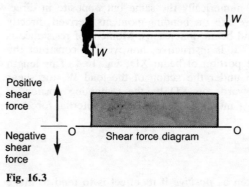

Positive shear force

Negative shear force

O — Shear force diagram — O

Fig. 16.3

16.3 Bending moment

The *bending effect* at any section X of a concentrated load W at D, Fig. 16.4, is measured by the applied moment Wx, where x is the perpendicular distance of the line of action of W from section X. This moment is called the *bending moment M* and is balanced by an equal and opposite moment exerted by the material of the beam at X, called

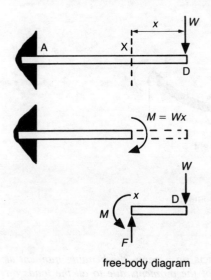

free-body diagram

Fig. 16.4

the *moment of resistance*. This resisting moment is due to the internal forces acting at the section. Note that the bending-moment and the moment of resistance are numerically the same but opposite in direction. The calculations here are on bending-moments derived directly from the external loads and beam dimensions. Moment of resistance is considered in Chapter 17. It is instructive, however, to construct the free-body diagram for the portion of beam XD, Fig. 16.4. This length of beam is in equilibrium under the action of the load W, the shear force $F = W$, exerted upwards on XD by the remaining portion of beam XA, and the *resisting* moment (M), due to the internal forces at the section X.

Sign convention

A bending moment is taken as *positive* if its effect is to tend to make the beam *sag* at the section considered, Fig. 16.5. If the moment tends to make the beam bend upward or *hog* at the section it is *negative*.

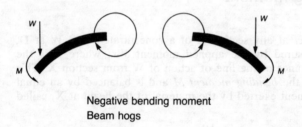

Negative bending moment
Beam hogs

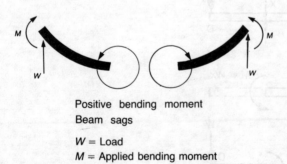

Positive bending moment
Beam sags

W = Load
M = Applied bending moment

Fig. 16.5

When more than one load acts on a beam *the bending moment at any section is the algebraic sum of the moments due to all the loads on one side of the section.* It does not matter which side of the section is considered but all loads to that side must be taken into account, including any moments exerted by fixings.

16.4 Bending moment diagram

The variation of bending moment along the beam is shown in a *bending moment diagram*. For the cantilever beam of Fig. 16.6 the bending moment at any section X is given by:

bending moment $= -Wx$ (negative, since the beam hogs at X)

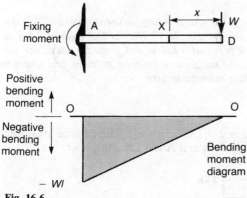

Fig. 16.6

Since there is no other load on the beam this expression for the bending moment applies for the whole length of beam from $x = 0$ to $x = l$. The moment is proportional to x and hence the bending moment diagram is a straight line. Hence the diagram can be drawn by calculating the moment at two points and joining the two corresponding points on the graph by a straight line.

At D, $x = 0$ and bending moment $= 0$

at A, $x = l$ and bending moment $= -Wl$

Since the bending moment is everywhere negative the graph is plotted below the line O–O of zero bending moment, Fig. 16.6. At the fixed end A the wall exerts a moment Wl anticlockwise *on* the beam; this is called a *fixing moment*.

16.5 Calculation of beam reactions

When a beam is fixed at some point, or supported by props, the fixings and props exert *reaction* forces on the beam. To calculate these reactions the procedure is:

(*a*) equate the net vertical force to zero;
(*b*) equate the total moment about any convenient point to zero.

Note: Distinguish carefully between 'taking moments' and calculating a 'bending moment':

1. The Principle of Moments states that the algebraic sum of the moments of all the forces about any point is zero, i.e. when forces *on both sides* of a beam section are considered.
2. The bending moment is the algebraic sum of the moments of forces *on one side* of the section about that section.

Example   *Draw the shear force and bending moment diagrams for the cantilever beam loaded as shown, Fig. 16.7. The vertical load of 2 kN at C is partly supported by a force of 3 kN at the prop B. State: (a) the reaction at the built-in end; (b) the greatest bending moment and where it occurs; (c) where the bending moment is zero.*

SOLUTION
Reaction
The net external load $= 3 - 2 = 1$ kN, upward. For balance therefore the vertical reaction at the built-in end A is **1 kN** *downward*.

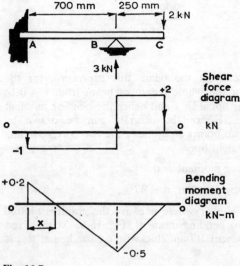

Fig. 16.7

Shear force diagram

The diagram is drawn by making use of the fact that on any unloaded portion of the beam the shear force is uniform and the graph between loads is a horizontal line. Starting at the left-hand end draw a line to scale from the zero line O–O of length and direction corresponding to that of the reaction at A, i.e. 1 kN downward.

Between A and B the shear force is uniform and of amount −1 kN (using the sign convention given). At B it changes by +3 kN. The shear force just to the right of B, in BC, is therefore:

$$-1 + 3 = 2 \text{ kN}$$

The shear force is uniform from B to C and changes by −2 kN at C. At C it is therefore zero, as it should be at a free end; (just to the left of C the shear force is of course +2 kN). The complete shear force diagram is shown in Fig. 16.7.

Bending moment diagram

The diagram is drawn by making use of the fact that on unloaded portions of the beam the bending moment is represented by straight lines. The bending moment is therefore calculated at the load and reaction points; the corresponding points on the diagram are then joined by straight lines.

At B, bending moment $= -2 \times 0.25 = -0.5$ kN-m

At A, bending moment $= -2 \times 0.95 + 3 \times 0.7 = 0.2$ kN-m

At C, bending moment $= 0$ (at the free end)

The greatest bending moment is at B and is **0.5 kN-m**. Note that the greatest bending moment occurs at a point where the shear force changes sign.

The bending moment is zero at section X, a distance x from the end A found by simple proportion from similar triangles in the bending moment diagram, Fig. 16.7.

i.e. $\dfrac{x}{0.2} = \dfrac{0.7 - x}{0.5}$

thus

$$x = 0.2 \text{ m} = \textbf{200 mm}$$

Since bending-moments of opposite sign indicate bending of opposite curvature the points of *change of sign* are important. Where the bending-moment changes sign and the value is zero as at section X the curvature of the beam changes and this is called a point of *contraflexure* or *inflexion*.

Example *The beam shown, Fig. 16.8, is simply supported at C and B, and loaded at A and D by concentrated masses of 1 tonne and 3 tonne, respectively. Draw the shear force and bending moment diagrams.*

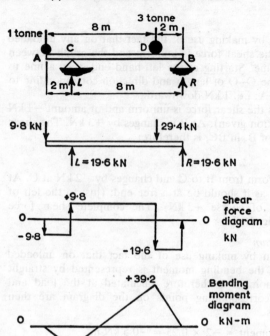

Fig. 16.8

SOLUTION

Reactions

A 'simple support' is one in which the beam is rested as on a knife-edge. The reaction exerted by the support is assumed to act at a point and is vertical. The reaction at B is found by taking moments about C for *all* the loads on the beam, i.e. equating clockwise and anticlockwise moments about C. The weight of 1 tonne = 9.8 kN and the weight of 3 tonne = 29.4 kN.

$$9.8 \times 2 + R \times 8 = 29.4 \times 6$$

therefore $R = 19.6$ kN

Similarly, taking moments about B:

$$L \times 8 = 29.4 \times 2 + 9.8 \times 10$$

therefore $L = 19.6$ kN

(Check

$$L + R = 19.6 + 19.6$$

$$= 39.2 \text{ kN}$$

$$= \text{net downward load})$$

Shear force diagram
The diagram is drawn by remembering

(a) the shear force changes abruptly at a concentrated load;
(b) the shear force is uniform on an unloaded portion of the beam.

Using the given sign convention we may start at the *left-hand end* and draw the diagram by following the arrows representing the loads and reactions, thus we draw:

at A, 9.8 kN down, then a horizontal line to C;

at C, 19.6 kN up, then a horizontal line to D;

at D, 29.4 kN down, then a horizontal line to B;

at B, 19.6 kN up to the zero line again.

Note that in this method we have followed the *changes* in shear force along the beam.

Bending moment diagram
At the free ends A and B, the bending moment is zero. At C, considering the left-hand portion AC:

$$\text{bending moment} = -9.8 \times 2 = -19.6 \text{ kN-m}$$

(negative, since the 9.8 kN load at A tends to make the beam hog at C). At D, considering the left-hand portion AD:

$$\text{bending moment} = -9.8 \times 8 + L \times 6$$

$$= -78.4 + 117.6, \text{ since } L = 19.6 \text{ kN}$$

$$= +39.2 \text{ kN-m}$$

This is the greatest *bending moment* in the beam. The values of the bending moment at A, B, C, and D are plotted in Fig. 16.8 and the bending moment diagram completed by joining the resulting points by straight lines. Note that the two points of greatest bending moment occur at C and D where the shear force changes sign.

Example *For the beam ABC, Fig. 16.9 find the vertical reaction at the pin-joint A and the bending moment at B. The beam is supported by a cable at B.*

SOLUTION
Reactions
The joint at the pinned end A may be assumed frictionless and therefore carries no bending moment. We are interested only in forces transverse to the horizontal beam, hence only *vertical* components of the tension in the cable and the reaction at A need be considered. The weight of the 2 Mg mass is $2 \times 9.8 = 19.6$ kN.

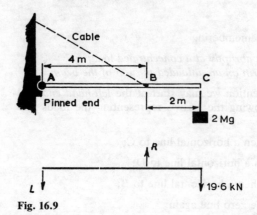

Fig. 16.9

Let L and R be the vertical forces at A and B, respectively. To find R take moments about A

$$R \times 4 = 19.6 \times 6$$

therefore $R = 29.4$ kN (upward)

Since the net vertical force is zero, assuming L to act downwards,

$$-L + R - 19.6 = 0$$

thus

$$L = +29.4 - 19.6$$

$$= +9.8 \text{ kN (downward)}$$

At B, considering the right-hand portion BC:

bending moment $= -19.6 \times BC$

$$= -19.6 \times 2$$

$$= -39.2 \text{ kN-m}$$

This is negative since the load at C tends to cause the beam to hog at B.

Problems

1. Draw the shear force and bending moment diagrams for the cantilever beams shown in Fig. 16.10. State in each case: (*a*) the shear force between points A and B; (*b*) the bending moment at points A and B.
   ((*a*) 60 kN; 540 kN-m; 240 kN-m; (*b*) 17.3 kN; 52 kN-m; 0; (*c*) 42.3 kN; 98 kN-m; 0; (*d*) 300 kN;; 0; 90 kN-m)

2. Draw the shear force and bending moment diagrams for the simply supported beams shown in Fig. 16.11. State in each case: (i) the reactions L and R; (ii) the greatest shear force in the beam; (iii) the bending moment at point A.
   ((*a*) L, 22.54 kN; R, 16.66 kN; 22.54 kN; 13.4 kN-m; (*b*) L, 6 kN; R, 24 kN; 14 kN; 30 kN-m; (*c*) L, 21.23 kN; R, 8.17 kN; 19.6 kN; 9 kN-m)

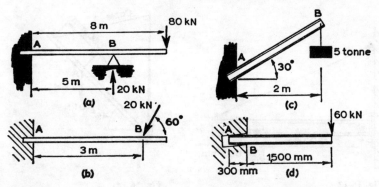

Fig. 16.10

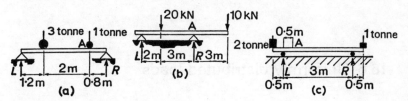

Fig. 16.11

3. For the cantilever beam shown in Fig. 16.12, determine the greatest value of the bending moment and the upward reaction at the built-in end. Draw the shear force and the bending moment diagrams.

(30 kN-m at built-in end; 20 kN)

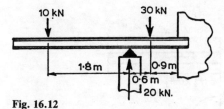

Fig. 16.12

4. Draw the shear force and bending moment diagrams for the pin-ended beam of Fig. 16.13. State the greatest value of bending moment and shear force. If the cable is at 30° to the beam what is the tension in the cable?

(4.2 kN-m; 7 kN; 14 kN)

5. Draw the shear force and bending moment diagrams for the pin-ended jib crane shown in Fig. 16.14. State the bending moment at the point A, and calculate the shear force in the beam at the pin-joint. If the cable is at 60° to the beam what is the tension in the cable?

(2.55 kN-m; 2.83 kN; 4.9 kN)

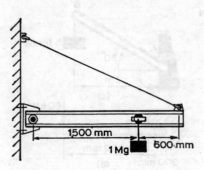

Fig. 16.13

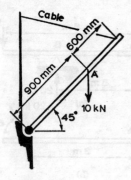

Fig. 16.14

16.6 Uniformly distributed loads

When a load is not concentrated at a point but spread uniformly over a portion of the beam the load is said to be *uniformly distributed*. The loading is usually given as an intensity of loading (force or mass) per unit length of the beam, i.e. N/m or kg/m. Suppose a beam to carry a load of *weight w* per unit length over a length x as shown in Fig. 16.15. Then the shear force at section X *due to this load* is the total load $w \times x$, i.e.

shear force at X = wx

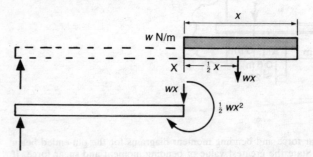

Fig. 16.15

The distance of the centroid of the load wx from section X is $\frac{1}{2}x$. Therefore the bending moment at X *due to this load* is

$$wx \times \tfrac{1}{2}x = \tfrac{1}{2}wx^2$$

Example *Draw the shear force and bending moment diagrams for the uniformly loaded, simply supported beam of Fig. 16.16. The load intensity is 10 Mg/m. State the value of the maximum bending moment and where it occurs.*

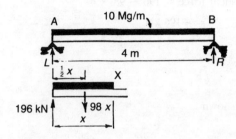

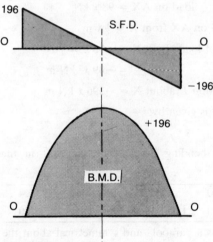

Fig. 16.16

SOLUTION
Reactions

$$w = 10 \times 9.8 = 98 \text{ kN/m}$$

Total load $= 98 \times 4 = 393$ kN

From symmetry

$$L = R = 196 \text{ kN}$$

Shear force diagram
For section X, distant x m from A:

shear force $= L -$ load on AX

$$= 196 - 98\,x \text{ kN}$$

which gives a straight line graph.

At A, $x = 0$, and shear force $= 196$ kN

At B, $x = 4$ m, and shear force $= 196 - 98 \times 4$

$$= -196 \text{ kN}$$

At centre of beam $x = 2$, thus

shear force $= 196 - 98 \times 2$

$$= 0$$

The shear force diagram is as shown.

Bending moment diagram
For section X:

$$\text{load on AX} = 98\,x \text{ kN}$$

distance of centroid of load on AX from $X = \frac{1}{2}\,x$ m

$$\text{moment of load about } X = -98\,x \times \tfrac{1}{2}\,x$$

$$= -49\,x^2 \text{ kN-m}$$

$$\text{moment of reaction } L \text{ about } X = -196\,x \text{ kN-m}$$

The total bending moment at X is given by:

$$196\,x - 49\,x^2 \text{ kN-m}$$

The calculated values of the bending moment are set out in the following table:

x (m)	0	1	1.5	2	2.5	3	4
Bending moment (kN-m) ..	0	147	184	196	184	147	0

The bending moment diagram is a parabola and symmetrical about the middle of the beam. The *maximum* bending moment is at the middle, the point at which the slope of the curve is zero and the shear force changes sign, therefore

maximum bending moment = **196 kN-m**

This is also the greatest numerical value of the bending moment.

16.7 Combined loading

When a beam carries both concentrated and uniformly distributed loads it is necessary to consider the beam and its loading in convenient portions and obtain expressions for the shear and bending moment in each portion separately. This is because the formula for bending moment and shear force changes at each concentrated load.

Example *Draw the shear force and bending moment diagrams for the beam shown, Fig. 16.17(a). The beam is simply supported at A and D.*

SOLUTION
Reactions
To determine the support reactions L and R the distributed load over AC may be taken as acting at its centroid, 1 m from A.

Total distributed load = $15 \times 2 = 30$ kN

Moments about A:

$$R \times 4 = 10 \times 1.5 + 30 \times 1 + 20 \times 4.5$$

thus $R = 33.75$ kN

Moments about D:

$$L \times 4 = 10 \times 2.5 + 30 \times 3 - 20 \times 0.5$$

therefore $L = 26.25$ kN

(Check:
$$L + R = 26.25 + 33.75$$
$$= 60 \text{ kN}$$
$$= \text{net downward load.})$$

Shear force diagram
For length AB, Fig. 16.17(*b*). Consider shear force at section X due to load on left-hand portion of beam AX:

load on AX = $15\,x$

The shear force at X is:

$$L - 15\,x = 26.25 - 15\,x \text{ kN}$$

This gives a straight line graph.

At A, $x = 0$,   shear force = 26.25 kN

At B, $x = 1.5$ m, shear force = $26.25 - 15 \times 1.5$
$$= 3.75 \text{ kN}$$

These values are plotted in Fig. 16.17(*c*). The shear force diagram is completed for portion AB by joining the plotted points by a straight line as shown.

Refer to Fig. 16.17(*d*). Consider shear force at X in BC due to loads on left-hand portion of beam AX. Shear force at X is:

$$26.25 - 10 - 15\,x$$
$$= 16.25 - 15\,x \text{ kN}$$

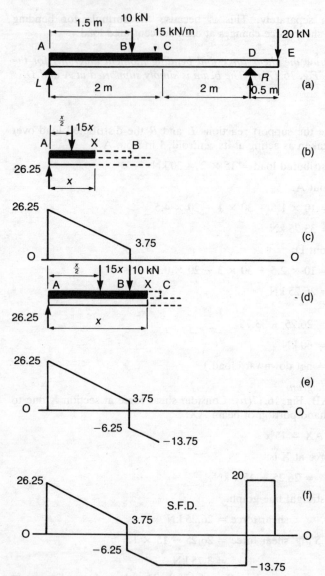

Fig. 16.17

At B, $x = 1.5$ m, shear force $= 16.25 - 15 \times 1.5 = -6.25$ kN

At C, $x = 2$ m, shear force $= 16.25 - 15 \times 2 = -13.75$ kN

These values are plotted in Fig. 16.17(*e*) and the shear force diagram completed as far as point C.

The portion CD is unloaded, hence the shear force is uniform at −13.75 kN.

DE, Fig. 16.17(f). At support D the shear force changes abruptly due to the reaction R upwards, i.e.

shear force to right of D = −13.75 + 33.75

$$= 20 \text{ kN}$$

The shear force is uniform at 20 kN along DE since this portion of the beam is unloaded. The complete shear force diagram is shown in Fig. 16.17(f). Since it is made up of straight lines it could have been drawn by calculating the shear force at the principal points and joining the plotted points by straight lines.

Bending moment diagram

Refer to Fig. 16.18(b). Consider bending moment at X due to load on portion AX:

bending moment due to reaction L = 26.25 x kN-m

bending moment due to load on AX = −15 $x \times \frac{1}{2} x$

$$= -7.5 \, x^2 \text{ kN-m}$$

Total bending moment at X = 26.25 x − 7.5 x^2 kN-m

Calculated values of the bending moment are tabulated below:

x (m)	0	0.5	1	1.25	1.5
Bending moment (kN-m)	0	11.25	18.75	21.1	22.5

The bending moment diagram for portion AB is shown in Fig 16.18(c).

Refer to Fig. 16.18(d). Consider the bending moment at X in BC due to load on portion AX.

Bending moment at X = 26.25 x − 15 $x \times \frac{1}{2} x$ − 10(x − 1.5)

$$= 15 + 16.25 \, x - 7.5 \, x^2 \text{ kN-m}$$

Values of the bending moment in BC are given below:

x (m)	1.5	1.75	2
Bending moment (kN-m)	22.5	20.44	17.5

The bending moment diagram for length BC is shown in Fig. 16.18(e).

In the lengths CD and DE there are no distributed loads and the bending moment diagram is therefore completed by calculating the bending moment at D and E, plotting these values and joining these points by straight lines. At D, considering portion DE:

bending moment = −20 × 0.5 = −10 kN-m

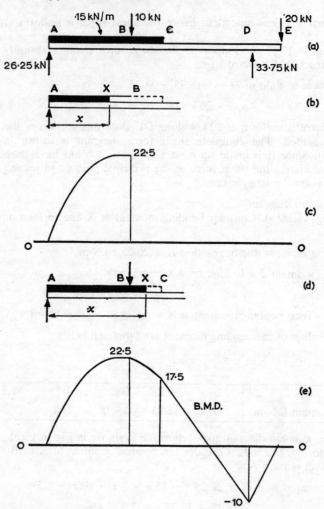

Fig. 16.18

At E,

bending moment = 0

The completed bending moment diagram is shown in Fig. 16.18(*e*). Often only a sketch may be required, in which case the values of the bending moment at the principal points should be marked as indicated. The greatest numerical value of the bending moment is at B (where the shear force changes sign) and is **22.5** kN-m. There is also another 'peak' value at D, of 10 kN-m, where the shear force again changes sign.

Problems

1. Draw the shear force and bending moment diagrams for the cantilever beams shown in Fig. 16.19. State the greatest value of shear force and bending moment in each case.

((a) 15 kN; 15.5 kN-m; (b) 50 kN; 39 kN-m)

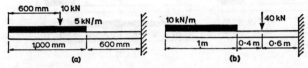

Fig. 16.19

2. The cantilever beam shown in Fig. 16.20 is partially supported by a load P exerted by a prop distant x m from the built-in end. Calculate the value of P if x is 1.6 m, in order that the bending moment at the built-in end shall be zero.

If the load in the prop is 10 kN and the bending moment at the built-in end is to be zero, what is the value of x required?

(1.764 kN; 282 mm)

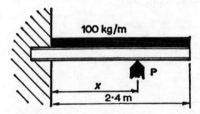

Fig. 16.20

3. Draw shear force and bending moment diagrams for the simply supported beams shown in Fig. 16.21. State the values of the reactions L and R and the bending moments at the points A and B, in each case.

((a) L, 16.86 kN; R, 22.74 kN; 14.11 kN-m; 11.37 kN-m; (b) L, 27.6 kN; R, 19.9 kN; 27.6 kN-m; 49.75 kN-m)

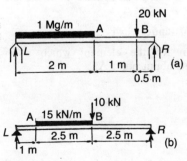

Fig. 16.21

4. For the simply supported beams shown in Fig. 16.22 determine the greatest value of the bending moment. State the values of the reactions L and R.

((*a*) 7.25 kN-m; L, 7.25 kN; R, 4.75 kN; (*b*) 20 kN-m; L, 27.2 kN; R, 14.8 kN)

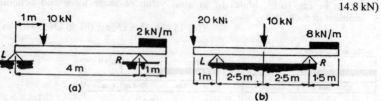

(a)

(b)

Fig. 16.22

16.8 Condition for a maximum bending moment

Consider an element of a beam cut off by transverse sections A and B (Fig. 16.23) such that $AB = dx$. For simplicity let AB be unloaded, although the remainder of the beam may be loaded.

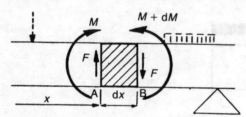

Fig. 16.23

Let shear force at A = F upwards. Since element AB is not loaded the shear force at B is F, downwards. These forces F are the forces exerted *on* the element by the other parts of the beam.

Let the bending moment at A = M, clockwise. Then owing to the effect of the length dx there must be a change in bending moment dM between A and B, thus:

bending moment at B = $M + dM$, anticlockwise

Fig. 16.23 is the free-body diagram for the element which is in equilibrium, therefore, taking moments about point A:

$$M + F \times dx = M + dM$$

i.e. $$dM = F \, dx$$

or $$\frac{dM}{dx} = F$$

Therefore, if the bending moment is expressed as a function of x and this expression differentiated with respect to x, the shear force is obtained. $\dfrac{dM}{dx}$ is the slope of the tangent to the bending moment curve. The point at which the tangent is horizontal corresponds to zero slope and a maximum or minimum bending moment. This point is a 'turning point' on the curve where the slope of the bending moment diagram changes sign and so accordingly does the shear force change sign passing through zero. Thus for maximum (or minimum) bending moment

$$\frac{dM}{dx} = 0$$

i.e. $\quad F = 0$

Hence the bending moment is a maximum when the shear force is zero.

This result remains true when the portion AB of the beam is loaded. It is true for any point at which the shear force has zero value. For example Fig. 16.24 shows a simply supported beam under a 'concentrated' load W which, in practice, is distributed over a small length of beam ab. If W is assumed uniformly distributed over ab then the shear force diagram for this region is the line $a'b'$, slightly inclined to the vertical, and the bending moment curve is $a''b''$. At the mid-point of ab, the shear force is zero, and the bending moment therefore takes a mathematical maximum value under the load. A convenient method of finding the greatest bending moment on a beam, particularly with complex loading, is to find *by inspection* where the shear force is zero or changes sign (*see* previous worked examples).

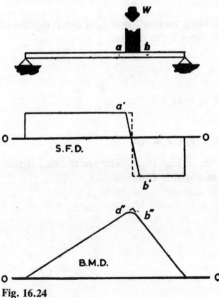

Fig. 16.24

Example *Calculate the position and magnitude of the maximum bending moment in the simply supported beam loaded as shown in Fig. 16.25.*

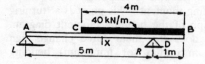

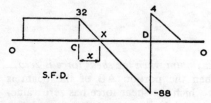

Fig. 16.25

SOLUTION

Moments about A:

$$R \times 5 = 40 \times 4 \times 4$$

therefore $R = 128$ kN

and $L + R = 160$ kN

thus $L = 32$ kN

To determine the maximum bending moment first find the positions of zero shear force by drawing the shear force diagram.

At C, shear force = 32 kN

At D (just to left-hand side):

shear force = $32 - 40 \times 3 = -88$ kN

At D (just to right-hand side):

shear force = $32 - 40 \times 3 + 128 = 40$ kN

The shear force diagram is shown in Fig. 16.25. The shear force is zero at D and at X. At D, bending moment is 20 kN-m.

At X, by simple proportion

$$\frac{CX}{CD} = \frac{32}{32 + 88} = 0.267$$

thus

$$x = CX$$

$$= 0.267 \times CD$$

$$= 0.267 \times 3$$
$$= 0.8 \text{ m}$$

Hence at X,

$$\text{bending moment} = L \times \text{AX} - \tfrac{1}{2} \times 40 \times x^2$$
$$= 32 \times 2.8 - \tfrac{1}{2} \times 40 \times 0.8^2$$
$$= 76.8 \text{ kN-m}$$

The greatest value of the bending moment therefore is at X, distant **0.8 m** from C between C and D, and its magnitude is **76.8 kN-m**.

Problems

1. Draw the shear force and bending moment diagrams for the beams shown in Fig. 16.26. State the maximum values of bending moment and shear force and where they occur.

   ((a) 2.5 kN-m, 1 m from l.h. end; 5 kN, 1 m from l.h. end; (b) 7.67 kN-m, 5 m from l.h. end; 5.17 kN, 6 m from r.h. end)

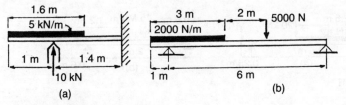

Fig. 16.26

2. For the simply supported beam shown in Fig. 16.27 find where the shear force is zero and hence obtain the maximum bending moment.

   (840 mm from l.h. end; 3.53 kN-m)

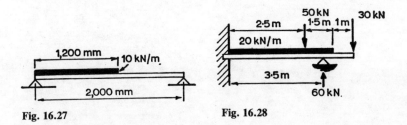

Fig. 16.27   **Fig. 16.28**

3. Fig. 16.28 shows a loaded cantilever beam propped at a point 3.5 m from its left-hand end. Draw to scale the shear force and bending moment diagrams. Find the maximum bending moment and state where it occurs.

   (225 kN-m, at the built-in end)

4. Draw to scale the shear force and bending moment diagrams for the pin-ended beam shown in Fig. 16.29. State the maximum values of the shear force and bending moment. If the cable is at 40° to the beam what is the tension in the cable? State the reaction at the pin joint.

(113.8 kN at 2.5 m from l.h. end; 46.6 kN-m at 1.077 m from l.h. end; 315 kN, 257 kN)

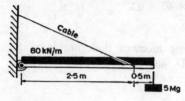

Fig. 16.29

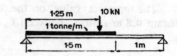

Fig. 16.30

5. Determine the maximum shear force and bending moment in the beam shown in Fig. 16.30. Where does the maximum bending moment occur?

(15.29 kN; 11.46 kN-m, at the 10-kN load)

6. State the position and magnitude of the maximum bending moment in each of the simply supported beams shown in Fig. 16.31.

((a) 1.86 m from l.h. end, 102 kN-m; (b) 2 m from l.h. end. 26.5 kN-m)

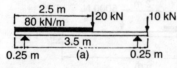

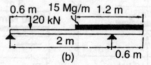

Fig. 16.31

2. The cross section of the beam is symmetrical about the plane of bending. If not symmetrical, the beam would twist as well as bend. In Fig.17.2, *xy* is the plane of bending and the section shown is symmetrical about *xy* and normal to the plane of bending.

CHAPTER 17

Bending of Beams

17.1 Pure bending of an elastic beam

The beam AB (Fig. 17.1) loaded by equal and opposite couples M at A and B, is said to be subjected to pure or *simple* bending, that is, bending without shear force. At any section X–X the bending moment is M. For equilibrium of any portion XB the couple must be balanced by an equal and opposite couple exerted by internal forces within the beam. Thus M at B is balanced by the couple represented by the pair of equal and opposite parallel forces F shown in the free-body diagram for XB. These are internal forces exerted by the portion AX on the portion XB at X–X. The upper force represents a tension in XB, the lower a corresponding compression. Thus the top layers are stretched and lower layers are compressed.

The internal forces are not actually point loads as shown in Fig. 17.1, but are distributed across the depth of the beam as tensile

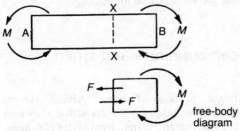

free-body
diagram

Fig. 17.1

and compressive stresses in the upper and lower portions of the section, respectively. These stresses are not uniform, however, and in order to calculate them it is necessary to make the following *assumptions*:

1. The beam is initially straight.
2. Bending takes place in the plane of the applied bending moment (the plane of the paper).

3. The cross-section of the beam is symmetrical about the plane of bending. If not symmetrical the beam would twist as well as bend. In Fig. 17.2 Y–Y is the plane of bending and the section shown is symmetrical about, and normal to, the plane of bending.

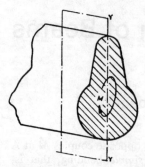

Fig. 17.2

4. The stresses are uniform across the width.
5. The material of the beam is elastic and obeys Hooke's law.
6. The stresses do not exceed the limit of proportionality.
7. The moduli of elasticity in tension and compression are the same.
8. A plane transverse section of the beam remains a plane section after bending.
9. Each layer of the beam is free to carry stress without interference from adjacent layers.

These assumptions are justified since they are found to give substantially correct values for the stresses due to bending.

17.2 Relation between curvature and strain

Consider the initially straight portion of beam ABCD shown, Fig. 17.3(*a*). After bending the beam takes up the arc shape shown in Fig. 17.3(*b*). Plane cross-sections remain plane, hence straight lines, BA, CD remain straight after bending and meet at some point O. Lines such as EF and SN form part of circular arcs with a common centre at O. Since the top layers are stretched and the bottom layers are compressed there is a layer, the *neutral surface*, which is neither stretched nor compressed. The lines S′N′ and N′N′ represent the trace of the neutral surface, the line N′–N′ normal to the plane of bending being known as the *neutral axis*. The *radius of curvature R* of the bent beam is measured from O to the neutral surface.

If θ is the angle in radians subtended by the arc S′N′ at O then, since the neutral layer remains unchanged in length:

line SN = arc S'N' = $R\theta$

Consider the thin layer EF, distant y from the neutral surface:

initial length of EF = SN = $R\theta$

final length of EF = E'F'

= $(R + y)\theta$

thus

extension of EF = $(R + y)\theta - R\theta$

= $y\theta$

therefore

strain in EF, $\varepsilon = \dfrac{y\theta}{R\theta}$

= $\dfrac{y}{R}$

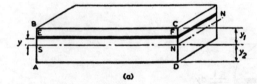

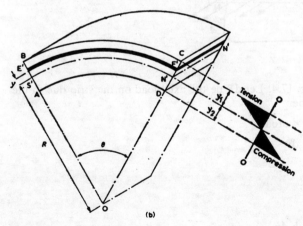

Fig. 17.3

The stress in the layer EF, normal to the beam section, is given by:

$\sigma = E \times \varepsilon$

$$= \frac{Ey}{R}$$

or $\quad \dfrac{\sigma}{y} = \dfrac{E}{R}$

Since E is constant for the beam and R is constant for the portion considered, the stress σ varies across the depth with the distance y from the neutral axis. The distribution of stress across the depth of the beam is sketched in Fig. 17.3(*b*), tensile stress being plotted to the left of the base O–O, compressive stress to the right. The *maximum* stress occurs at the *outside* surfaces such as AD and BC where y takes its largest values y_2 and y_1.

17.3 Position of the neutral axis

The position of the neutral axis is found from the fact that there is no net axial force on the beam. Consider a small strip of width b,

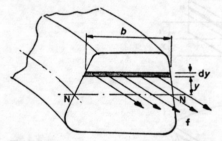

Fig. 17.4

thickness dy, Fig. 17.4. Let dF be the axial load on the strip due to the bending stress, then

$$dF = \text{stress} \times \text{area}$$

$$= \sigma \times b \, dy$$

but $\quad \sigma = \dfrac{Ey}{R}$

thus

$$dF = \frac{Ey}{R} \times b \, dy$$

The total load on the section is the sum of the elementary loads, therefore

$$F = \int dF$$

$$= \int \frac{Ey}{R} \times b \, dy$$

and, since the total axial load is zero

$$\int \frac{Ey}{R} \times b \, dy = 0$$

and thus

$$\int y \times b \, dy = 0$$

since E and R are constants. But $y \times b \, dy$ is the moment of area $b \, dy$ about the neutral axis N–N. Hence $\int yb \, dy$ represents the moment of the whole area about the neutral axis and *this is zero only if the neutral axis passes through the centroid of the cross-section.* This result is true for any shape of cross-section.

Note: (a) The above results do not apply completely to a strip which is wide compared with its depth; (b) In a symmetrical section $y_1 = y_2$, and the maximum tensile and compressive stresses are therefore equal.

In a non-symmetrical section the *numerically largest* bending stress will occur at the outer layer most distant from the neutral axis, and may be tensile *or* compressive.

Example *Calculate the maximum stress in a coil of steel rod 4 mm diameter due to coiling it on a drum of 2 m, diameter.* E = 200 GN/m².

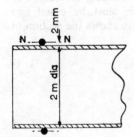

Fig. 17.5

SOLUTION

Radius of curvature = 1 m approx.

The greatest distance of outside surface from the neutral axis N–N is 2 mm. The maximum stress due to bending is

$$\sigma = \frac{E}{R}y$$

$$= \frac{200 \times 10^9}{1} \times 0.002$$

$$= 400 \times 10^6 \text{ N/m}^2 = \textbf{400 MN/m}^2$$

Problems

1. A steel strip forming a bandsaw is wrapped round a 400-mm diameter drum. If the strip thickness is 1 mm, calculate the maximum stress due to bending. $E = 200$ kN/mm².

   (500 MN/m²)

2. Aluminium alloy tube of 20 mm diameter is wound on a drum of 3 m diameter. Calculate the maximum bending stress in the tube. $E = 70$ GN/m².

   (467 MN/m²)

3. Calculate the minimum diameter of drum on which copper strip 2.5 mm thick may be wound if the maximum bending stress is not to exceed 350 MN/m². $E = 100$ GN/m².

   (714 mm)

4. A steel strip of rectangular cross-section 10 mm thick is bent to the arc of a circle until the steel just yields at the top surface. Find the radius of curvature of the neutral surface if the yield stress of the material is 270 MN/m². $E = 195$ kN/mm².

   (3.61 m)

17.4 Moment of resistance

The *moment of resistance* of a beam is the moment about the neutral axis of the internal forces resisting the applied bending moment. For equilibrium, the internal moment of resistance must be equal and opposite to the applied bending moment. Fig. 17.6 shows the section of

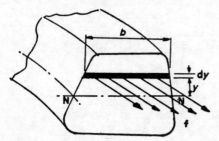

Fig. 17.6

a beam on which the bending moment is M. Let σ be the stress produced in a thin strip at a distance y from the neutral axis N–N then, if dy is the thickness of the strip

   load on strip = dF = stress × area

   $= \sigma \times b \, dy$

The moment of this load about the neutral axis is:

$$dM = dF \times y$$
$$= \sigma b \, dy \times y$$

but $\sigma = \dfrac{E}{R}y$

thus $dM = \dfrac{E}{R}y \times by \, dy$

$$= \dfrac{E}{R}by^2 \, dy$$

Hence total moment of resistance is:

$$M = \int dM$$
$$= \int \dfrac{E}{R}by^2 \, dy$$
$$= \dfrac{E}{R}\int by^2 \, dy$$
$$= \dfrac{E}{R}I$$

where $I = \int by^2 \, dy$, is the *second moment of area of the section about the neutral axis*. Hence

$$\dfrac{M}{I} = \dfrac{E}{R}$$

and since

$$\dfrac{\sigma}{y} = \dfrac{E}{R}$$

then

$$\dfrac{M}{I} = \dfrac{E}{R} = \dfrac{\sigma}{y}$$

17.5 *I* of rectangular and circular sections

For a rectangular section of width b mm, depth d mm (Fig. 17.7) the second moment of area about an axis through the centroid parallel to the width is

$$I_G = \dfrac{bd^3}{12} \ \text{mm}^4$$

For a circular section, diameter d mm, the second moment of area about any axis through the centre, i.e. a diametral axis, is

$$I_G = \frac{\pi d^4}{64} \text{ mm}^4$$

 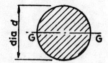

Fig. 17.7

Example *A rectangular section beam has a depth of 100 mm, width 24 mm, and is subjected to a bending moment of 25 kN-m. Calculate (a) the maximum stress in the beam and (b) the radius of curvature of the neutral surface. E = 206 GN/m².*

SOLUTION

(a) The axis of bending is the axis through the centroid parallel to the 24-mm side, therefore

$$I = \frac{bd^3}{12} = \frac{0.024 \times 0.1^3}{12} = 2 \times 10^{-6} \text{ m}^4$$

$$\sigma = \frac{M}{I}y = \frac{25 \times 10^3}{2 \times 10^{-6}} \times y = 12.5 \times 10^9 \, y \text{ N/m}^2$$

where y is in metres. Now the bending stress σ takes its greatest value at the outside surface where

$$y = \pm \frac{d}{2} = \pm 0.05 \text{ m}$$

hence

$$\sigma_{max} = \pm 12.5 \times 10^9 \times 0.05 = \pm 625 \times 10^6 \text{ N/m}^2$$

$$= \pm 625 \text{ MN/m}^2$$

The positive answer denotes a tensile stress at one outer surface, the negative a compressive stress at the other outer surface.

(b) $\qquad R = \dfrac{I}{M}E = \dfrac{2 \times 10^{-6} \times 206 \times 10^9}{25 \times 10^3}$

$$= 16.48 \text{ m}$$

Example *A beam of symmetrical I-section has the following dimensions: flange 150 mm wide, 30 mm thick; web 30 mm thick; total depth of beam 200 mm. Calculate the second moment of area of the beam section about an axis through the centroid parallel to the flange face. If the beam is simply supported over a length of 2 m and carries a uniformly distributed load of 6 tonne/m run, calculate the maximum bending stress in the beam.*

SOLUTION

The beam section is symmetrical about the centroidal axis G–G, Fig. 17.8. Hence I is given by the difference between I for the rectangle ABCD and I for the two cross-hatched rectangles shown. For whole rectangle ABCD:

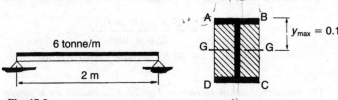

Fig. 17.8

$$I = \frac{bd^3}{12}$$

$$= \frac{150 \times 200^3}{12}$$

$$= 10^8 \text{ mm}^4$$

For each cut-out rectangle

$$b = 60 \text{ mm}$$

$$d = 140 \text{ mm}$$

therefore

$$I = \frac{60 \times 140^3}{12}$$

$$= 0.137 \times 10^8 \text{ mm}^4$$

For the I-section, by difference

$$I_G = 10^8 - 2 \times 0.137 \times 10^8$$

$$= \mathbf{72.6 \times 10^6 \text{ mm}^4}$$

Note: This method of calculation does not apply when the I-section is not symmetrical about G–G (*see* paragraph 17.7)

6 tonne/m = $6 \times 9.8 = 58.8$ kN/m

Total load on beam = $58.8 \times 2 = 117.6$ kN

thus

$$\text{reaction at each support} = \frac{117.6}{2} = 58.8 \text{ kN}$$

The maximum bending moment occurs at the middle of the beam; therefore

maximum bending moment $M_{max} = 58.8 \times 1 - 58.8 \times 1 \times 0.5$

$$= 29.4 \text{ kN-m}$$

The maximum bending stress occurs at the outside surface of the section of maximum bending moment, therefore

$$\text{maximum bending stress, } \sigma_{max} = \frac{M}{I} y_{max}$$

$$= \frac{29.4 \times 10^3 \times 0.1}{72.6 \times 10^{-6}}$$

$$= 40 \times 10^6 \text{ N/m}^2 = 40 \text{ MN/m}^2$$

Example   *The cantilever beam shown in Fig. 17.9 is loaded by a single concentrated load W kN at its free end. It is of hollow section 90 mm outside diameter, 80 mm inside diameter. Calculate the maximum value of W if the bending stress is not to exceed 60 MN/m².*

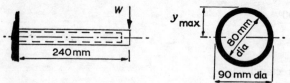

Fig. 17.9

SOLUTION
Since the beam cross-section is symmetrical about the neutral axis the second moment of area is found by subtracting that of the inner circle area from that of the second moment for the whole area. For a circular section:

$$I = \frac{\pi d^4}{64}$$

therefore for hollow section

$$I = \frac{\pi \times 90^4}{64} - \frac{\pi \times 80^4}{64}$$

$$= 1.21 \times 10^6 \text{ mm}^4$$

$$= 1.21 \times 10^{-6} \text{ m}^4$$

$$\text{Allowable moment, } M = \frac{\sigma}{y_{max}} I$$

$$= \frac{60 \times 10^6}{0.045} \times 1.21 \times 10^{-6}$$

$$= 1,610 \text{ N-m} = 1.61 \text{ kN-m}$$

The maximum bending moment on the cantilever is $W \times 0.24$ kN-m.

Therefore

$$W \times 0.24 = 1.61$$

i.e. $\qquad W = \textbf{6.71 kN}$

Problems

1. A light-alloy beam, rectangular section 5 mm deep, 12 mm wide rests on supports 200 mm apart and carries a central load of 70 N. Calculate the maximum bending stress.

$$(70 \text{ MN/m}^2)$$

2. A light wooden bridge is supported by six parallel timber beams, each 300 mm deep, and 200 mm wide. Each beam may be considered as simply supported over a 4.5 m span. If the allowable bending stress in the timber is 5.60 MN/m² calculate the greatest uniformly distributed load the bridge can support.

$$(179.2 \text{ kN or } 18.3 \text{ Mg})$$

3. The crane beam shown in Fig. 17.10 is made up of two 25-mm thick steel plates each 400 mm deep at the middle section. Calculate the maximum allowable central load W if the span is 4.5 m and the ultimate tensile stress of the steel is 370 MN/m². Allow a factor of safety of 6.

$$(73 \text{ kN or } 74.5 \text{ tonne})$$

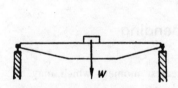

Fig. 17.10

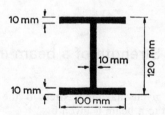

Fig. 17.11

4. Calculate the maximum bending moment which may be applied to the cast-iron section shown in Fig. 17.11 if the ultimate tensile stress of the material is 280 MN/m² and a factor of safety of 10 is to be used.

$$(I = 6.9 \times 10^{-6} \text{ m}^4; 3.22 \text{ kN-m})$$

5. Calculate the maximum allowable uniformly distributed load, in kilograms, on a channel which is used as a cantilever of length 3 m. The channel is to be loaded along its whole length and the maximum bending stress permitted is 72 MN/m². Centroid, 150 mm below top face; relevant I of section about centroid = 86×10^6 mm⁴.

$$(935 \text{ kg/m})$$

6. The bar of section shown in Fig. 17.12 is simply supported over a span of 1 m and carries a central load of 20 kN. Find the maximum bending moment and the maximum bending stress in the material.

$$(I = 77.44 \times 10^4 \text{ mm}^4; 5 \text{ kN-m}; 194 \text{ MN/m}^2)$$

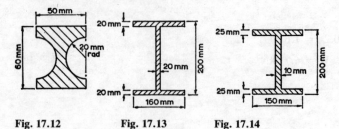

Fig. 17.12   **Fig. 17.13**   **Fig. 17.14**

7. Fig. 17.13 shows a section of a light alloy cantilever beam 1.2 m long which is to carry a load of 2 tonne at its free end. Calculate the second moment of area of the section about its axis of bending and find the maximum bending stress in the material.

$$(I = 58.9 \times 10^6 \text{ mm}^4; \; 39.9 \text{ MN/m}^2)$$

8. The steel joist shown in section in Fig. 17.14 is simply supported over a span of 3 m. Calculate the maximum bending stress in the joist and the stress at the inside face of each flange due to a load of 10 Mg mid-way between the supports.

$$(I = 60.63 \times 10^6 \text{ mm}^4; \; 121.3 \text{ MN/m}^2; \; 91 \text{ MN/m}^2, \text{ one tensile, other compressive})$$

9. A cast-iron pipe is carried over a span of 6 m and may be considered as simply supported at each end. The pipe is 200 mm bore, 10 mm thick and full of water. Calculate the maximum bending stress in the pipe. Density of water = 1 Mg/m³. Density of cast iron = 7.2 Mg/m³.

$$(10.5 \text{ MN/m}^2)$$

17.6 Strength of a beam in bending

For a given maximum stress σ the greatest moment which may be applied to a beam is given by

$$M_{\max} = \frac{\sigma}{y_{\max}} \times I$$

i.e. $M_{\max}$ is proportional to I and inversely proportional to $y_{\max}$, the distance of the extreme fibres from the neutral axis. For a rectangular section

$$I = \frac{bd^3}{12}, \text{ and } y_{\max} = \tfrac{1}{2}d$$

Hence

$$M_{\max} = \frac{\sigma \times bd^3/12}{\tfrac{1}{2}d}$$

$$= \frac{\sigma}{6} \times bd^2$$

which is proportional to the square of the depth of the beam.

For the same area of section (or weight per unit length) the *I*-value for various beams varies widely with the *shape* of section. Thus Fig. 17.15 shows four typical sections, each having an area of about 8,750 mm². The table indicates the corresponding *I* and maximum moment which can be carried for the same maximum bending stress of 75 MN/m².

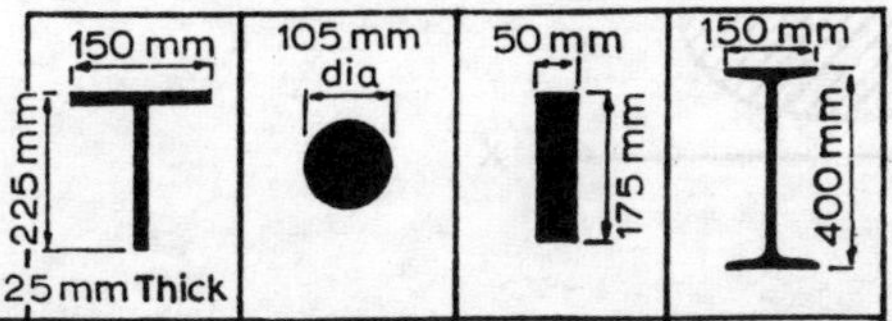

	150 mm 225 mm 25 mm Thick	105 mm dia	50 mm 175 mm	150 mm 400 mm
I (mm⁴)	44 × 10⁶	6 × 10⁶	22·3 × 10⁶	240 × 10⁶
M_{max} (kN-m)	22·2	8·6	19·2	90

Fig. 17.15

Comparison of these figures shows that the standard rolled steel joist is by far the strongest in bending. Evidently in building up a beam section it is advantageous to place the greatest area at the largest possible distance from the centroid.

For steel having the same strength in tension as in compression a symmetrical section is usually adequate. For cast iron, however, which has a lower strength in tension than in compression, the centroid of the cross-section should lie nearer to the tension flange so that the greatest stress occurs at the compression flange.

17.7 Calculation of *I* for complex sections

In dealing with complex and asymmetrical sections the previous method of calculating *I* is insufficient. Also the position of the centroid and thus the neutral axis may be unknown. It is necessary therefore first to locate the centroid and then obtain the second moment of area about the neutral axis. The calculation of the second moment requires the theorem of parallel axes. The *theorem of parallel axes* states that if I_G is the second moment of area of a section about an axis G–G through the centroid and I_X is the second moment about an axis X–X parallel to G–G, then

$$I_X = I_G + Ah^2$$

where A is the area of the section and h the perpendicular distance between the axes G–G and X–X, Fig. 17.16.

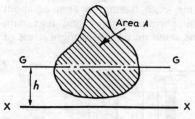

Fig. 17.16

The method of calculation for various sections is illustrated in the following examples.

Example *Calculate the second moment of area of the T-section shown, Fig. 17.17, about a line X–X through the centroid parallel to the flange face.*

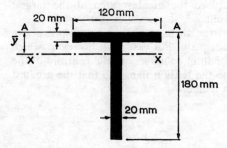

Fig. 17.17

SOLUTION
To find position of centroid
    Area of section = 5,600 mm^2

    moment of flange area about edge A–A = $120 \times 20 \times 10$

$$= 24{,}000 \text{ mm}^3$$

    moment of web area about A–A = $160 \times 20 \times 100$

$$= 320{,}000 \text{ mm}^3$$

    total moment = 344,000 mm^3

Let distance of centroid from A–A be $\bar{y}$, then

$$\text{(total area)} \times \bar{y} = 344{,}000$$

i.e. $5{,}600 \times \bar{y} = 344{,}000$

thus $\bar{y} = 61.4 \text{ mm}$

Calculation of $I_{\text{X-X}}$

Flange: second moment of flange about its own centroid:

$$I_\text{G} = \frac{bd^3}{12}$$

$$= \frac{120 \times 20^3}{12}$$

$$= 8 \times 10^4 \text{ mm}^4$$

Distance of centroid of flange from axis X–X through centroid of section is

$$61.4 - 10$$

$$= 51.4 \text{ mm}$$

Second moment of flange about axis X–X is

$$I_\text{flange} = I_\text{G} + Ah^2$$

$$= 8 \times 10^4 + 120 \times 20 \times 51.4^2$$

$$= 6.42 \times 10^6 \text{ mm}^4$$

Web: I for web about its own centroid is:

$$I_\text{G} = \frac{bd^3}{12} = \frac{20 \times 160^3}{12} = 6.83 \times 10^6 \text{ mm}^4$$

Distance of web centroid from axis X–X is

$$100 - 61.4 = 38.6 \text{ mm}$$

$$I_\text{web} \text{ about X–X} = I_\text{G} + Ah^2$$

$$= 6.83 \times 10^6 + 160 \times 20 \times 38.6^2$$

$$= 11.6 \times 10^6 \text{ mm}^4$$

Therefore,

$$\text{total } I \text{ of section about X–X} = I_\text{flange} + I_\text{web}$$

$$= 6.42 \times 10^6 + 11.6 \times 10^6$$

$$\mathbf{= 18.02 \times 10^6 \text{ mm}^4}$$

The solution can be conveniently set out in tabular form; thus, in millimetre units:

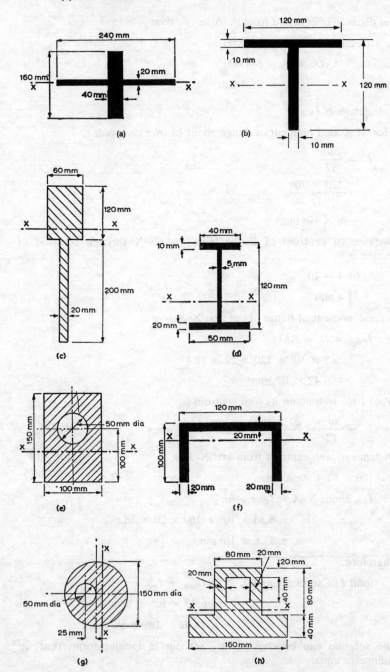

Fig. 17.18

Part	Flange	Web
b	120	20
d	20	160
A	2,400	3,200
I about own centroid $bd^3/12$	$\dfrac{120 \times 20^3}{12} = 0.08 \times 10^6$	$\dfrac{20 \times 160^3}{12} = 6.83 \times 10^6$
h	51.4	38.6
Ah^2	6.34×10^6	4.77×10^6
I about X–X	$0.08 \times 10^6 + 6.34 \times 10^6$ $= 6.42 \times 10^6$	$6.83 \times 10^6 + 4.77 \times 10^6$ $= 11.6 \times 10^6$
	Total $I_X = \mathbf{18.02 \times 10^6\ mm^4}$	

Problems

Calculate the second moment of area of each of the sections shown in Fig. 17.18 about an axis X–X through the centroid.

Answers:

(a) $13.8 \times 10^6\ mm^4$

(b) Centroid 33.7 mm from top flange face; $I = 3.19 \times 10^6\ mm^4$

(c) Centroid 117.2 mm from top edge; $I = 87.5 \times 10^6\ mm^4$

(d) Centroid 75 mm from top flange face; $I = 3.686 \times 10^6\ mm^4$

(e) Centroid 78.8 mm from top edge; $I = 26.4 \times 10^6\ mm^4$

(f) Centroid 38.6 mm from top edge; $I = 5.42 \times 10^6\ mm^4$

(g) Centroid 3.13 mm from centre of 150-mm circle; $I = 23.3 \times 10^6\ mm^4$

(h) Centroid 45.8 mm from base; $I = 13.9 \times 10^6\ mm^4$

Example *The beam section shown in Fig. 17.19 is subjected to a bending moment M kN-m acting in the sense shown. If the maximum tensile and compressive stresses are limited to 30 MN/m² and 45 MN/m², respectively, calculate the maximum allowable value of M.*

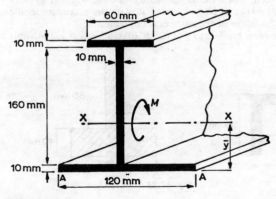

Fig. 17.19

SOLUTION

Using the method of the previous example it will be found that $\bar{y}$, the distance of the centroid from AA is 75 mm and $I_X = 15.66 \times 10^6$ mm⁴.

Using the formula: $M = \dfrac{\sigma}{y}I$

In tension,

$$y_{max} = 0.075 \text{ m and } \sigma_{max} = 30 \text{ MN/m}^2$$

therefore

$$M = \frac{30 \times 10^6 \times 15.66 \times 10^{-6}}{0.075} = 6{,}270 \text{ N-m}$$

$$= 6.27 \text{ kN-m}$$

In compression,

$$y_{max} = 0.105 \text{ m and } \sigma_{max} = 45 \text{ MN/m}^2$$

therefore

$$M = \frac{45 \times 10^6 \times 15.66 \times 10^{-6}}{0.105} = 6{,}720 \text{ N-m}$$

$$= 6.72 \text{ kN-m}$$

The maximum allowable bending moment must be the smaller of the two values found, i.e. **6.27 kN-m**.

Problems

1. Fig. 17.20 shows the section of a cantilever beam 3 m long, which is to carry a concentrated load of 500 kg at its free end. Calculate the maximum tensile stress in the beam if the flange of the channel is uppermost.

   (Centroid 27.2 mm from top face; $I = 3.025 \times 10^6$ mm⁴; stress 132 MN/m²)

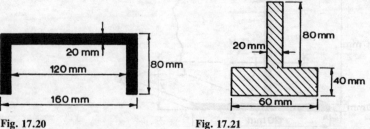

Fig. 17.20   Fig. 17.21

2. Calculate the maximum bending moment which can be carried by a cast-iron bracket having the T-section shown in Fig. 17.21, if the maximum bending stress permitted is 14 MN/m².

   (Centroid 44 mm from bottom face; $I = 4.63 \times 10^6$ mm⁴; moment 853 N-m)

3. A cantilever 5 m long has the section shown in Fig. 17.22. The short flange is uppermost and the beam carries a uniformly distributed load of 21 kN/m run. Calculate the stresses at the two faces of the short flange at the section where the beam is fixed, stating whether they are tensile or compressive.

   (Centroid 120.4 mm from bottom face; $I = 106.5 \times 10^6$ mm⁴; stresses, 295 and 196 MN/m² tensile)

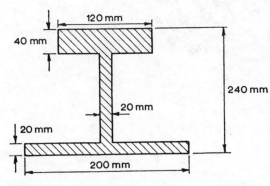

Fig. 17.22

4. A light bridge is to be supported by a number of beams of the section shown in Fig. 17.23 in which the top flange is uppermost. The beams are simply supported on a span of 7 m and carry a load of 20 kN at the mid-point. The maximum allowable stress in the beam material is 125 MN/m². Calculate the minimum number of beams required.

   (Centroid 56.4 mm from bottom face; $I = 8.56 \times 10^5$ mm⁴; number of beams, 18.5, say nineteen)

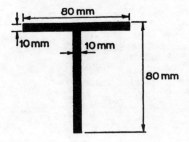

Fig. 17.23

5. In the cast-iron bracket lever shown in Fig. 17.24 the pull P is perpendicular to the bracket arm. Considering section X–X find the maximum allowable value of P if the stress due to bending is not to exceed 15 MN/m².

$(I = 56.2 \times 10^3$ mm⁴; 175 N)

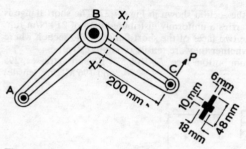

Fig. 17.24

CHAPTER 18

Combined Bending and Direct Stress

18.1 Principle of superposition

The stress at any point of a structure, beam or strut carrying several loads may be found by considering each load separately *as if it acted alone*. The total stress is then the algebraic sum of the stresses due to each separate load. This is the *method of superposition*.

A particular case is that of combined bending and direct stress due to a single load. The cantilever shown in Fig. 18.1(*a*) is subject to the

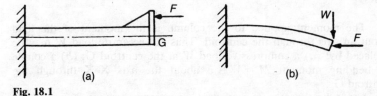

(a)   (b)

Fig. 18.1

axial load F offset from the centroid of cross-section G. This produces two effects: (*a*) a simple compression; (*b*) a bending moment about an axis through G. The stress due to each effect may be calculated separately and added to give the total stress. The method of superposition does not apply if one load alters the *character* or effect of another. For example, the cantilever, Fig. 18.1(*b*), subject to an axial load F and a transverse load W may deflect sufficiently under the transverse load so as to increase the moment due to F. Only when this deflexion is negligible does the principle of superposition apply.

18.2 Combined bending and direct stress of a loaded column

A short concrete column is loaded in compression by a concentrated load W at point A on the axis of symmetry, distant e from the centroid G of the cross-section, Fig. 18.2. It is required to find the maximum eccentricity of the load if there is to be no tensile stress in the concrete and to obtain the value for a rectangular section.

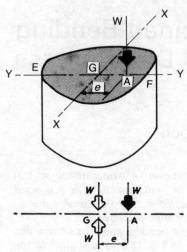

Fig. 18.2

The moment applied to the column is the moment of the load about an axis through the centroid. Thus the eccentric load W may be replaced by: (*a*) a compressive load W at the centroid G; (*b*) a couple or bending moment $M = W \times e$ about the axis X–X through the centroid G.

Let A be the area of section, I the second moment of area about X–X, y the distance of any point in the plane of the section from X–X:

direct compressive stress $= -\dfrac{W}{A}$

bending stress $= +\dfrac{M}{I}y$, tensile in EG when y is positive,

compressive in GF, when y is negative

total stress at any point
distant y from X–X $= -\dfrac{W}{A} + \dfrac{M}{I}y$

Fig. 18.3 shows the variation of direct, bending and total stress across the section. The effect of the eccentricity of loading is that the line of zero stress (neutral axis) is shifted to the left of the centroid. The form of the total stress diagram depends on the magnitudes of the direct and bending stresses.

The maximum tensile stress is at E. For no tension at any point

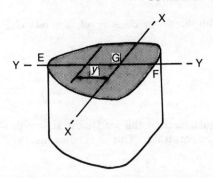

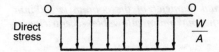

Direct stress

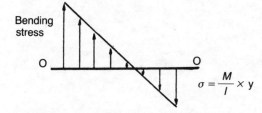

$\dfrac{W}{A}$

Bending stress

$\sigma = \dfrac{M}{I} \times y$

Total stress

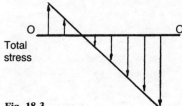

Fig. 18.3

the stress at E must be just zero. Now $M = W \times e$ and at E, $y = y_1$; therefore

$$\text{stress at E} = -\frac{W}{A} + \frac{W \times e}{I} y_1$$

$$= 0, \text{ for no tensile stress}$$

hence

$$\frac{Wey_1}{I} = \frac{W}{A}$$

or $\quad e = \dfrac{I}{Ay_1}$

For a rectangular section, breadth b, depth d, $A = bd$, $I = bd^3/12$, $y_1 = \frac{1}{2}d$. Therefore

$$e = \frac{bd^3/12}{bd \times \frac{1}{2}d}$$

$$= \frac{d}{6}$$

Thus for no tension in a rectangular section the load must act on or within the distance $d/6$ from the centroid. This is known as the *middle-third rule*.

Example   *A short hollow cast-iron column (Fig. 18.4) is to support a vertical load of 1 MN. The external diameter of the column is 250 mm and the thickness 25 mm. Find the maximum allowable eccentricity of*

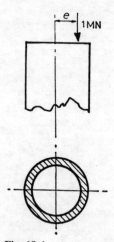

Fig. 18.4

this load if the maximum tensile stress is not to exceed 30 N/mm². What is then the value of the maximum compressive stress?

SOLUTION
(working throughout in N and mm; 1 MN = 10⁶ N).

$$A = \frac{\pi}{4}(250^2 - 200^2) = 17.7 \times 10^3 \text{ mm}^2$$

$$I = \frac{\pi}{64}(250^4 - 200^4) = 113.5 \times 10^6 \text{ mm}^4$$

$$\text{direct stress} = -\frac{1 \times 10^6}{17.7 \times 10^3} = -56.5 \text{ N/mm}^2 \text{ (compressive)}$$

bending moment = $10^6 e$, where e is the eccentricity of the load from the centroid in millimetres.

maximum bending stress $= \dfrac{My}{I} = \pm \dfrac{10^6 \times e \times 125}{113.5 \times 10^6}$

$$= \pm 1.1 \, e \text{ N/mm}^2$$

maximum tensile stress $= 1.1 \, e - 56.5 = 30 \text{ N/mm}^2$

therefore

$e = \textbf{78.5 mm}$

Maximum compressive stress $= -1.1 \, e - 56.5$

$$= -1.1 \times 78.5 - 56.5$$

$$= 143 \text{ N/mm}^2 = \textbf{143 MN/m}^2$$

Example *A screw clamp is tightened on a proving ring as shown, Fig. 18.5. From measurement of the deflexion of the ring the clamping force is estimated as 11 kN. Find the maximum tensile and compressive*

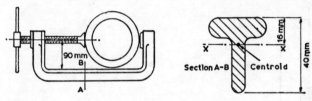

Fig. 18.5

stresses in the material at section A–B due to bending and direct loading. Area of section = 480 mm², $I_{XX} = 6.4 \times 10^4$ mm⁴.

SOLUTION

The resisting force exerted by the ring puts section A–B of the clamp under a direct tensile load of 11 kN and a bending moment $= 11 \times 10^3(0.09 + 0.016) = 1{,}166$ N-m about the axis X–X, through the centroid of section. Since the effect of the resisting force exerted by the ring is to tend to open out the clamp the top portion of the section is in tension and the bottom in compression.

Direct stress $= \dfrac{11 \times 10^3}{480 \times 10^{-6}} = 22.9 \times 10^6 \text{ N/m}^2$

$$= 22.9 \text{ MN/m}^2 \text{ tension}$$

bending stress at top face $= \dfrac{M}{I}y = \dfrac{1{,}166}{(6.4 \times 10^4 \times 10^{-12})} \times 0.016$

$$= 291.5 \times 10^6 \text{ N/m}^2$$

$$= 291.5 \text{ MN/m}^2 \text{ (tension)}$$

bending stress at bottom face $= -\dfrac{1{,}166}{6.4 \times 10^4 \times 10^{-12}} \times 0.024$

$$= - 437.3 \times 10^6 \text{ N/m}^2$$

$$= - 437.3 \text{ MN/m}^2 \text{ (compression)}$$

at top face, maximum stress $= 22.9 + 291.5$

$$= \textbf{314.4 MN/m}^2 \text{ (tension)}$$

at bottom face, maximum stress $= 22.9 - 437.3$

$$= - \textbf{414.4 MN/m}^2 \text{ (compression)}$$

Example *A uniform masonry chimney of outside diameter 3 m, inside diameter 2.4 m is subjected to a horizontal wind of load 1.8 kN/m of height. The weight of masonry is 17.5 kN/m³. Calculate the maximum chimney height to avoid tensile stress at the base section.*

SOLUTION

Total wind load on height h m (Fig. 18.6) $= 1,800h$ N

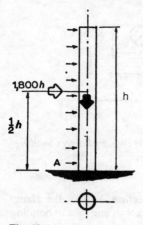

Fig. 18.6

bending moment at base section due to this load $= 1,800 \, h \times \frac{1}{2} \, h$

$$= 900 \, h^2 \text{ N-m}$$

I of section $= \dfrac{\pi}{64}(3^4 - 2.4^4) = 2.35 \text{ m}^4$

$y_{\max} = 1.5 \text{ m}$

maximum tensile stress at base $= \dfrac{My}{I} = \dfrac{900 \, h^2 \times 1.5}{2.35}$

$$= 575 \, h^2 \text{ N/m}^2 \text{ at point A}$$

total weight of chimney

$$= \text{volume} \times \text{specific weight}$$

= area of section × height × specific weight

= Ah × 17,500

compressive stress at base due to dead load = $\dfrac{Ah \times 17{,}500}{A}$

= 17,500 h N/m²

For no tensile stress at base, total stress at A = 0. Therefore

575 h^2 − 17,500 h = 0 or h = **30.4 m**

Problems

1. A 50-mm diameter tie bar carries a pull of 80 kN offset a distance of 3 mm from the axis of the bar. Calculate the maximum and minimum tensile stresses in the bar.

    (21.25, 60.35 MN/m²)

2. The cranked tie bar shown in Fig. 18.7 carries a load of F kN. Calculate the maximum value of F if the tensile stress in section X–X is limited to 75 MN/m².

    (90.3 kN)

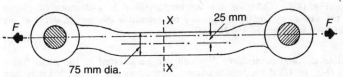

Fig. 18.7

3. A short cast-iron column of rectangular section 50 mm × 30 mm carries a load of F kN as shown in Fig. 18.8. Calculate the greatest value of F if the maximum *tensile* stress is limited to 15 MN/m².

    (2.8 kN)

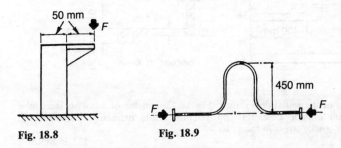

Fig. 18.8   **Fig. 18.9**

4. Fig. 18.9 shows an expansion loop in a steam pipe. The pipe has an external diameter of 100 mm and internal diameter of 90 mm. If the end load F is 5 kN calculate the maximum tensile stress in the pipe.

    (63.3 MN/m²)

5. The aircraft undercarriage shown in Fig. 18.10 is constructed of alloy tube. Calculate the maximum tensile and compressive stresses in the tube A if it is 50 mm outside diameter and 10 mm thick.

(416.2 MN/m² tensile, 480 MN/m² compressive)

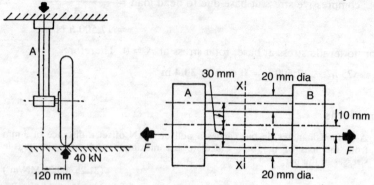

Fig. 18.10   Fig. 18.11

6. Two blocks A and B are connected by two short 20-mm diameter bars as shown, Fig. 18.11. Calculate the maximum pull F at a distance of 10 mm from the axis of symmetry which may be exerted if the tensile stress at the section X–X is limited to 150 N/mm².

(65.8 kN)

7. Calculate the maximum force F which can be exerted in the press frame shown (Fig. 18.12) if the tensile stress is limited to 100 N/mm². What is then the maximum compressive stress?

(1.3 MN; 50 MN/m²)

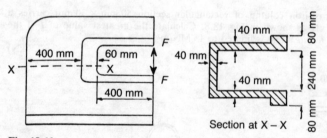

Fig. 18.12

8. The pillar of the radial drill shown, Fig. 18.13, is made of a hollow steel tube of 160 mm outside diameter and 120 mm inside diameter. Calculate the maximum tensile stress in the pillar.

(22.35 MN/m²)

9. A steel chimney is 36 m high, 1.5 m external diameter, 20 mm thick. It is rigidly fixed at the base and is acted upon by a horizontal wind pressure of intensity 1.1 kN/m² of projected area. Calculate the maximum stress in the steel at the base if steel weighs 74 kN/m².

(34.1 MN/m²)

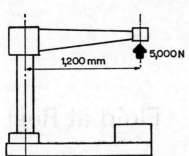

Fig. 18.13

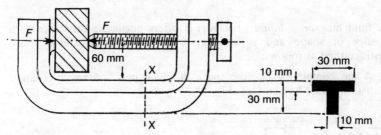

Fig. 18.14

10. The clamp shown, Fig. 18.14, exerts a force of F N on the workpiece. If the section X–X is to carry a maximum tensile stress of 65 N/mm² find the maximum clamping force.

(2.77 kN; centroid of section is 11 mm from top face, I about centroid = 36,200 mm⁴)

CHAPTER 19

Fluid at Rest

19.1 Fluid

A fluid may be a *liquid* or a *gas*; it offers negligible resistance to a change of shape and is capable of flowing. Liquid and gas are distinguished as follows:

1. A gas completely fills the space in which it is contained; a liquid usually has a free surface, Fig. 19.1.

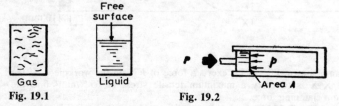

Fig. 19.1   **Fig. 19.2**

2. A gas is a fluid which can be compressed relatively easily; a liquid can be compressed only with difficulty.

Hydrostatics, or the statics of a fluid, is the study of force and pressure in a fluid at rest.

19.2 Pressure

A fluid in a closed cylinder (Fig. 19.2) may be put into a state of pressure by applying a force P to the piston shown. Neglecting the weight of the fluid, the pressure p in the fluid is the ratio:

$$\frac{\text{force } P \text{ on piston}}{\text{area } A \text{ of piston}}$$

or $p = \dfrac{P}{A}$

The basic SI unit of pressure is the same as that for stress, i.e. *newtons per square metre* (**N/m²**). The name *pascal* (**Pa**) is also given to this unit but we shall use only the form N/m² and the following multiples: kN/m² = 10³ N/m², MN/m² = 10⁶ N/m², GN/m² = 10⁹ N/m². For pressure (but not for stress) we shall use also the *bar* and its multiples. The bar is the name given to a special multiple of the unit N/m². Thus:

$$1 \ bar \qquad\qquad = 10^5 \ \text{N/m}^2$$
$$1 \ millibar \ (\text{mbar}) \ = 10^{-3} \ \text{bar} = 100 \ \text{N/m}^2$$
$$1 \ hectobar \ (\text{hbar}) \ = 100 \ \text{bar} \ = 10^7 \ \text{N/m}^2$$

Pressure in a fluid has the following important features:

1. The pressure at a point is the same in all directions.
2. The pressure exerted at a point on any surface is normal to the surface.

Fig. 19.3

Fig. 19.3 shows equal pressures acting on all surfaces of a very small body immersed in a fluid. The pressure is everywhere normal to the surface. Similarly the pressure exerted by a fluid on its container is everywhere normal to the vessel wall, Fig. 19.3.

19.3 Transmission of fluid pressure

A simple hydraulic press is shown in Fig. 19.4. A load of weight W is supported by a piston of area A in the cylinder D. Cylinder D is connected by a pipe to another cylinder E which, in turn, contains a

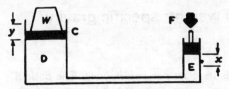

Fig. 19.4

piston of area a. The cylinders are filled with a liquid which is assumed incompressible. It is required to find the force F on the piston in cylinder E in order to hold the load W in equilibrium. To do this we make use of the *principle of work*.

Let the piston in cylinder D move down a distance x causing the piston in cylinder E to rise a distance y, then

volume of liquid leaving E $= a \times x$

volume of liquid entering D $= A \times y$

Since the liquid is incompressible

$$Ay = ax$$

or $\quad y = \dfrac{ax}{A}$

work done by force $F = F \times x$

work done on load $W = W \times y$

Neglecting friction, the work done by F must be equal to the work done on W, therefore

$$Fx = Wy$$

$$= W \times \frac{ax}{A}$$

or $\quad \dfrac{F}{a} = \dfrac{W}{A}$

But F/a is the pressure in cylinder E, and W/A is the pressure in cylinder D. Hence these two pressures are equal.

This is a demonstration of the *principle of transmission of pressure*, which states that the pressure intensity at any point of a fluid at rest is transmitted without loss to all other points of the fluid. The principle does not depend on frictionless pistons. The effect of friction is merely to increase the effort, F, required to hold the load W, above the theoretical value $W \times a/A$. This effect is true of machines in general.

Similarly, the principle of transmission of pressure does not depend on the fluid being incompressible. It applies to gases provided the forces are applied sufficiently slowly. (Owing to the compression of gas the work principle does not apply in a simple form.)

19.4 Density; specific weight; specific gravity; relative density

The *density* ρ of a substance is its *mass per unit volume*. The basic SI units of mass and volume are respectively the kilogram and the (metre)^3, so the basic units of density are *kilograms per cubic metre* (**kg/m³**). Other forms may be used, e.g. tonne/m^3, kg/litre, g/millilitre. Note that

$$1 \text{ litre} = 1 \text{ dm}^3 = \quad 10^{-3} \text{ m}^3$$
$$1 \text{ Mg/m}^3 = 1 \text{ tonne/m}^3 = 1 \text{ kg/litre} = 1 \text{ kg/dm}^3$$
$$1 \text{ kg/m}^3 = 1 \text{ g/litre} = \quad 1 \text{ g/dm}^3$$

The density of water is most important and should be remembered, thus,

density of water = 1,000 kg/m³ = 1 Mg/m³ = 1 tonne/m³

= 1 kg/litre

The *specific weight w* of a substance is its *weight per unit volume*. Since the weight of a substance is mass multiplied by the acceleration due to gravity, g, the relation between specific weight and density is

specific weight = density $\times g$

$$w = \rho g$$

Thus for water

specific weight of water = 1,000 × 9.8 = 9,800 N/m³ = 9.8 kN/m³

The *specific gravity s* of a substance is the ratio:

$$\frac{\text{weight of substance}}{\text{weight of equal volume of pure water}}$$

It may also be expressed as the ratio of the masses and is thus called, alternatively, *relative density*. It is a pure number and has no units. For example, the statement that mercury has a specific gravity of 13.59 means that it has a weight or mass 13.59 times as great as an equal volume of water.

In general, the specific weight w of a substance is given in terms of its specific gravity s, and the specific weight of water, by the expression

$$w = s \times \textit{specific weight of water}$$

Thus, if the specific gravity or relative density of petrol is 0.8 its specific weight is 0.8 × 9,800, or 7,840 N/m³, and its density is 0.8 × 1,000 = 800 kg/m³.

19.5 Pressure in a fluid due to its own weight

Consider a vertical tube of fluid, Fig. 19.5, of height h, uniform cross-sectional area A, specific weight w, and density ρ.

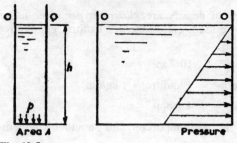

Fig. 19.5

Total weight of fluid column = specific weight × volume

$$= w \times Ah$$

pressure p at depth $h = \dfrac{\text{weight of column}}{\text{area of base}}$

$$= \frac{wAh}{A}$$

$$p = wh = \rho gh$$

It follows, therefore, that the pressure in a fluid due to its own weight is proportional to the depth h below the free surface O–O. Fig. 19.5 shows the vertical variation with depth of the pressure in an open tank of liquid. If the *atmospheric pressure* on the free surface of the liquid is p_a (Fig. 19.6) the total pressure p at depth h in the liquid is

$$p = p_a + wh$$

The pressure of the atmosphere is usually about 101.3 kN/m².

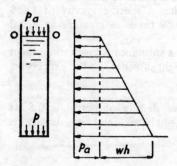

Fig. 19.6

19.6 Measurement of pressure

A container at zero *absolute pressure* is one which is completely empty.

The absolute pressure of a fluid is measured above this zero. The absolute pressure of the atmosphere is measured by a *barometer*, which consists of a tube sealed at the top and standing with its open end in a mercury bath open to the atmosphere, Fig. 19.7. Let the mercury rise to a height h above its free surface at C. By the principle of transmissibility of pressure, the upward pressure exerted by the mercury in the tube at D is equal to the downward pressure p_a exerted by the atmosphere on the free surface C. But

pressure at D $= wh$

where w is the specific weight of mercury. Hence at D

$p_a = wh$

or   atmospheric pressure = specific weight of barometer fluid

$\times$ 'height' h of barometer

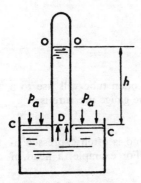

Fig. 19.7

For example, the height of a mercury barometer corresponding to an atmospheric pressure of 101.3 kN/m² is given by:

$$h = \frac{p_a}{w}$$

$$= \frac{p_a}{s \times specific\ weight\ of\ water}$$

$$= \frac{101.3 \times 10^3}{13.59 \times 9,800}$$

$$= 0.76 \text{ m or } 76 \text{ cm or } 760 \text{ mm}$$

Thus the pressure of the atmosphere is 101.3 kN/m² or 760 mm of mercury.

19.7 Measurement of gauge pressure

Gauge pressure is pressure measured above that of the atmosphere by a *manometer* or pressure gauge. Hence the absolute pressure is the sum of the gauge and atmospheric pressures,* thus:

absolute pressure = gauge pressure + atmospheric pressure

The simplest type of manometer is the *piezometer tube*, which is an open tube fitted into the top of the vessel containing liquid, whose pressure is to be measured, Fig. 19.8. If the vessel contains liquid under

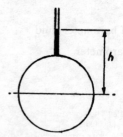

Fig. 19.8   Piezometer tube

pressure, the free surface of the liquid in the open tube will rise to a height h above the centre line of the vessel. The gauge pressure is:

$$p = wh = \rho gh$$

The height h is known as the *pressure head* and is often expressed in metres, centimetres, or millimetres of water. For example, a head of 10 mm of water corresponds to a pressure:

$$p = wh = 9,800 \times 10 \times 10^{-3} = 98 \text{ N/m}^2$$

Atmospheric pressure, 101.3 kN/m³, is equivalent to a head of water given by

$$h = \frac{p}{w} = \frac{101.3 \times 10^3}{9,800} \simeq 10.4 \text{ m}$$

A pressure head is often expressed as an equivalent *head of water*. Thus a head of 2 cm of oil of specific gravity 0.8 is equivalent to a head of $2 \times 0.8 = 1.6$ cm of water.

---

* All pressures stated in this book denote absolute pressures unless otherwise stated. Atmospheric pressure is normally taken as 101.3 kN/m² (1.013 bar). For large pressures the unit used for atmospheric pressure is the *atmosphere* = 101.3 kN/m².

19.8 Measurement of pressure differences

The pressure difference between two pipes containing a liquid may be measured by an inverted U-tube, formed by two piezometer tubes combined, Fig. 19.9. The pressure difference is measured by the distance h between the two liquid levels. For example, if the pipes and tube contain oil of specific gravity 0.8, a difference in head h of 10 mm corresponds to a pressure difference of

$$wh = 0.8 \times 9,800 \times 0.01 = 78.4 \text{ N/m}^2$$

The upper part of the tube is fitted with an air vent. By adjusting this vent the levels in the arms of the tube may be brought to a convenient part of the scale, whatever the pressure difference may be.

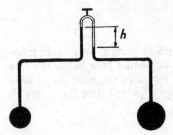

Fig. 19.9   Inverted U-tube

19.9 Total thrust on a vertical plane surface

Consider a plane surface of area A immersed vertically in a liquid of specific weight w. The pressure on one side of the surface is normal to the surface and gives rise to a resultant force or *thrust* on that side, Fig. 19.10.

Pressure p on one side of thin strip DE at depth $x = wx$

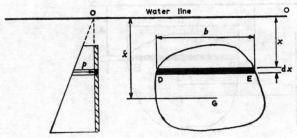

Fig. 19.10

area of strip of breadth b, thickness $dx = b \times dx$

$$\text{force on strip} = p \times b \ dx$$
$$= wx \times b \ dx$$
$$= wbx \ dx$$

total force F on area $A = \int wbx \ dx = w \int bx \ dx$

But $\int bx \ dx$ is the total moment of area A about an axis through O in the water surface:

$$= A \times \bar{x}$$

where $\bar{x}$ is the depth of the centroid G of the plane surface below the water line. Hence

$$F = w \times A \bar{x}$$
$$= wA \bar{x} = \rho g A \bar{x}$$

Thus the total thrust on an immersed plane vertical surface is proportional to the depth of the centroid of the wetted area below the free surface. Note, however, that the line of action of the total thrust does not pass through the centroid but through a point called the *centre of pressure*, which has yet to be found.

19.10 Centre of pressure

The centre of pressure is the point of application of the resultant force due to fluid pressure on one face of an immersed surface. To determine the depth $\bar{y}$ of the centre of pressure we use the *principle of moments*. The sum of the moments about the water surface O–O (Fig. 19.11) of the forces on all the thin strips such as DE must equal the moment $F \times \bar{y}$ of the resultant thrust F about O–O.

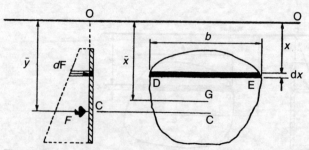

Fig. 19.11

Force on strip DE = pressure × area of strip

or $$\mathrm{d}F = p \times b \, \mathrm{d}x$$

$$= wx \times b \, \mathrm{d}x, \text{ since } p = wx$$

Moment of this force about O–O is:

$$\mathrm{d}F \times x$$

$$= wxb \, \mathrm{d}x \times x$$

$$= wx^2b \, \mathrm{d}x$$

total moment about O–O $= \int wx^2b \, \mathrm{d}x$

and this is equal to the moment $F \times \bar{y}$ of the resultant force F about O–O. Hence

$$F \times \bar{y} = w \int x^2b \, \mathrm{d}x, \text{ since } w \text{ is a constant}$$

But $\int x^2b \, \mathrm{d}x$ is the total second moment of area I_O of the area A about O–O, thus

$$F \times \bar{y} = w \times I_O$$

Now $F = wA\bar{x}$, where $\bar{x}$ is the distance of the centroid G below O–O. Thus

$$\bar{y} = \frac{wI_O}{P}$$

$$= \frac{wI_O}{wA\bar{x}}$$

$$= \frac{I_O}{A\bar{x}}$$

$$= \frac{\text{second moment of area about O–O}}{\text{first moment of area about O–O}}$$

Note that this refers to *wetted* area only. This last expression is conveniently re-written as follows:

Let I_G = second moment of area A about axis through G parallel to water surface O–O

$$= Ak^2$$

where k is the corresponding radius of gyration. Then the parallel axis theorem for second moments of area states:

$$I_O = I_G + A\bar{x}^2$$

Hence

$$\bar{y} = \frac{I_O}{A\bar{x}}$$

$$= \frac{I_G + A\bar{x}^2}{A\bar{x}}$$

$$= \frac{Ak^2 + A\bar{x}^2}{A\bar{x}}$$

$$= \frac{k^2 + \bar{x}^2}{\bar{x}}$$

$$= \frac{k^2}{\bar{x}} + \bar{x}$$

Thus the distance of the centre of pressure C below the centroid G is

$$GC = \bar{y} - \bar{x}$$

$$= \frac{k^2}{\bar{x}}$$

This expression tends to zero as the distance $\bar{x}$ becomes very large. Hence the centre of pressure is always below the centroid of the wetted area, but tends to coincide with the centroid at very great depth.

The second moment of area I_G and the corresponding k^2 for rectangular and circular areas are shown in Fig. 19.12.

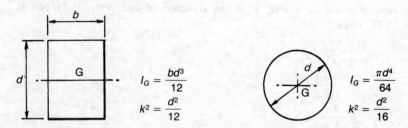

Fig. 19.12

Example *A lock gate has sea-water to a depth of 3.6 m on one side and 1.8 m on the other. Find (a) the resultant thrust per metre width on the gate, (b) the resultant moment per metre width tending to overturn the gate at its base, Fig. 19.13. Specific weight of sea-water = 10 kN/m³.*

SOLUTION
On left-hand side:

depth of centroid, $\bar{x} = 1.8$ m

total force F_1 from left to right $= wA\bar{x}$

$$= 10 \times 3.6 \times 1 \times 1.8$$

$$= 64.8 \text{ kN/m width}$$

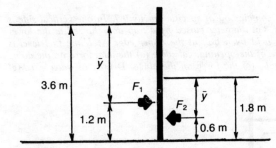

Fig. 19.13

    depth of centre of pressure, $\bar{y} = \bar{x} + \dfrac{k^2}{\bar{x}}$

where

$$k^2 = \frac{d^2}{12}$$

$$= \frac{3.6^2}{12} = 1.08 \text{ m}^2$$

therefore

$$\bar{y} = 1.8 + \frac{1.08}{1.8}$$

$$= 2.4 \text{ m}$$

Thus the total thrust on the left-hand side acts at 1.2 m from the base. Moment of total force about the base is:

$$64.8 \times 1.2 = 77.76 \text{ kN-m/m width}$$

Similarly, on right-hand side:

$$\bar{x} = 0.9 \text{ m}$$

$$F_2 = 10 \times 1.8 \times 1 \times 0.9 = 16.2 \text{ kN/m width}$$

$$\bar{y} = 0.9 + \frac{1.8^2}{12} \times \frac{1}{0.9} = 1.2 \text{ m}$$

moment of force F_2 about base = 16.2×0.6

$$= 9.72 \text{ kN-m/m width}$$

resultant thrust (from left to right) = $F_1 - F_2$

$$= 64.8 - 16.2$$

$$= \textbf{48.6 kN/m}$$

net overturning moment = $77.76 - 9.72$

$$= \textbf{68.04 kN-m/m width, clockwise}$$

Example   *A fuel tank contains oil of specific gravity 0.7. In one vertical side is cut a circular opening 1.8 m diameter closed by a trap door hinged at the lower end B (Fig. 19.14) and held by a bolt at the upper edge A. If the fuel level is 1.8 m above the top edge of the opening, calculate (a) the total force on the door, (b) the force P in the bolt, (c) the force on the hinge. Density of water = 1 Mg/ m³.*

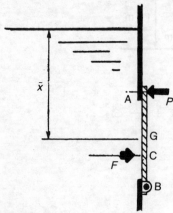

Fig. 19.14

SOLUTION

(a) Specific weight of fuel, w = density of water $\times g \times s$

$$= 1 \times 9.8 \times 0.7$$

$$= 6.86 \text{ kN/m}^3$$

depth of centroid of door, $\bar{x} = 1.8 + 0.9 = 2.7$ m

total force F on door $= wA\bar{x}$

$$= 6.86 \times \frac{\pi}{4} \times 1.8^2 \times 2.7$$

$$= \mathbf{47.2 \text{ kN}}$$

(b) Depth of centre of pressure C below centroid G is:

$$\frac{k^2}{\bar{x}} = \frac{d^2}{16\,\bar{x}}$$

therefore

$$\text{GC} = \frac{1.8^2}{16 \times 2.7} = 0.075 \text{ m}$$

Moments about hinge B:

$$P \times \text{AB} = F \times \text{CB}$$

$$P \times 1.8 = 47.2 \times (0.9 - 0.075)$$

$$P = \mathbf{21.6 \text{ kN}}$$

(*c*) Resultant horizontal force on hinge

$$= \text{water thrust} - \text{bolt force}$$

$$= F - P$$

$$= 47.2 - 21.6$$

$$= \mathbf{25.6\ kN}$$

Problems

1. A tank 1.2 m high, 0.9 m wide and 2.4 m long is filled with water. Find the total force and the position at which it acts for (*a*) an end, (*b*) a side, (*c*) the base. Density of water = 1 Mg/m³.

   ((*a*) 6.35 kN at 0.4 m above base; (*b*) 16.95 kN at 0.4 m above base; (*c*) 25.4 kN at mid-point)

2. A lock gate is of rectangular section 7.2 m wide. The depth of water on the lower side is 2.4 m, the depth on the opposite side is *h* m. The maximum allowable resultant thrust is 1.5 MN. Calculate the least value of *h* if this thrust is not to be exceeded.

   (6.95 m)

3. A sluice gate 0.9 m wide, 1.8 m deep, weighs 3.5 kN and works in vertical guides. The water surface on one side of the gate is 0.3 m below the upper edge of the gate. If the coefficient of friction between the gate and the guides is 0.4 find the force required just to lift the gate. (Specific weight of water 9,800 N/m³.)

   (7.47 kN)

4. A dock gate 18 m wide has sea-water of specific weight 10 kN/m³ to a depth of 6 m on one side and 3 m on the other. Find (*a*) the resultant thrust on the gate, (*b*) the resultant moment tending to overturn the gate about its lower edge.

   (2.43 MN; 5.67 MN-m)

5. A dock gate 6 m high and 9 m wide is pivoted at its base A and held vertical by a cable at the top of the gate, Fig. 19.15. There is a total depth of 3 m of salt water of density 1.03 Mg/m³ on one side of the gate. Find the tension in the cable.

   (79 kN)

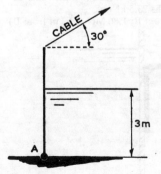

Fig. 19.15

6. A tank of water has vertical sides and a rectangular opening in one end. The opening is covered by a door hinged along its top edge A (Fig. 19.16) and held by four bolts at the lower end B. The opening is 0.6 m wide and 1.5 m deep and the hinge at A is 1.2 m below the water surface. Find (*a*) the depth at which the resultant water thrust acts on the door, (*b*) the load on each bolt, (*c*) the load on the hinge at A. Density of water = 1 kg/dm³.

(2.046 m; 2.42 kN; 7.52 kN)

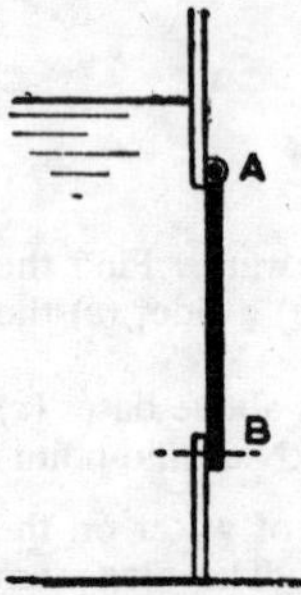

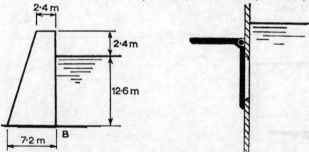

Fig. 19.16

7. A tank with vertical sides contains oil of specific gravity 0.8 to a total depth of 1.8 m. A hole 0.6 m diameter is covered by a trap door, the lowest point of which is level with the base of the tank. Find (*a*) the total thrust on the door, (*b*) the depth below the surface of the line of action of this thrust. Density of water = 1 Mg/m³.

(3.33 kN; 1.515 m)

8. A vertical dock gate of rectangular section 2.4 m deep and 1.2 m wide pivots about a hinge at its lower edge. It is held in position by four 12-mm diameter cables attached to its upper edge. The cables are at 45° to the horizontal. If sea-water weighs 10 kN/m³ calculate (*a*) the maximum thrust on the dock gate, (*b*) the stress in the cables, (*c*) the vertical and horizontal forces on the hinge.

(34.56 kN; 36.1 MN/m²; 11.55 kN; 23 kN)

9. A concrete reservoir dam has a vertical face on the water side and dimensions as shown, Fig. 19.17. The top water level reaches 2.4 m below the crest of the dam. Find the magnitude of the resultant thrust per metre width on the base (*a*) when the reservoir is empty, (*b*) when the reservoir is full. State in each case where the line of action of the resultant thrust cuts the base. Water weighs 9.8 kN/m³, concrete weighs 23.5 kN/m³.

(1.69 MN; 2.6 m from B; 1.86 MN, 4.53 m from B)

Fig. 19.17   **Fig. 19.18**

10. Fig. 19.18 shows a pivoted sluice gate of mass 1,450 kg covering a rectangular opening 0.9 m by 0.9 m. The pivot of the gate is 1.2 m above the base of the opening. The gate is just about to open when the head of water is 1.35 m above the base of the opening. Find the distance of the centre of gravity of the gate from the pivot. Specific weight of water = 9.8 kN/m³.

(415 mm)

19.11 Inclined surface

The method of finding the total force on an inclined surface, and the depth of the centre of pressure, is similar to that for the vertical surface. The variable distance x is now measured along the plane of the incline however, Fig. 19.19.

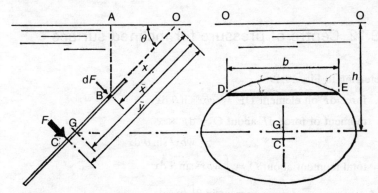

Fig. 19.19

Consider element DE of the inclined plane, distant x from the water line O–O, x being measured along the incline. The pressure p on the element at vertical depth AB is given by:

$$p = w \times AB$$

$$= w \times x \sin \theta$$

where θ is the acute angle made by the plane with the water surface.

Force dF on the element = $p \times$ area of element

$$= wx \sin \theta \times b \, dx$$

$$= wbx \sin \theta \, dx$$

where b is width of element and dx = thickness of element.

Total force F on surface = $\int dF$

$$= \int wbx \sin \theta \, dx$$

$$= w \sin \theta \int bx \, dx$$

But $\int bx \, dx$ = total moment of area A about O–O = $A \times \bar{x}$

where $\bar{x}$ is the distance OG of centroid G from O–O and A is the area of *wetted* surface. Therefore

$$F = wA \bar{x} \sin \theta$$

Now $\bar{x} \sin \theta$ is the *vertical* depth h of centroid G below O–O. Thus

$$F = wAh = \rho gAh$$

The total force on the inclined surface is determined by the vertical depth h of the centroid. The force is, however, normal to the surface at C, the centre of pressure.

19.12 Centre of pressure for inclined surface

Referring to Fig. 19.19:

force dF on element DE = $wbx \sin \theta \, dx$

moment of force dF about O = $dF \times x$

$$= wbx^2 \sin \theta \, dx$$

total moment about O = $\int wbx^2 \sin \theta \, dx$

$$= w \sin \theta \int bx^2 \, dx$$

$$= w \sin \theta \times I_O$$

where $I_O = \int bx^2 \, dx$ = total second moment of area A about water line O–O. The total moment exerted by the elementary forces is equal to the moment of the resultant force F about O, i.e.

$$wI_O \sin \theta = F \times \bar{y}$$

where $\bar{y}$ is the distance OC of centre of pressure C from O, measured along the inclined surface. But

$$F = wA \bar{x} \sin \theta$$

hence

$$wI_O \sin \theta = wA \bar{x} \sin \theta \times \bar{y}$$

thus

$$\bar{y} = \frac{I_O}{A \bar{x}}$$

From the parallel axis theorem

$$I_O = I_G + A\bar{x}^2 = A(k^2 + \bar{x}^2)$$

where I_G is the second moment of area about the centroid G, and k the radius of gyration about G. Therefore

$$\bar{y} = \frac{A(k^2 + \bar{x}^2)}{A\bar{x}}$$

or $\quad \bar{y} = \bar{x} + \dfrac{k^2}{\bar{x}}$

which is the same as for a vertical surface. Note, however, that $\bar{x}$ and $\bar{y}$ are now measured *along the inclined surface*. It follows that the distance GC of the centre of pressure from the centroid of area is given by

$$GC = \frac{k^2}{\bar{x}}$$

Example *In Fig. 19.20 QR represents a trap-door, 3 m wide and 2.4 m deep, in the side of a water tank. It is pivoted at Q and held against water pressure by eight bolts at R. Calculate the force in each bolt. The water level is 1.8 m vertically above R. Density of water = 1,000 kg/m³.*

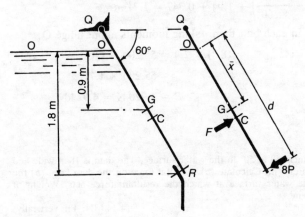

Fig. 19.20

SOLUTION
The vertical depth of centroid G of *wetted* surface OR is 0.9 m and the area of wetted surface is

$$A = \frac{1.8}{\sin 60°} \times 3 = 6.23 \text{ m}^2$$

total force F on surface OR $= \rho g A h$

$$= 1,000 \times 9.8 \times 6.23 \times 0.9$$
$$= 55,200 \text{ N}$$

The distance $\bar{x}$ of centroid G of *wetted* surface from water level O–O is

$$\frac{0.9}{\sin 60°} = 1.04 \text{ m}$$

For wetted area

$$k^2 = \frac{d^2}{12}$$
$$= \frac{1}{12}\left(\frac{1.8}{\sin 60°}\right)^2$$
$$= 0.36 \text{ m}^2$$

Distance GC of centre of pressure C from centroid G is

$$\frac{k^2}{\bar{x}} = \frac{0.36}{1.04} = 0.347 \text{ m}$$
$$QC = QO + OG + GC$$
$$= \left(2.4 - \frac{1.8}{\sin 60°}\right) + 1.04 + 0.347 = 1.71 \text{ m}$$

If P is the force in each bolt then, taking moments about hinge Q;

$$(8P) \times 2.4 = 55,200 \times QC$$
$$= 55,200 \times 1.71$$

hence $\qquad P = 4,920 \text{ N} = \textbf{4.92 kN}$

Problems

1. A dam face is inclined at 50° to the water surface. The dam is 18 m wide and the depth of water 6 m. Calculate (*a*) the total force on the dam face, (*b*) the depth below the water surface at which the resultant force acts. Weight of water = 9.8 kN/m³.

(4.14 MN; 4 m vertically)

2. The side of a tank makes 45° with the water surface. A trap-door 0.3 m diameter is hinged at a point 0.9 m below the water surface and bolted against the water pressure at the lowest point of the door. Calculate (*a*) the force in the bolt, (*b*) the load on the hinge. Density of water = 1 Mg/m³.

(356 N; 340 N)

3. A tank of oil of specific gravity 0.8 contains a rectangular trap-door 3 m wide and 2.4 m deep. The door makes an angle of 110° with the line of the oil surface and is hinged at a point 0.6 m vertically below the surface. If the door is held against oil pressure by four bolts along its lower edge find the force in each bolt. Specific weight of water = 9,800 kN/m³.

(14.85 kN)

CHAPTER 20

Fluid in Motion

The science of a fluid in motion is most conveniently studied using energy methods. It will be found that, in addition to potential and kinetic energy, a fluid may possess *pressure energy* by virtue of the work done in introducing it into a container under pressure. The fluid will be assumed to be an incompressible liquid; the flow of a compressible gas is not considered.

20.1 Pressure energy

Consider a tank of liquid with free surface at B, Fig. 20.1. We calculate the work done in introducing an additional volume of liquid at level A

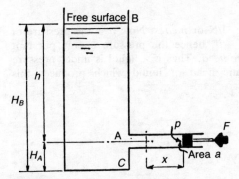

Fig. 20.1

into the tank against the pressure p at A. The liquid is to be forced into the tank by means of a small piston of area a; the head h of liquid (and thus pressure p) is assumed constant while this is done. The area a is assumed small enough for pressure p to be considered uniform across the face of the piston. Friction between the piston and cylinder is neglected. The force on the piston is:

$$F = p \times a$$

The work done in small slow displacement x is:

$$F \times x = pa \times x$$

Since, if the piston is released, both fluid and piston would be forced out by the fluid pressure this work is not lost but is recoverable. It therefore represents energy possessed by the fluid forced into the tank, i.e. pressure energy.

Volume of liquid entering tank = area of cylinder × length

$$= a \times x$$

Mass of liquid entering tank = density × volume

$$= \rho \times ax$$

Hence, energy possessed by *unit mass* of liquid is:

$$\frac{\text{work done on liquid}}{\text{mass of liquid}} = \frac{pax}{\rho ax}$$

$$= \frac{p}{\rho}$$

This is the pressure energy of unit mass of liquid and since p is in newtons per square metre (N/m^2), ρ in kilograms per cubic metre (kg/m^3), the units are newton-metres per kilogram (N-m/kg), i.e. *joules per kilogram* (**J/kg**).

The pressure energy *per unit weight* is

$$\frac{p}{\rho g} = \frac{p}{w}$$

and the units are N-m/N, i.e. J/N or *metres*. Now p is the pressure of liquid at depth h, and $p/w = h$, hence the pressure energy per unit weight is termed the *pressure head*. Thus if a liquid is under pressure p, h is the equivalent static head of liquid which produces this pressure.

20.2 Potential energy

Since pressure at depth h is

$$p = wh = \rho gh$$

therefore

$$h = \frac{p}{w} \text{ and } gh = \frac{p}{\rho}$$

Now, since $\dfrac{p}{\rho}$ is the pressure energy per unit mass, the quantity gh must also represent energy. It is in fact equal in magnitude to the work done against gravity in raising unit mass of the liquid slowly from A to B, Fig. 20.1. In rising from A to B a particle of liquid of unit mass would gain *potential energy* of amount $1 \times gh$. In falling from B to A it would lose potential energy and gain a corresponding amount of pressure energy. Similarly, since $\dfrac{p}{w}$ is the pressure energy per unit weight, the head h also represents energy per unit weight, i.e. h is the *potential energy per unit weight* of liquid at B. Thus potential and pressure energy may be converted one into the other, i.e. *in a liquid at rest*:

potential energy + pressure energy = constant

i.e. $h + \dfrac{p}{w} =$ constant, per unit weight

and $gh + \dfrac{p}{\rho} =$ constant, per unit mass

In any change, a gain in one term will be balanced by a corresponding loss in the other, to keep the sum of the two constant. A datum level must be fixed for measuring potential energy. We have measured h from level A so that we have assumed level A as datum. Thus at level B, the potential energy (per unit weight) is h and the pressure energy is zero; at level A the potential energy is zero and the pressure energy is $p/w = h$. If however we take a datum level for measuring potential energy at some other level C, Fig. 20.1, then the potential energy per unit weight of liquid at level A is H_A where H_A is the height of level A above the datum position. Similarly at B the potential energy is H_B. Thus, relative to datum level C,

total energy at A $= H_A + \dfrac{p_A}{w}$

total energy at B $= H_B = H_A + h = H_A + \dfrac{p_A}{w}$

In general we may write, for any point *in a liquid at rest*:

$H + \dfrac{p}{w} =$ **constant, per unit weight**

and $gH + \dfrac{p}{\rho} =$ **constant, per unit mass**

where H is the potential energy per unit weight (i.e. height) measured above a convenient datum position.

The units of potential energy are the same as those of pressure energy, i.e. per unit mass, J/kg, and per unit weight, J/N or m.

20.3 Kinetic energy

Now let the piston of Fig. 20.1 be removed so that liquid may flow freely from the opening (orifice) at A. Consider a particle of unit mass falling freely from the free surface at B to level A and then escaping with velocity v. The potential energy lost by the particle in falling is balanced by the kinetic energy gained in attaining a velocity v just outside the tank at A.

potential energy lost per unit mass $= gh$

kinetic energy gained per unit mass $= \frac{1}{2}v^2$

potential energy lost $=$ kinetic energy gained

i.e.
$$gh = \frac{v^2}{2}$$

$$h = \frac{v^2}{2g}$$

$$v = \sqrt{2gh}$$

Hence there is now an interchange of potential and kinetic energy. h is the potential energy per unit weight therefore $\frac{v^2}{2g}$ is the *kinetic energy per unit weight* and is termed the *velocity head*. Note that if a liquid has a velocity v, then h is the equivalent head of static liquid to produce this velocity at an opening. The units of kinetic energy per unit mass are J/kg. The units of kinetic energy per unit weight are J/N or m.

We have assumed level A as datum for potential energy. If we take level C as datum, Fig. 20.1, then per unit weight:

potential energy lost $=$ kinetic energy gained

$$(H_B - H_A) = \frac{v^2}{2g}$$

i.e.
$$h = \frac{v^2}{2g}, \text{ as before}$$

20.4 Interchange of pressure and kinetic energy

It is not necessary for any particle to have actually fallen from B to A before flowing from the orifice. If already at the level A it will possess pressure energy $\frac{p}{w}$ and thus in turn may be converted into kinetic energy on escaping. Thus we may write (per unit weight):

pressure energy lost = kinetic energy gained

or

$$\frac{p}{w} = \frac{v^2}{2g}$$

This assumes that p is the pressure measured above that of the atmosphere outside the jet of fluid at A, i.e. the gauge pressure. This is usually the case.

20.5 Bernoulli's equation (conservation of energy)

We have seen that, for an incompressible liquid in motion, there are three forms of energy to be considered:

(a) pressure energy, $\dfrac{p}{w}$;

(b) kinetic energy, $\dfrac{v^2}{2g}$;

(c) potential energy, H, per unit weight;

and there may be an interchange between any of these forms of energy. This is expressed by *Bernoulli's equation* which states that, for a liquid in motion:

pressure energy + kinetic energy + potential energy = constant

or, for unit weight of liquid

$$\frac{p}{w} + \frac{v^2}{2g} + H = constant$$

and, for unit mass of liquid

$$\frac{p}{\rho} + \frac{v^2}{2} + gH = constant$$

where the potential energy per unit weight (height) H is measured above an arbitrary datum level. In calculations we shall use the first form of the equation since engineers find it useful to deal with 'heads', i.e. pressure head, p/w, velocity head, $v^2/2g$, and head (height of liquid) H. Note that p is the *gauge* pressure.

Bernoulli's equation shows that a 'loss' or reduction in one term is always balanced by an increase in one or both of the other energy terms. Thus a drop in pressure p may accompany a corresponding increase in height H, or in velocity v. The equation holds in the above form provided:

(*a*) there is no loss of energy by friction or leakage;
(*b*) the motion is not turbulent or fluctuating.

Bernoulli's equation is a statement of the *principle of conservation of energy* for the particular case of a liquid in steady motion.

20.6 Pipe flow: equation of continuity

If suffices 1, 2 denote two points A, B, respectively, in a pipe, Fig. 20.2, then for a liquid flowing in the pipe from A to B, Bernoulli's equation is

$$\frac{p_1}{w} + \frac{v_1^2}{2g} + H_1 = \frac{p_2}{w} + \frac{v_2^2}{2g} + H_2 = \text{constant}$$

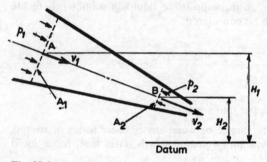

Fig. 20.2

This equation alone is usually insufficient to solve problems on pipe flow. We must find a second equation using the assumption that the liquid is incompressible.

If the liquid is incompressible the volume of liquid passing A per second must be the same as the volume passing B per second. Denoting the cross-sectional areas of the pipe by A_1 and A_2:

volume passing A per second = area of pipe × velocity of liquid

$$= A_1 \times v_1$$

volume passing B per second $= A_2 \times v_2$

Equating:

$$A_1 \times v_1 = A_2 \times v_2$$

or
$$\frac{v_1}{v_2} = \frac{A_2}{A_1}$$

This is known as the *equation of continuity* and, when the pipe dimensions are known, gives the ratio of the velocities at any two points in the pipe. The equation states that the velocity of flow is inversely proportional to the area of pipe section.

20.7 Flow-rate

The volumetric flow-rate Q is the quantity of liquid flowing per second. Thus:

Q = volume per second

$\quad = Av$

If A is in square metres (m^2), v in metres per second (m/s), the flow-rate Q is in cubic metres per second (m^3/s).

The corresponding mass flow-rate per second is

$$\dot{m} = \rho \times Q = \rho Av = s \times \text{density of water} \times Av$$

where ρ is the density of the liquid and s is the relative density (specific gravity) of the liquid.

The *weight* of liquid flowing per second is

$$wAv = \rho gAv = s \times \text{specific weight of water} \times Av$$

where w is the specific weight of the liquid. Note that for water

$$Q(m^3/s) = Q \times 10^3 \ dm^3/s$$

and $\quad \dot{m} = Q \times 10^3 \ kg/s = Q \ Mg/s$

since the density of water is $1 \ kg/dm^3$.

20.8 Variation in pressure head along a pipe

Fig. 20.3 shows the pressure head at two points A and B of an inclined pipe connected to an open tank and containing fluid *at rest*. The *pressure head* of the liquid is measured by the head h in the piezometer tubes, since

$$h = \frac{p}{w}$$

The potential energy per unit weight of the fluid at the free surface in the tank is H, and at point A the potential energy is H_1. Since the fluid is at rest the difference $H - H_1$ corresponds to the pressure head, i.e.

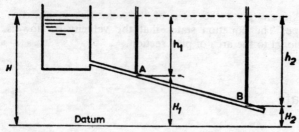

Fig. 20.3

$$H - H_1 = h_1 = \frac{p_1}{w}$$

Similarly for point B, we have

$$H - H_2 = h_2 = \frac{p_2}{w}$$

hence

$$H = H_1 + h_1 = H_1 + \frac{p_1}{w}$$

$$= H_2 + h_2 = H_2 + \frac{p_2}{w}$$

Fig. 20.4 shows the corresponding piezometer head levels for a *flowing* liquid in a pipe of uniform section. Since the pipe has constant

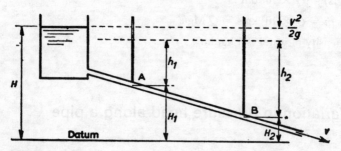

Fig. 20.4

area of section the velocity v and, therefore, the velocity head $v^2/2g$, is constant along the pipe. Neglecting friction and losses, the total energy is the same at all points along the pipe. If the liquid in the tank is assumed at rest, then Bernoulli's equation is

energy at free surface = energy at A = energy at B

i.e. $H = H_1 + \dfrac{p_1}{w} + \dfrac{v^2}{2g} = H_2 + \dfrac{p_2}{w} + \dfrac{v^2}{2g}$

Fig. 20.5 shows the variation in head along a horizontal diverging pipe connected to a cylinder of liquid at constant pressure p. Since the area at B is greater than that at A, the velocity head $v^2/2g$ decreases between A and B and the pressure head h shows a corresponding increase. As before, the total energy remains constant along the pipe.

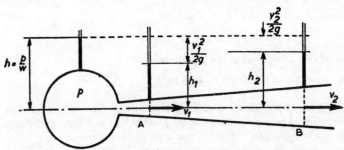

Fig. 20.5

Example *Oil flows along a horizontal pipe which varies uniformly in section from 100 mm diameter at A to 150 mm diameter at B. At A the gauge pressure is 126 kN/m² and at B 140 kN/m². Find the flow rate in litres per second and kilograms per second. The density of water is 1,000 kg/m³ and the relative density of the oil is 0.8.*

SOLUTION

Area of section at A $= \dfrac{\pi}{4} \times 0.1^2 = 7.854 \times 10^{-3}$ m²

area of section at B $= \dfrac{\pi}{4} \times 0.15^2 = 17.7 \times 10^{-3}$ m²

From equation of continuity:

$$\frac{v_1}{v_2} = \frac{A_2}{A_1}$$

therefore

$$v_1 = \frac{17.7}{7.854} \times v_2 = 2.25 v_2$$

Specific weight of oil, $w = 0.8 \times 1,000 \times 9.8 = 7,840$ N/m³

Applying Bernoulli's theorem to points A and B:

$$\frac{p_1}{w} + \frac{v_1^2}{2g} = \frac{p_2}{w} + \frac{v_2^2}{2g}$$

i.e. $\dfrac{126 \times 10^3}{7,840} + \dfrac{v_1^2}{2 \times 9.8} = \dfrac{140 \times 10^3}{7,840} + \dfrac{v_2^2}{2 \times 9.8}$

thus

$$v_1^2 - v_2^2 = 35$$

Substituting for v_1 in terms of v_2:

$$(2.25v_2)^2 - v_2^2 = 35$$

hence $\qquad v_2 = 2.94 \text{ m/s}$

therefore $\qquad Q = A_2 v_2$

$$= 17.7 \times 10^{-3} \times 2.94$$

$$= 0.052 \text{ m}^3\text{/s} = \textbf{52 litres/s}$$

The mass of one litre of oil is 0.8 kg, therefore

mass flow rate $= 52 \times 0.8 = \textbf{41.6 kg/s}$

Example　*Water flows down a sloping pipe which has one end 1.3 m above the other. The pipe section tapers from 0.9 m diameter at the top end A to 0.45 m diameter at the lower end B. The flow of water is 9 tonne/min. Find the difference in pressure between A and B in kilonewtons per square metre. Density of water = 1 kg/litre.*

SOLUTION

Area of pipe at A $= \dfrac{\pi}{4} \times 0.9^2 = 0.637 \text{ m}^2$

area of pipe at B $= \dfrac{\pi}{4} \times 0.45^2 = 0.159 \text{ m}^2$

$\dot{m} = 9{,}000 \text{ kg/min}$

$Q = 9{,}000 \text{ litres/min}$

$\quad = \dfrac{9{,}000 \times 10^{-3}}{60} \text{ m}^3\text{/s}$

$\quad = 0.15 \text{ m}^3\text{/s}$

From equation of continuity:

$$Q = A_1 v_1 = A_2 v_2$$

therefore

$$0.15 = 0.637 v_1 = 0.159 v_2$$

and $v_1 = 0.235 \text{ m/s}$

$\quad v_2 = 0.944 \text{ m/s}$

Applying Bernoulli's equation:

$$H_1 + \frac{p_1}{w} + \frac{v_1^2}{2g} = H_2 + \frac{p_2}{w} + \frac{v_2^2}{2g}$$

thus

$$1.3 + \frac{p_1}{9{,}800} + \frac{0.235^2}{2 \times 9.8} = 0 + \frac{p_2}{9{,}800} + \frac{0.944^2}{2 \times 9.8}$$

and $p_2 - p_1 = 12,330 \text{ N/m}^2$

$\qquad\qquad = \textbf{12.3 kN/m}^2$

Problems

1. A tank contains oil of specific gravity 0.85 to a depth of 2.4 m. It discharges through a 25-mm diameter straight pipe at a point 6 m below the bottom of the tank. Calculate the discharge in litres per second and tonnes per hour and also find the oil pressure at a point half-way along the pipe. Specific weight of water = 9.8 kN/m³.

   (6.28 litres/s; 19,25 tonnes/h; −25 kN/m² gauge or 76.3 kN/m² abs.)

2. The diameter of a pipe tapers gradually in the direction of water flow as the level drops 9 m from point A to point B. At A the gauge pressure is 210 kN/m² and the pipe diameter 200 mm; at B the diameter is 100 mm. What is the pressure at B when the flow rate is 72 litres per second? Specific weight of water = 9.8 kN/m³.

   (259 kN/m² gauge)

3. A horizontal pipe tapers gradually from 150 mm to 300 mm diameter in the direction of flow. At the narrow section a pressure gauge reads 140 kN/m². At the wide section the pressure is 280 kN/m². Neglecting losses calculate the flow rate of water in cubic metres per second, litres per second, and tonnes per hour. Density of water = 1 Mg/m³.

   (0.306 m³/s; 306 litres/s; 1,102 tonnes/h)

4. Oil of specific weight 8.8 kN/m³ flows through a horizontal pipe which reduces smoothly from 75 mm to 50 mm diameter. If the gauge pressure at these points is 70 kN/m² and 49 kN/m², respectively, find the velocity at the larger diameter and the flow rate in tonnes per minute. Specific weight of water = 9,800 N/m³.

   (3.41 m/s; 0.813 tonnes/min)

5. Oil of relative density 0.8 flows at the rate of 216 dm³/s through a falling pipe which tapers gradually in the direction of flow. The diameter at a point A is 0.6 m and at a point B, 3.6 m vertically below A, it is 0.3 m. The gauge pressure at A is 84 kN/m². Calculate the pressure at B. Density of water = 1,000 kg/m³.

   (109 kN/m² gauge)

20.9 The flow of real fluids

So far we have assumed the fluid to be perfectly frictionless and the pipe walls to be perfectly smooth. A frictionless fluid would flow as in Fig. 20.6, each layer travelling in a smooth path without interference

Fig. 20.6

from adjacent layers. Such a smooth regular flow is called *laminar* or *streamline*. In practice we have to consider the effects of both *fluid friction* and pipe wall friction.

20.10 Viscosity

The *viscosity* of a fluid is the internal resistance to a change of shape. Typically viscous fluids are treacle, glycerine and thick oils; all fluids are viscous in some degree.

20.11 Flow at low velocities

Consider now a viscous fluid entering smoothly and slowly into a pipe. Owing to wall friction fluid sticks to the wall surface, forming a *boundary layer* which is at rest relative to the pipe, Fig. 20.7. Owing to the viscosity of the fluid a drag force or shear stress is exerted on the remainder of the moving fluid by the boundary layer. The velocity of the fluid outside the boundary layer is, however, roughly uniform across the pipe, decreasing within the boundary layer to zero at the pipe wall.

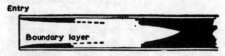

Fig. 20.7

Fig. 20.8

As the fluid continues down the pipe the boundary layer thickens until it completely fills the pipe. The flow is now said to be fully developed; this occurs at a distance equal to a few pipe diameters from the entry. The distribution of velocity across the pipe is now parabolic in form, Fig. 20.8. Note, however, that viscous flow is still regular, i.e. streamline or laminar. The viscous drag forces in the fluid involve a loss of pressure and thus a drop in pressure head along the pipe. This drop in pressure:

(*a*) is proportional to the mean flow velocity;
(*b*) is proportional to the length of pipe;
(*c*) varies inversely as the square of the pipe diameter;
(*d*) is greater with more viscous fluids;
(*e*) is independent of pipe roughness.

Viscous flow of this nature occurs with very viscous oils at low speeds and with ordinary fluids such as water when the pipe diameter is very small indeed.

20.12 Onset of turbulence

At high velocities the fluid flow loses its regular streamline form and takes on an irregular motion. Fig. 20.9 shows the effect of flow velocity on the motion of a thin stream of dye injected into water flowing in a pipe. In Fig. 20.9(*a*) the velocity is low, the flow laminar, and the dye flows as a thin thread. At a higher velocity the fluid takes on a sinuous wavy motion as shown by the dye, Fig. 20.9(*b*). Finally, at sufficiently high velocities the dye thread breaks up and takes on the irregular motion of the main flow. This irregular motion is called *turbulence*, Fig. 20.9(*c*). For a given fluid and pipe diameter there is a *critical velocity* above which turbulence sets in. This critical velocity increases with the viscosity and density of the fluid and decreases with the pipe diameter. That is, turbulence is more likely with fluids of low viscosity in large diameter pipes. For water this critical velocity would be about 0.15 m/s in a 25-mm diameter pipe.

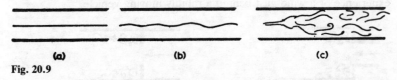

| (a) | (b) | (c) |

Fig. 20.9

Turbulence arises from the initial presence of a boundary layer; fluid near the boundary tends to drag behind the main stream to disturb a uniform flow. Once turbulence has set in the viscosity of the fluid is no longer of great importance. Layers of fluid near the pipe wall still adhere to it however, even though the fluid is turbulent; a laminar boundary layer remains but is thinner than in viscous flow. When the pipe surface roughness is such that the irregularities are larger than the boundary layer thickness then pipe roughness becomes important.

20.13 Pressure loss in turbulent flow

The drop in pressure head due to turbulence does not obey the same laws as in viscous laminar flow. The loss is now:

(*a*) proportional to the *square* of the mean velocity;

(*b*) proportional to the length of pipe, as before;
(*c*) inversely proportional to the pipe diameter;
(*d*) nearly independent of fluid viscosity, but may depend on pipe roughness.

In most practical pipeline applications turbulence may usually be assumed to occur. This is almost always the case for water flow. Note that the pressure loss in turbulent flow is much greater than that for viscous laminar flow.

20.14 Eddy formation

The retarded boundary layer formed at the pipe wall as a result of pipe friction and fluid viscosity can–under certain conditions–give rise to the formation of *eddies* or *vortices*. Eddies occur at a discontinuity in the pipe surface, such as a sudden enlargement or contraction, rapid increases in pipe diameter, sharp bends and valves. At a discontinuity of section or obstruction the boundary layer breaks away from the pipe surface to form the eddy, with a consequent further loss of pressure at the obstruction, Fig. 20.10. The reader may compare this effect with his experience of a wind gust behind a rapidly moving vehicle.

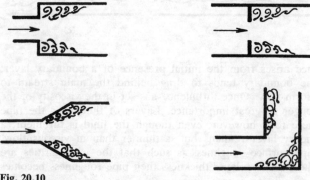

Fig. 20.10

Although eddies may be formed whether the flow is initially laminar or turbulent, turbulence of any sort in the oncoming fluid tends to promote the onset of eddies. The latter may perhaps be thought of as a turbulence on a larger scale, but localized near the obstruction. The formation of eddies involves a further source of pressure loss which is proportional to the square of the mean fluid velocity. Eddy formation at an enlargement in a pipe is prevented by allowing the pipe to open out only very gradually. Similarly the provision of smooth changes of section (streamlining) at the *downstream* side of an obstruction helps to reduce eddies and the loss of pressure.

20.15 Energy of a fluid and pressure loss

Fig. 20.11 shows the effect of pressure loss due to friction on the pressure head, for liquid flow through a uniform horizontal pipe

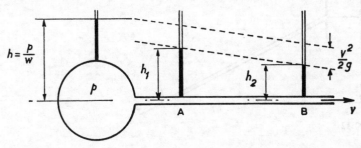

Fig. 20.11

connected to a reservoir at constant pressure p. Note that, since the area is uniform, the velocity head $v^2/2g$ is uniform along the pipe. The head h_f lost in friction is given by:

$$h_f = h_1 - h_2$$

or $\quad h_1 - h_f = h_2$

More generally we can say that:

$$[\text{total energy at A}] - \begin{bmatrix} \text{energy lost in friction} \\ \text{between A and B} \end{bmatrix} = \text{total energy at B}$$

i.e. $\quad H_1 + \dfrac{p_1}{w} + \dfrac{v_1{}^2}{2g} - h_f = H_2 + \dfrac{p_2}{w} + \dfrac{v_2{}^2}{2g}$

or $\quad H_1 + h_1 + \dfrac{v_1{}^2}{2g} - h_f = H_2 + h_2 + \dfrac{v_2{}^2}{2g}$

This is Bernoulli's equation per unit weight of liquid modified to allow for friction loss in the pipe-line. The mechanical energy 'lost' reappears, of course, as heat. Note that the energy quantities in this equation refer to *unit weight of liquid*. The actual energy per second is found by multiplying each term by the weight of liquid flowing per second. Thus, for example, h_f is the head lost in friction in metres and if $\dot{m}$ is the mass rate of flow of liquid per second, then the weight of liquid flowing per second is $\dot{m}g$, and hence the energy lost in friction per second, i.e. the loss of power, is given by

$$\dot{m}gh_f \text{ J/s} \quad \text{or} \quad \dot{m}gh_f \text{ W}$$

Example   *A 50-mm diameter pipe-line falls a vertical distance of 30 m from an open oil reservoir and discharges into an open tank Fig. 20.12. The head of oil above the pipe entrance is 6 m and the loss of head due to pipe friction is 3.6 m. Calculate the discharge in litres per second and in tonnes per hour. Specific gravity of oil is 0.8. Specific weight of water = 9.8 kN/m³.*

What is the loss of power due to friction, in kilowatts?

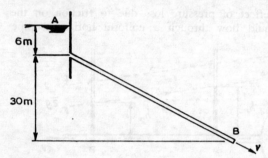

Fig. 20.12

SOLUTION

Total head available, measured above pipe exit B:

$$= \text{head at entry} + \text{fall of pipe} - \text{friction head}$$

$$= 6 + 30 - 3.6$$

$$= 32.4 \text{ m}$$

$$\text{velocity head at exit} = \frac{v^2}{2g}$$

The pressure at exit and at the free surface of the reservoir is atmospheric. Therefore there is no change in pressure energy between A and B. Thus:

kinetic energy at B = potential energy at A

and $\dfrac{v^2}{2g} = 32.4$

i.e. $v = \sqrt{2 \times 9.8 \times 32.4}$

$$= 25.2 \text{ m/s}$$

(This result may be arrived at directly by applying Bernoulli's equation.)

The volumetric flow-rate is given by:

$$Q = Av$$

$$= \frac{\pi}{4} \times 0.05^2 \times 25.2$$

$$= 0.0495 \text{ m}^3/\text{s}$$

$$= \textbf{49.5 litres/s}$$

The mass flow-rate is

$$\dot{m} = 49.5 \times 0.8 \text{ kg/s}$$

$$= \frac{49.5 \times 0.8 \times 3,600}{1,000} \text{ tonne/h}$$

$$= \textbf{142.5 tonne/h}$$

The friction loss is 3.6 m, i.e. the loss is 3.6 J/N = 3.6 × 9.8 J/kg, since unit weight of oil, one newton, has a mass of 1/9.8 kg. Thus:

$$\text{power loss} = \text{kg of oil/s} \times \text{loss of energy/kg}$$

$$= (49.5 \times 0.8) \times 3.6 \times 9.8 \text{ J/s}$$

$$= 1,395 \text{ W}$$

$$= \textbf{1.395 kW}$$

Problems

1. A horizontal pipe of 50 mm diameter connected to a cylinder of water at 210 kN/m² gauge pressure discharges freely to the atmosphere. If the head lost in friction in the pipe is 4.2 m calculate the discharge in litres per minute.

(2,172)

2. Water is pumped up from a level A to level B through a vertical height of 11 m, through a pipe tapered in diameter from 100 mm at A to 150 mm at B. The pressure head at A is 24 m of water and at B 13.5 m. The friction loss of head between A and B is 1.5 m. Find the discharge at B in cubic metres per second and the energy loss due to friction in watts. Density of water = 1 Mg/m³.

(0.055 m³/s; 809 W)

20.16 Measurement of pipe flow-rate: Venturi meter

The flow-rate Q of liquid in a closed pipe is measured by a *Venturi meter*. This consists of a constriction in the pipe line, Fig. 20.13. The pipe converges in the direction of flow from the flange A to the throat B, and then diverges gradually to the full pipe diameter at C. Manometer tubes are inserted in the pipe at A and at the throat B. The rate of flow of liquid in the pipe is then proportional to the square root of the difference in pressure or manometric head h between the pipe and the throat. This is proved as follows:

Applying Bernoulli's equation to points A and B, assuming no loss:

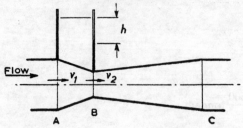

Fig. 20.13

$$H_1 + \frac{p_1}{w} + \frac{v_1^2}{2g} = H_2 + \frac{p_2}{w} + \frac{v_2^2}{2g}$$

and, since for a horizontal pipe, $H_1 = H_2$, then

$$\frac{v_2^2}{2g} - \frac{v_1^2}{2g} = \frac{p_1}{w} - \frac{p_2}{w}$$

But $\dfrac{p_1}{w} - \dfrac{p_2}{w} = h$

the measured difference in pressure; hence

$$\frac{v_2^2}{2g} - \frac{v_1^2}{2g} = h \qquad \qquad \dots [1]$$

The velocity v_2 at the throat B is obtained from the equation of continuity:

$$v_2 A_2 = v_1 A_1$$

i.e. $v_2 = \dfrac{A_1}{A_2} v_1$

Substituting for v_2 in equation [1]:

$$\frac{\{(v_1)(A_1/A_2)\}^2 - v_1^2}{2g} = h$$

or $v_1^2\left(\dfrac{A_1^2}{A_2^2} - 1\right) = 2\,gh$

i.e. $v_1 = \sqrt{\left(\dfrac{2\,gh}{(A_1/A_2)^2 - 1}\right)}$

The required theoretical volumetric flow-rate in the pipe is

$$Q_t = A_1 \times v_1$$

$$= A_1 \sqrt{\left\{\frac{2\,gh}{(A_1/A_2)^2 - 1}\right\}}$$

or $Q_t = C\sqrt{h}$

where

$$C = A_1 \sqrt{\left\{ \frac{2g}{(A_1/A_2)^2 - 1} \right\}}$$

This quantity C is a constant for a given meter. The flow-rate Q_t is therefore seen to be proportional to the square root of the difference in head h between pipe and throat.

20.17 Coefficient of discharge for a Venturi meter

In practice, owing to friction in the convergent portion, the discharge from the pipe is less than $C\sqrt{h}$. The *coefficient of discharge C_d* for the meter is defined as the ratio:

$$\frac{\text{actual discharge } Q}{\text{theoretical discharge } Q_t}$$

i.e.

$$C_d = \frac{Q}{C\sqrt{h}}$$

and, actual discharge is

$$Q = C_d \times C\sqrt{h}$$

C_d is always less than unity and usually about 0.97; for very small meters it may be as low as 0.9. The discharge coefficient C_d for a meter may be found experimentally by weighing the actual discharge

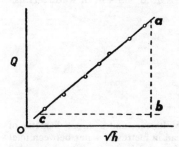

Fig. 20.14

and hence calculating the flow-rate Q, and at the same time measuring the difference of head h. If Q is plotted against $\sqrt{h}$ a straight line graph is obtained, Fig. 20.14:

$$\text{slope} = \frac{\mathbf{ab}}{\mathbf{bc}} = \frac{Q}{\sqrt{h}}$$

hence

$$C_d = \frac{Q}{C\sqrt{h}}$$

$$= \frac{1}{C} \times \frac{ab}{bc}$$

The constant C is determined from the dimensions of the meter.

Example   *The flow in a 600-mm diameter horizontal water main is measured by means of a Venturi meter with a throat diameter of 300 mm. The difference in pressure between pipe and throat corresponds to 250 mm of mercury. Find the flow in cubic metres per second if the discharge coefficient for the meter is 0.99, and the relative density of mercury is 13.6.*

SOLUTION
At pipe,

$$A_1 = \frac{\pi}{4} \times 0.6^3 = 0.283 \text{ m}^2$$

At throat,

$$A_2 = \frac{\pi}{4} \times 0.3^2 = 0.0707 \text{ m}^2$$

$$\text{Constant } C = A_1 \sqrt{\left\{ \frac{2g}{(A_1/A_2)^2 - 1} \right\}}$$

$$C = 0.283 \sqrt{\left\{ \frac{2 \times 9.8}{(4^2 - 1)} \right\}}$$

$$= 0.323 \text{ (m and s units)}$$

250 mm of mercury corresponds to $0.25 \times 13.6 = 3.4$ m of water. Thus

$$Q = C_d \times C\sqrt{h}$$

$$= 0.99 \times 0.323 \times \sqrt{3.4}$$

$$= \textbf{0.59 m}^3\textbf{/s}$$

Problems

1. A Venturi meter has an inlet diameter of 100 mm and a throat diameter of 50 mm. What will be the difference of head in metres of water between inlet and throat if the flow rate is 15 dm³/s of water? If the flow-rate is doubled what would then be the difference in head? Specific weight of water = 9.8 kN/m³.

   (2.78 m; 11.12 m)

2. A Venturi meter is to be designed to measure a maximum flow-rate of 400 tonne/h in a 150-mm diameter pipe line, with a maximum difference of head between flange and throat of 3.6 m. Calculate the corresponding throat diameter required, assuming no losses. If the throat diameter chosen is 100 mm what would be the flow-rate for a head of 3.6 m? Density of water = 1 Mg/m³.

   (116 mm; 0.074 m³/s or 266 tonne/h)

3. The measured discharge of water through a Venturi meter is 78 Mg/h. The inlet and throat diameters are 120 mm and 55 mm, respectively. The pressure drop between inlet and throat is 42 kN/m². Find the discharge coefficient for the meter.

(0.975)

20.18 Discharge through a small orifice

Liquid under a static head h is allowed to flow through an orifice, whose diameter is small compared with the head, Fig. 20.15. The

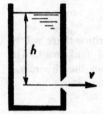

Fig. 20.15

velocity v of the issuing jet is then obtained by equating the pressure energy or head of the liquid in the tank to its kinetic energy or velocity head at the jet. Thus, neglecting energy losses the theoretical velocity of flow v is given by:

$$\frac{v^2}{2g} = h$$

or $v = \sqrt{(2gh)}$

If A is the area of the orifice, the theoretical flow-rate Q_t, is given by

$$Q_t = A \times \sqrt{(2gh)}$$

20.19 Coefficient of discharge for a small orifice

In practice this flow-rate is never achieved and we define a *coefficient of discharge C_d* by the ratio:

$$\frac{\text{actual discharge } Q}{\text{theoretical discharge } Q_t}$$

i.e. $C_d = \dfrac{Q}{A\sqrt{(2gh)}}$

Thus the actual flow-rate is

$$Q = C_d \times A \sqrt{(2\,gh)}$$

The value of C_d is about 0.6–0.7. The value depends slightly on the head h and on the shape and condition of the orifice. C_d is increased by the use of a sharp-edged orifice. Failure to attain the full theoretical discharge is due mainly to two reasons:

1. The theoretical velocity is not achieved due to losses.
2. The full area of the orifice is not utilized.

20.20 Coefficient of velocity

Since the liquid is in motion near the inside of the orifice there will be a loss of head due to friction between liquid and tank wall. The velocity of the jet will therefore be slightly less than the theoretical value and the *coefficient of velocity* C_v is defined by the ratio:

$$\frac{\text{actual velocity of jet}}{\text{theoretical velocity of jet}}$$

i.e. $C_v = \dfrac{v}{\sqrt{(2\,gh)}}$

20.21 Vena contracta: coefficient of contraction

Fig. 20.16 shows a jet issuing from a sharp-edged orifice. The liquid in the tank streams into the orifice as shown, with the result that each

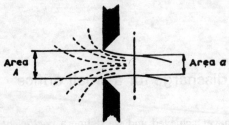

Fig. 20.16

particle of liquid has a component of velocity perpendicular to the jet axis at the opening. The effect is to cause the jet to contract just after leaving the orifice. The section of the jet where it first becomes parallel, and the area is least, is known as the *vena contracta*. The jet velocity therefore reaches its greatest value at the vena contracta. The ratio of the area of the vena contracta to the actual orifice is termed the *coefficient of contraction* C_c, i.e.

$$C_c = \frac{\text{area } a \text{ of jet at vena contracta}}{\text{area } A \text{ of orifice}}$$

$$= \frac{a}{A}$$

Hence the effective area of the jet is

$$a = C_c \times A$$

The coefficient of contraction for small sharp-edged circular orifices is usually about 0.63–0.65.

20.22 Relation between the coefficients

The actual velocity of the jet is $C_v\sqrt{(2\,gh)}$, hence the actual discharge is:

$$Q = \text{effective area} \times \text{actual jet velocity}$$

$$= a \times v$$

$$= C_c A \times C_v\sqrt{(2\,gh)}$$

$$= C_c C_v \times A\sqrt{(2\,gh)}$$

By comparison with the equation,

$$Q = C_d \times A\sqrt{(2\,gh)}$$

it is seen that

$$C_d = C_c \times C_v$$

20.23 Power of a jet

If a jet has a velocity v then the kinetic energy–the energy of motion–of a particle of mass m is $\frac{1}{2}mv^2$. If $\dot{m}$ is the mass rate of flow of fluid per second then the energy per second or *power* of the jet is $\frac{1}{2}\dot{m}v^2$. Thus:

$$\text{power} = \tfrac{1}{2}\dot{m}v^2$$

If ρ is the density of the fluid and a the area of cross-section of the jet, then volume of fluid per second = area of section × velocity

and mass of fluid per second, $\dot{m} = \rho \times$ volume per second

$$= \rho a v$$

Hence

power $= \frac{1}{2}\dot{m}v^2 = \frac{1}{2}\rho a v^3$

The units of power (energy per second) are joules per second, i.e. *watts*, when the density ρ is in kg/m³, mass flow $\dot{m}$ in kg/s, and velocity v in m/s.

When the jet flows from an orifice the actual velocity v is given by $C_v\sqrt{2gh}$ and the actual discharge $\dot{m}$ by

$$\rho C_d A \sqrt{(2gh)} = \rho(C_c A) \times C_v \sqrt{(2gh)}$$

where A is the area of the orifice.

Example   *A hydraulic machine is driven by a jet from a nozzle of 25 mm diameter in a water main under a gauge pressure of 700 kN/m². Neglecting any loss of energy find the power supplied to the machine. For water, density = 1,000 kg/m³, specific weight = 9,800 N/m³.*

Solution

Pressure head across machine, $h = \dfrac{700 \times 10^3}{9,800} = 71.4$ m

velocity of jet, $v = \sqrt{(2\,gh)}$

$= \sqrt{(2 \times 9.8 \times 71.4)}$

$= 37.4$ m/s

$Q = Av$

$= \dfrac{\pi}{4} \times 0.025^2 \times 37.4$

$= 0.01835$ m³/s

Mass of water per second, $\dot{m} = 0.01835 \times 10^3$ kg/s

kinetic energy of jet $= \frac{1}{2}\dot{m}v^2$ J/s

$= \frac{1}{2} \times 0.01835 \times 10^3 \times 37.4^2$

$= 12,850$ J/s

power supplied $= \dfrac{12,850}{1,000} = \mathbf{12.85\ kW}$

Example   *Water flows from an orifice in the side of a tank. The head of water above the centre line of the orifice is 12 m and the opening is 30 mm diameter. The coefficient of velocity for the orifice is 0.95 and the coefficient of discharge 0.6. Find the power of the jet.*

SOLUTION

Theoretical velocity = $\sqrt{(2gh)} = \sqrt{(2 \times 9.8 \times 12)} = 15.35$ m/s

Actual velocity, $v = 0.95 \times 15.35 = 14.55$ m/s

Actual discharge, $Q = C_d A \sqrt{(2gh)}$

$$= 0.6 \times \frac{\pi}{4} \times 0.03^2 \times 15.35$$

$$= 0.0065 \text{ m}^3/\text{s}$$

(Note that C_d includes C_v)

Mass flow-rate, $\dot{m} = 0.0065 \times 10^3$ kg/s

$$= 6.5 \text{ kg/s}$$

Energy per second of the jet $= \frac{1}{2}\dot{m}v^2$

$$= \frac{1}{2} \times 6.5 \times 14.55^2 \text{ J/s}$$

i.e. power = **690 W**

20.24 Experimental determination of orifice coefficients

The most easily obtained coefficient is C_d, the coefficient of discharge. It is found directly by weighing the liquid discharged in a given time while the head h is kept constant. The coefficient of velocity C_v is found from the geometry of the jet.

Fig. 20.17 shows a jet issuing horizontally from an orifice. At a point distant y below the centre of the orifice the distance of the centre of the jet from the vena contracta is x. We may assume each particle of liquid to act as a projectile, travelling without interference from other particles.

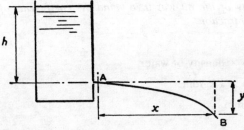

Fig. 20.17

If v is the actual horizontal velocity at the orifice, then the distance x is given by:

$$x = vt \quad \text{or} \quad v = \frac{x}{t}$$

where t is the time of flight from A to B. t is also the time taken for a particle to fall freely a distance y from rest, hence y is given by:

$$y = \tfrac{1}{2} gt^2 \quad \text{or} \quad t = \sqrt{\left(\frac{2y}{g}\right)}$$

hence

$$v = \frac{x}{t} = \frac{x}{\sqrt{(2y/g)}}$$

$$= x\sqrt{\left(\frac{g}{2y}\right)}$$

The theoretical velocity $= \sqrt{(2\,gh)}$

hence

$$C_v = \frac{\text{actual velocity}}{\text{theoretical velocity}}$$

$$= \frac{v}{\sqrt{(2\,gh)}} = \frac{x\sqrt{(g/2y)}}{\sqrt{(2\,gh)}}$$

$$= \frac{x}{2\sqrt{(yh)}}$$

Finally, the coefficient of contraction is obtained from the relation:

$$C_c = \frac{C_d}{C_v}$$

C_c may be obtained by direct measurment of the jet diameter at the vena contracta but this is neither an easy nor an accurate method.

Example *A tank of oil discharged through an orifice of 10 mm diameter. The measured discharge was 13.6 kg/min when the head of oil in the tank was 1.8 m measured from the centre line of the orifice. The jet issued horizontally, falling a distance of 330 mm in a distance of 1.5 m. The relative density of the oil was 0.76. Find the coefficients of discharge, velocity and contraction.*

SOLUTION

Density of oil $= 0.76 \times$ density of water

$$= 0.76 \times 1 \text{ kg/dm}^3$$

Discharge $= 13.6$ kg/min

therefore

$$Q = \frac{13.6}{0.76 \times 60} \times 10^{-3} \text{ m}^3/\text{s}$$

$$= 0.000298 \text{ m}^3/\text{s}$$

$$Q = C_d \times A \sqrt{(2gh)}$$

where h is head of oil in metres. Thus:

$$0.000298 = C_d \times \frac{\pi}{4} \times 0.01^2 \sqrt{(2 \times 9.8 \times 1.8)}$$

i.e. $C_d = \mathbf{0.637}$

$$C_v = \frac{x}{2\sqrt{(yh)}}$$

$$= \frac{1.5}{2\sqrt{(0.33 \times 1.8)}}$$

$$= \mathbf{0.975}$$

$$C_d = C_c \times C_v$$

therefore

$$C_c = \frac{0.637}{0.975}$$

$$= \mathbf{0.653}$$

Problems

(Density of water = 1,000 kg/m³; specific weight of water = 9,800 N/m³)

1. A tank of oil, of relative density 0.8, discharges through a 12-mm diameter orifice. The head of oil above the centre line of the orifice is kept constant at 1.2 m and the measured discharge rate is 16 kg/min. Calculate the coefficient of discharge for the orifice.

   (0.61)

2. A large tank contains water to a depth of 0.9 m. Water issues from a sharp-edged orifice of 25 mm diameter and is collected in a circular tank of 0.9 m diameter. The water level in the cylinder rises 600 mm in 5 min. Calculate the discharge coefficient for the orifice.

   (0.62)

3. A square section sharp-edged orifice is to discharge 90 kg of water per minute under a constant head of 750 mm. Assuming a coefficient of discharge of 0.63 find the length of the side of the orifice.

   (24.9 mm)

4. A tank of water on level ground has a sharp-edged orifice in the side 400 mm from the bottom. The coefficient of velocity of the orifice is 0.98. Find how far from the orifice the jet will strike the ground when the head of water in the tank is 2.4 m above the orifice.

   (1.92 m)

5. In an experiment on an orifice the following results were obtained: orifice diameter, 6.5 mm; discharge of water, 23.4 kg in 4 min; fall of jet, 300 mm in a horizontal distance of 1,175 mm; head of water above centre line of orifice, 1.2 m. Calculate the coefficients of discharge, velocity and contraction.

(0.605; 0.98; 0.617)

6. A jet issues horizontally from an orifice under a head of 1.2 m. The ordinates of its path, measured from the vena contracta, are 1.5 m horizontally and 525 mm vertically. Obtain, from first principles, the coefficient of velocity for the orifice.

(0.944)

7. A pressure vessel contains water at 2.1 MN/m^2 gauge pressure. A jet issues from a nozzle of 50 mm diameter, having a coefficient of velocity 0.95 and coefficient of discharge 0.65. Find the power of the jet.

(157 kW)

8. Water flows from an orifice in the side of a tank. The head of water above the level of the orifice is 3.6 m and the opening is 25 mm diameter. If the coefficient of discharge for the orifice is 0.64 calculate the power of the jet.

(93 W)

20.25 Impact of jets. Dynamic pressure

When a jet of fluid impinges on a plate or vane or passes a bend in a pipe the effect of the sudden change in the magnitude and/or direction of its velocity is to cause a *dynamic pressure* or *force* to be set up. The *average* force exerted on the vane is found by applying Newton's second law which states that the rate of change of momentum (or momentum per second) is equal to the applied force and takes place in the direction of the force.

In considering the impact of jets on single vanes we shall assume steady flow and that the area of section of the jet is small compared to that of the vane. The cases considered where a jet strikes a single moving vane, although impossible in practice because the jet must continually lengthen as the vane moves away, are important as an introduction to more advanced work on Pelton wheels and turbines.

20.26 Jet normal to vane

Stationary vane, Fig. 20.18

The jet moving with velocity v has momentum $\dot{m}v$ where $\dot{m}$ is the mass rate of discharge of fluid. This momentum in the normal direction is destroyed by the vane, hence the normal force *on* the vane is given by the *loss* of momentum per second, i.e.

$$F = \dot{m}v = \rho a v^2$$

since $\dot{m} = \rho a v$.

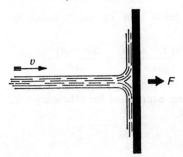

Fig. 20.18

Vane moving in initial direction of jet, Fig. 20.19

The jet is assumed to overtake the vane and only the fluid which

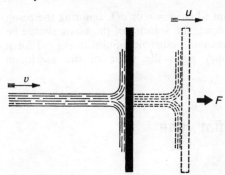

Fig. 20.19

actually strikes the vane per second has its momentum affected. The amount of fluid striking the vane per second depends on the relative velocity between the jet and the vane. If u is the blade speed then the relative velocity is $v - u$ and the fluid overtaking (and leaving) the vane per second is

$$\dot{m}_O = \rho a(v - u) = \dot{m}\frac{(v - u)}{v}$$

where $\dot{m}$ is the *total* mass flow rate.

The initial momentum per second of this quantity $\dot{m}_O$ of fluid is $\dot{m}_O v$ and its final momentum per second is $\dot{m}_O u$, hence the normal force *on* the vane is

$$F = \dot{m}_O v - \dot{m}_O u = \dot{m}_O(v - u) = \rho a(v - u)^2$$

This force moves through a distance u in each second therefore

work done per second (power) $= Fu = \rho a(v - u)^2 u$

The total energy available in the jet is $\frac{1}{2}\dot{m}v^2$ or $\frac{1}{2}\rho a v^3$, hence the efficiency of the operation is

$$\eta = \frac{\text{work done per second}}{\text{energy input per second}} = \frac{\rho a(v - u)^2 u}{\frac{1}{2}\rho a v^3} = \frac{2(v - u)^2 u}{v^3}$$

For a given jet velocity v the value of the blade speed u for maximum efficiency is found by differentiating the expression for efficiency with respect to u and equating to zero, thus:

$$\frac{\mathrm{d}\eta}{\mathrm{d}u} = 0$$

$$\frac{\mathrm{d}}{\mathrm{d}u}\{(v - u)^2 u\} = 0$$

$$\frac{\mathrm{d}}{\mathrm{d}u}(v^2 u - 2u^2 v + u^3) = 0$$

$$v^2 - 4uv + 3u^2 = 0$$

i.e. the efficiency is a maximum when $u = v$ or $v/3$. Ignoring the result $u = v$ which gives zero work done, the velocity of the blade should be *one-third* that of the jet for maximum efficiency. Substituting $v/3$ for u in the expression for efficiency gives the value of the maximum efficiency as 29.6 per cent.

20.27 Jet inclined to flat vane

Stationary vane, Fig. 20.20
The jet strikes the plate at an angle θ to the normal. If there is no friction or splash-back the jet divides into two streams to flow up and down the plate surface, both streams moving with velocity v, but having *different* flow rates. The component of the jet velocity normal to the plate is $v \cos \theta$ and the corresponding momentum $\dot{m}v \cos \theta$ is destroyed producing a normal force on the plate. The component of the initial velocity parallel to the plate, $v \sin \theta$, is left unchanged and hence there is no change of momentum parallel to the plate. The only force therefore is the normal force. Thus:

$$F = \dot{m}v \cos \theta = \rho a v^2 \cos \theta$$

The total thrust F may be resolved into two components, $F_p = F \cos \theta$, parallel to the plate, and $F_n = F \sin \theta$, normal to the jet. These component forces are shown in Fig. 20.20.

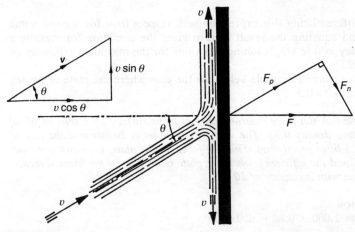

Fig. 20.20

Moving vane, Fig. 20.21

Suppose the blade to be moving away from the jet with velocity u in the same direction as the jet. The mass of fluid striking the vane per second is $\dot{m}_O = \rho a(v - u)$. the momentum per second of the jet before impact is $\dot{m}_O v$ and the component of this normal to the plate is $\dot{m}_O v \cos \theta$. After impact the momentum normal to the plate is $\dot{m}_O u \cos \theta$ hence the normal (resultant) force on the plate is

$$F = \dot{m}_O v \cos \theta - \dot{m}_O u \cos \theta = \dot{m}_O(v - u) \cos \theta = \rho a(v - u)^2 \cos \theta$$

Since we have neglected friction there is no force parallel to the plate.

The work done per second (power) is $F \times u \cos \theta$ and the kinetic energy of the jet per second initially is $\frac{1}{2}\rho a v^3$ hence the efficiency is

$$\eta = \frac{Fu \cos \theta}{\frac{1}{2}\rho a v^3} = \frac{2(v - u)^2 u \cos^2 \theta}{v^3}$$

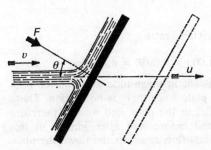

Fig. 20.21

Differentiating this expression with respect to u for a given value of v and equating the result to zero gives the condition for maximum efficiency as $u = v/3$, resulting in a value for the maximum efficiency of $\frac{8}{27} \cos^2 \theta$.

(*See* worked example below for the case where the plate is moving parallel to itself.)

Example *A flat plate is struck normal to its surface by a thin jet of oil of relative density 0.82. The velocity of the jet is 20 m/s and the rate of flow is 5 litres per second. Find the force on the plate, the work done per second and the efficiency when the plate is moving in the same direction as the jet with a velocity of 10 m/s.*

SOLUTION

$\rho = 1{,}000 \times 0.82 = 820 \text{ kg/m}^3$

$Q = av = 0.005 \text{ m}^3/\text{s}$

Total discharge, $\dot{m} = \rho Q = 820 \times 0.005 = 4.1 \text{ kg/s}$

Oil striking plate per second is

$$m_O = \dot{m} \times \frac{(v - u)}{v} = 4.1 \times \frac{(20 - 10)}{20}$$

$$= 2.05 \text{ kg/s}$$

$F = \dot{m}_O(v - u) = 2.05 \times (20 - 10) = \textbf{20:5 N}$

Work done per second (power) $= Fu = 20.5 \times 10 = \textbf{205 W}$

Initial energy of the jet $= \frac{1}{2}\dot{m}v^2 = \frac{1}{2} \times 4.1 \times 20^2 = 820 \text{ J/s}$

Efficiency $= \dfrac{205}{820} \times 100 = \textbf{25 per cent}$

Example *A jet of water of 40 mm diameter, moving at 7 m/s, impinges on a fixed flat plate inclined at 60° to the axis of the jet as shown, Fig. 20.22. Find the mass of water per second flowing upwards along the plate. Neglect all losses.*

SOLUTION

$$Q = av = \frac{\pi}{4} \times 0.04^2 \times 7 = 0.0088 \text{ m}^3/\text{s}$$

Total flow rate, $\dot{m} = \rho Q = 1{,}000 \times 0.0088 = 8.8 \text{ kg/s}$

The jet velocity remains constant in magnitude at 7 m/s for both streams of fluid up and down the plate since there is no friction. There is no change of momentum parallel to the plate and we can therefore equate the momentum per second before and after impact in this direction. If $\dot{m}_u$ and $\dot{m}_d$ are the mass flow rates up and down the plate respectively then

$$\dot{m}_u + \dot{m}_d = \dot{m} = 8.8 \text{ kg/s}$$

and $\dot{m}v \cos 60° = \dot{m}_u v - \dot{m}_d v$

i.e. $8.8 \times 7 \times 0.5 = (\dot{m}_u - \dot{m}_d) \times 7$

$$\dot{m}_u - \dot{m}_d = 4.4$$

hence

$$\dot{m}_u = \mathbf{6.6 \text{ kg/s}}$$

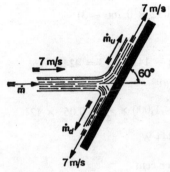

Fig. 20.22

Example *A jet of water 60 mm in diameter, moving with a velocity of 12 m/s, strikes a smooth flat plate inclined to the axis of the jet at an angle of 50° as shown in Fig. 20.23. The plate is moving away from the jet with a velocity of 3 m/s parallel to itself and in the direction of the normal to its surface. Find the work done per second and the efficiency.*

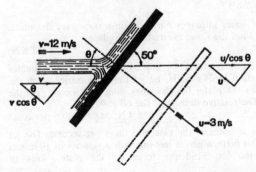

Fig. 20.23

SOLUTION
The angle between the axis of the jet and the normal to the plate is
$\theta = 40°$.

Mass of water striking plate per second $= \dot{m}_0 = \rho a\left(v - \dfrac{u}{\cos \theta}\right)$

Normal force,

$F =$ loss of momentum per second in direction normal to plate

$= \dot{m}_0 v \cos \theta - \dot{m}_0 u$

$= \rho a\left(v - \dfrac{u}{\cos \theta}\right)(v \cos \theta - u)$

$= 1,000 \times \dfrac{\pi \times 0.06^2}{4}\left(12 - \dfrac{3}{0.766}\right)(12 \times 0.766 - 3)$

$= 141.5 \text{ N}$

Work done per second $= Fu = 141.5 \times 3 = \mathbf{424.5 \text{ W}}$

Initial kinetic energy of the jet $= \frac{1}{2}\rho a v^3$

$$= \tfrac{1}{2} \times 1,000 \times \dfrac{\pi}{4} \times 0.06^2 \times 12^3$$

$$= 2,444 \text{ W}$$

Efficiency $= \dfrac{424.5}{2,444} \times 100 = \mathbf{17.36 \text{ per cent}}$

(The student should confirm that this is almost the maximum possible efficiency for the given jet speed.)

Problems

(Density of water $= 1,000 \text{ kg/m}^3$)

1. A 40-mm diameter jet of water impinges normally on a stationary flat plate. The jet velocity is 7 m/s. Find the force exerted on the plate.

(61.6 N)

2. A flat plate moving at 8 m/s is struck by a jet of oil of 20 mm diameter moving in the same direction at 24 m/s. If the jet impinges normally on the plate find (a) the force on the plate, (b) the work done per second, (c) the efficiency of the action. The relative density of the oil is 0.85.

(68.4 N; 547 J/s; 29.6 per cent)

3. A nozzle discharges a jet of water at the rate of 3 litres per second. The jet impinges normally on a flat plate which is moving with a velocity of 10 m/s in the same direction as the jet. Find the force on the plate. Area of jet $= 80 \text{ mm}^2$.

(60 N)

4. A jet of water of 200 mm diameter issues from a nozzle at the rate of 150 dm³/s. The jet impinges normally on a flat plate moving in the same direction as the jet with a velocity of 0.5 m/s. Find the rate of working of the jet.

(286 W)

5. A jet of oil 10 mm in diameter moving at 50 m/s impinges on a stationary flat plate at an angle of 60° to its surface. The relative density of the oil is 0.9. Find the normal force on the plate and the force in the direction of the jet. If the plate has a velocity of 10 m/s in the same direction as the jet find the normal force and the work done per second.

(153 N; 132.5 N; 97.8 N; 848 W)

6. A nozzle discharges a jet of water of cross-sectional area 1,000 mm² on to a flat plate which is moving parallel to itself with a velocity of 5 m/s in the direction of the normal to its surface. The jet axis is at 30° to the normal to the plate which is moving away from the jet. If the head of water serving the nozzle is 40 m, find the normal force on the plate, the work done per second, and the efficiency.

(428 N; 2.14 kW; 19.45 per cent)

20.28 Jet impinging tangentially on a curved vane

Stationary Vane, Fig 20.24

Neglecting friction and shock losses the velocity of the jet may be assumed constant in magnitude while passing over the blade. The initial momentum per second (in the horizontal direction) is $\dot{m}v$, shown by vector **oa**, and the final momentum per second is also $\dot{m}v$ but in a direction tangential to the blade at exit, shown by vector **ob**. The resultant force F on the blade is given by vector **ba** since vector **ab** represents the rate of change of momentum of the fluid. Alternatively,

horizontal force on blade,

$$F_h = \dot{m}v - \dot{m}v\cos\phi = \rho av^2(1 - \cos\phi) \qquad \cdots [1]$$

vertical force on blade,

$$F_v = 0 - \dot{m}v\sin\phi = -\rho av^2\sin\phi \qquad \cdots [2]$$

The minus sign indicates that the force is opposite in direction to the velocity v, i.e. F_v is downwards.

Resultant force on blade, $F = \sqrt{(F_h{}^2 + F_v{}^2)}$

and the direction of this force is given by

$$\tan\alpha = \frac{F_v}{F_h}$$

The effect of friction, shock at entry, and turbulent flow along the blade is to cause a reduction in the jet velocity at exit. If v_e is the actual velocity of the jet at exit equations [1] and [2] become

$$F_h = \dot{m}v - \dot{m}v_e\cos\phi$$

and $F_v = \dot{m}v_e\sin\phi$

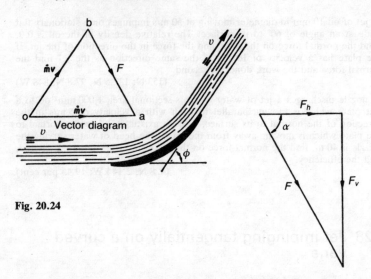

Fig. 20.24

In the vector diagram, Fig. 20.24, vector **ob** would represent $\dot{m}v_e$.

When $\phi = 180°$, $F_v = 0$, $F_h = 2\dot{m}v$. Thus, for a semi-circular (bucket) wheel the thrust is twice that which the same jet can develop against a flat plate.

Vane moving in initial direction of jet, Fig. 20.25

If the vane speed is u then the velocity of the jet relative to the vane at entry is $v - u$. If there is no friction or shock at entry the relative velocity of the jet to the blade is constant in magnitude as the jet flows across the vane but its direction changes. Thus the relative velocity of the jet to vane at exit is $v - u$ and its direction is tangential to the vane. Only the fluid which actually overtakes the vane is deflected and has its momentum affected. The amount of fluid overtaking (and leaving) the vane per second is

$$\dot{m}_O = \rho a(v - u)$$

and the corresponding momentum per second at entry is $\dot{m}_O v$.

To find the momentum per second of the fluid at exit we require the absolute velocity v_e of the fluid at exit. This is obtained from a *vector diagram* or *velocity diagram*. The velocity of the vane u may be represented by vector **ob**, Fig. 20.25, and the velocity of the jet relative to the vane, $v - u$, by vector **oc**. The absolute velocity of the fluid at exit is then the vector sum of $v - u$ and u, i.e. $v_e = $ **oa**. The horizontal component of the absolute velocity is $u + (v - u) \cos \phi$ and the vertical component is $(v - u) \sin \phi$.

At exit

momentum per second in horizontal direction

$$= \dot{m}_O[u + (v - u) \cos \phi]$$

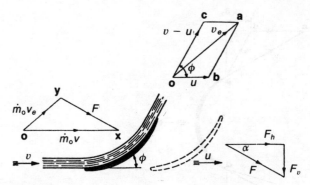

Fig. 20.25

momentum per second in vertical direction $= \dot{m}_O[v - u) \sin\phi$

Therefore

horizontal force on blade,

$$F_h = \dot{m}_O v - \dot{m}_O[u + (v - u)\cos\phi]$$
$$= \rho a(v - u)\{v - [u + (v - u)\cos\phi]\}$$

i.e. $F_h = \rho a(v - u)^2(1 - \cos\phi)$

and vertical force on blade,

$$F_v = 0 - \dot{m}_O(v - u)\sin\phi$$

i.e. $F_v = \rho a(v - u)^2 \sin\phi$

The minus sign indicates that the force is in the opposite direction to the relative velocity $v - u$ at exit, i.e. downwards.

Total force on blade, $F = \sqrt{(F_h{}^2 + F_v{}^2)}$

and its direction is obtained from $\tan\alpha = F_v/F_h$.

Alternatively the vector diagram can be drawn for the initial and final values of the momentum per second. The initial momentum per second is represented by vector $\mathbf{ox} = \dot{m}_O v$ and the final momentum per second by vector $\mathbf{oy} = \dot{m}_O v_e$. Force F is given by vector $\mathbf{yx}$.

The work done per second (power) is $F_h \times u$. The vertical force F_v produces an *end thrust* but does no work since the blade has no velocity in the vertical direction.

The effect of friction is to reduce the magnitude of the relative velocity at exit. Thus if the relative velocity at exit is k times that at entry then vector $\mathbf{oc} = k(v - u)$ and the above equations are modified accordingly. For example, the force F_h becomes

$$F_h = \dot{m}_O v - \dot{m}_O[u + k(v - u)\cos\phi]$$
$$= \rho a(v - u)^2(1 - k\cos\phi)$$

If the blade is not moving in the same direction as the jet at entry

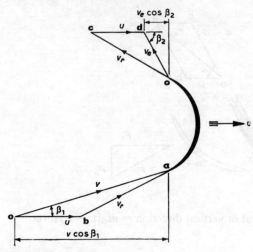

Fig. 20.26

but as shown in Fig. 20.26 the problem becomes more involved. At entry vector **oa** represents the jet velocity v and vector **ob** represents the blade velocity u. The relative velocity at entry is v_r, given by the vector difference **ba**. It is usually arranged that v_r is tangential to the blade at entry to minimize shock. The component of the absolute velocity of the jet at entry in the direction of the blade speed is $v \cos \beta_1$ (β_1 is the *nozzle angle*). At exit **oc** is drawn to represent v_r (assuming no friction) tangential to the blade, and **cd** is drawn to represent u. The absolute velocity of the jet at exit, v_e, is then given by vector **od**. The component of v_e in the direction of the blade speed is $v_e \cos \beta_2$. Hence the force F in the direction of motion of the blade is

$$F = \dot{m}_O v \cos \beta_1 - \dot{m}_O v_e \cos \beta_2$$

where $\dot{m}_O$ is the mass of fluid striking the blade per second, i.e. $\rho a v_r$. The sign of the second term is determined by the direction of velocity v_e. If velocity v_e is in the opposite direction to that of the blade speed as shown in Fig. 20.26 the second term is positive since the reaction from the leaving jet pushes the blade in the direction of its motion. If v_e has its component velocity in the same direction as the blade speed then the second term is negative since the reaction from the leaving jet opposes the blade motion.

Work done per second (power) = Fu

If friction reduces the relative velocity by a factor k then the relative velocity at exit is kv_r and vector **oc** is accordingly modified.

Having found the absolute velocity v_e at exit the work done per second may be found also from the change of kinetic energy per second of the fluid striking the blade. Thus

work done per second = $\frac{1}{2}\dot{m}_O v^2 - \frac{1}{2}\dot{m}_O v_e^2$

and the efficiency is

$$\eta = \frac{\frac{1}{2}\dot{m}_O(v^2 - v_e{}^2)}{\frac{1}{2}\dot{m}v^2} = \frac{\dot{m}_O}{\dot{m}}\left[1 - \left(\frac{v_e}{v}\right)^2\right] = \frac{v_r}{v}\left[1 - \left(\frac{v_e}{v}\right)^2\right]$$

Note: It is important to remember that the above work is on the basis of a jet impinging on a single blade. In practice there is a series of blades and the full discharge per second $\dot{m}$ would therefore be used in calculating the forces and the power.

Example *A nozzle under a head of 40 m directs a jet of water of 40 mm diameter horizontally on to a single curved fixed blade which deflects the jet through an angle of 150°. The force of the jet is measured as 1.7 kN in the horizontal direction. If the coefficient in velocity for the nozzle is 0.98 find the percentage reduction in velocity of the jet caused by friction as the water passes over the blade.*

SOLUTION
Referring to Fig. 20.24

$\qquad F_h = 1,700$ N, $\phi = 150°$; $\cos\phi = -0.866$

$\qquad v = C_v\sqrt{(2gh)} = 0.98 \times \sqrt{(2 \times 9.8 \times 40)} = 27.44$ m/s

$\qquad \dot{m} = \rho av = 1,000 \times \dfrac{\pi}{4} \times 0.04^2 \times 27.44 = 34.5$ kg/s

If the velocity of the jet leaving the blade is v_e then

$\qquad F_h = \dot{m}v - \dot{m}v_e\cos\phi$

$\qquad 1,700 = 34.5(27.44 + 0.866v_e)$

$\qquad v_e = 25.2$ m/s

$\qquad$ Reduction in velocity $= \dfrac{(27.44 - 25.2)}{27.44} \times 100 = $ **8.2 per cent**

Example *A 30-mm diameter jet of water is discharged from a nozzle under a head of 10 m. The jet is deflected through an angle of 135° by a stationary curved vane. When the vane is moving at 5 m/s in the same direction as the jet find the work done per second and the efficiency.*

SOLUTION
Referring to Fig. 20.25,

$\qquad \phi = 135°$; $\cos\phi = -0.707$; $\quad v = \sqrt{(2g \times 10)} = 14$ m/s;

$\qquad a = \pi \times \dfrac{0.03^2}{4} = 0.000707$ m²

Force in direction of jet,

$\qquad F_h = \rho a(v - u)^2(1 - \cos\phi)$

$\qquad\quad = 1,000 \times 0.000707(14 - 5)^2(1 + 0.707)$

$$= 97.6 \text{ N}$$

Work done per second $= F_h u = 97.6 \times 5 = \textbf{488 W}$

Initial kinetic energy of jet $= \frac{1}{2}\rho a v^3$

$$= \frac{1}{2} \times 1,000 \times 0.000707 \times 14^3$$

$$= 970 \text{ W}$$

Efficiency $= \dfrac{488}{970} \times 100 = \textbf{50.5 per cent}$

Alternatively, by drawing the velocity diagram for exit conditions the absolute velocity at exit v_e will be found to be 6.44 m/s. Hence using the expression derived on page 481.

$$\text{Efficiency} = \frac{(v-u)}{v}\left[1 - \left(\frac{v_e}{v}\right)^2\right]$$

$$= \frac{9}{14}\left[1 - \left(\frac{6.44}{14}\right)^2\right] \times 100$$

$$= \textbf{50. 5 per cent}$$

Example   *The curved blade shown in Fig. 20.27 has an exit angle 30° to the horizontal and is moving horizontally at 10 m/s. The nozzle angle at which a jet is directed on to the blade is 15° and the jet flows tangentially on to the blade without shock. The effect of friction is to reduce the relative velocity of the water across the blade by 20 per cent. The total discharge from the nozzle is 20 kg/s. Find the force on the blade in the direction of its motion, the work done per second and the efficiency for a jet velocity of 25 m/s.*

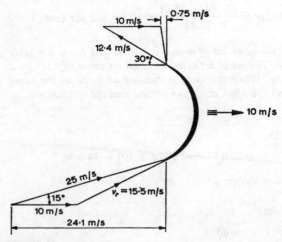

Fig. 20.27

SOLUTION

By drawing or calculation the relative velocity at entry, $v_r = 15.5$ m/s, hence the relative velocity at exit is $0.8 \times 15.5 = 12.4$ m/s.

Total discharge $\dot{m} = 20$ kg/s

Mass of water striking blade per second,

$$\dot{m}_O = \frac{v_r}{v} \times \dot{m} = \frac{15.5}{25} \times 20 = 12.4 \text{ kg/s}$$

By drawing or calculation the various velocities are found as indicated in Fig. 20.27. The force in the direction of motion of the blade is the loss of momentum in this direction, therefore

$$F_h = \dot{m}_O \times 24.1 + \dot{m}_O \times 0.75$$

$$= 12.4 \times 24.85$$

$$= \textbf{308 N}$$

Work done per second $= F_h \times 10 = 308 \times 10$

$$= \textbf{3,080 W}$$

Initial kinetic energy per second $= \frac{1}{2}\dot{m}v^2 = \frac{1}{2} \times 20 \times 25^2$

$$= \textbf{6,250 W}$$

$$\text{Efficiency} = \frac{\text{work done per second}}{\text{initial kinetic energy per second}}$$

$$= \frac{3,080}{6,250} \times 100$$

$$= \textbf{49.3 per cent}$$

Problems

(Density of water $= 1,000$ kg/m³)

1. A 20-mm diameter nozzle discharges a jet of water on to a fixed vane. The water impinges tangentially and is deflected through 160°. If the head of water above the nozzle is 30 m and the coefficient of velocity is 0.97 find the force on the vane in the direction of the jet and the total force on the vane. Assume that friction reduces the velocity of the water across the vane by 15 per cent.

(311 N; 317 N)

2. A jet of water has velocity 20 m/s and impinges on a curved vane moving at 8 m/s in a direction making 20° to that of the jet. Find the entry angle of the vane so that the water enters without shock.

(32° 20′)

3. A jet of water is deflected through 90° by a curved vane. The area of section of the jet is 1,000 mm² and its velocity is 10 m/s. If the vane is moving at 4 m/s in the same direction as the jet initially find the resultant force on the bend.

(51 N)

4. A jet of water 80 mm in diameter moving with a velocity of 20 m/s impinges tangentially on a stationary curved blade which deflects the jet through 120°. Find the resultant force on the blade in magnitude and direction. If the blade has a velocity of 5 m/s in the same direction as the jet, what is the work done per second and the efficiency? Neglect friction.

(3.48 kN, 30° to jet; 8.48 kW; 42.2 per cent)

5. A jet of water 30 mm in diameter has a velocity of 40 m/s. The jet impinges tangentially on a semicircular 'bucket' blade which is moving in the same direction as the jet at 10 m/s. Assuming the jet velocity relative to the blade is reduced 10 per cent by friction during passage across the blade, find the absolute velocity of the jet leaving the blade, the work done per second, and the efficiency.

(17 m/s; 12.15 kW; 53.6 per cent)

6. A jet of water impinges tangentially on a curved vane which deflects it through an angle of 45°. The jet velocity is 30 m/s. If the vane is moving with a velocity of 20 m/s in the same direction as the jet and the effect of friction is to reduce the velocity of the jet relative to the blade by 10 per cent, find the magnitude and inclination of the absolute velocity of the jet leaving the vane.

(27.2 m/s, 13° 33' to initial direction of jet)

7. A jet of water of cross-sectional area 2,000 mm² moving at 25 m/s impinges tangentially on a stationary vane which changes the direction of the jet through an angle of 145°. If the vane moves at 5 m/s in the same direction as the jet find the magnitude of the resultant force on the vane, the work done per second, and the efficiency. Assume that friction reduces the relative velocity of the jet to the vane by 10 per cent.

(1.405 kN; 6.55 kW; 41.9 per cent)

8. A jet of water 50 mm in diameter strikes a curved blade making an angle of 30° with the direction of motion of the blade. The absolute velocity of the jet is 30 m/s and the blade speed is 14 m/s. Find the blade angle at entry. If the blade angle at exit is the same as the blade angle at entry and friction reduces the relative velocity across the blade by 10 per cent, find (a) the absolute velocity of the jet at exit; (b) the force on the blade in the direction of motion of the blade; (c) the work done per second and the efficiency.

(51° 20'; 13.85 m/s; 865 N; 12.1 kW; 45.5 per cent)

CHAPTER 21

Experimental Errors and the Adjustment of Data

21.1 Experiment

The object of a student's experiment may be one or more of the following:

1. To verify a textbook theory.
2. To carry out a standard industrial test, such as a hardness or tensile test.
3. To determine the performance of a machine.
4. To determine a physical constant, such as the acceleration due to gravity, the discharge coefficient for an orifice, or the modulus of elasticity of a metal.

The object of the experienced investigator, however, might be to carry out an experiment when an adequate theory is not known, to verify or reject a new theory, or to provide data on which a theory may be based. Whether student or experienced investigator, however, the *scientific method* used is fundamentally the same, i.e.

1. To alter only one variable at a time.
2. To test the experimental method to show that it is valid, i.e. actually measures the effect it is designed to measure.
3. To test the reliability of the experiment, i.e. that the results are repeatable by any competent investigator and free from errors.

The scientist who subjects a theory to experimental test may often try to devise an experiment to show that his theory is false rather than to show it is correct. The engineer or the student will not usually go so far in expressing doubt. Nevertheless it is the *discrepancy* between theory and experiment that is often of greatest interest, and the *errors* that are of greatest importance in testing the reliability of the experiment. For example, a knowledge of the errors and of their source will often show how the experiment may be improved.

21.2 Error and discrepancy

We distinguish between error and discrepancy as follows:

Error is the difference between a measured quantity and the true value. Since the true value is often unknown the term 'error' usually refers to the estimated uncertainty in the result. If Δx is the *absolute error* in measurement of a quantity of magnitude x, the *relative error* is defined as the ratio $\Delta x/x$. The percentage error is given by $\Delta x/x \times 100$ per cent.

Discrepancy is the difference between two measured values when errors have been minimized, corrected or taken into account. For example, an experimental determination of the ultimate tensile strength of a steel will often differ from that given in a handbook. Nevertheless, since the properties of a steel may vary from batch to batch, the experimental value may be the more reliable for the batch from which the specimen was taken. Similarly a discrepancy may exist between an experimental and a theoretical result. For example, the period of vibration of a spring-supported light mass may differ from that calculated. A suitable graphical procedure may show that a more advanced theory is required to take into account the mass of the spring.

Note, however, that we are not justified in suggesting a discrepancy between theory and experiment, unless the sources of error have been fully investigated.

21.3 Classification of errors

Errors may be of four kinds, each of which requires different treatment; they are:

   (*a*) mistakes;
   (*b*) consistent or systematic errors;
   (*c*) accidental or random errors;
   (*d*) errors of calculation.

(*a*) Mistakes

Mistakes are usually avoidable and are due to inexperience, inattentiveness and faulty use of the apparatus.

Doubtful results should be repeated immediately if possible. For this reason a graph of measured values should be plotted as the test proceeds; mistakes can then be seen immediately. Where a physical disturbance occurred or an obvious mistake was made the measurement should be rejected. If there is no evident reason why a doubtful result should occur this result should be *retained*, but repeated if possible.

Sometimes it may be possible to repeat the measurement several times, then the doubtful value will have only a small effect on the average value. Mistakes in calculations should not be tolerated.

(b) Constant or systematic errors

Constant or systematic errors may be due to: (i) the instrument; (ii) the observer; (iii) the experimental conditions.

(i) The instrument

An instrument may read consistently high or low; the error involved is constant and may be allowed for by calibration against a standard. This type of error is vividly illustrated by comparing the scales of a number of rules made of different materials. A difference of length over a few centimetres is often visible to the naked eye.

Constant errors are usually *determinate*, i.e. they may be allowed for, or a *correction* made. For example, a spring balance may read 0.1 kg when unloaded. This *zero* error may be allowed for by subtracting 0.1 kg from all readings. Note, however, that it is sometimes necessary to check whether an error is uniform along the scale or varies with the reading. A complete calibration of an instrument involves checking every major scale reading against an accurate standard.

We have to distinguish now between *accuracy* and *precision*. A precision instrument will give consistent readings, perhaps to several significant figures, but will be accurate only if calibrated. For example, a micrometer may be read more precisely, to thousandths of a millimetre, by using a rotating drum and a vernier scale. However, only if the screw is accurately made and the micrometer correctly calibrated can we regard it as accurate.

A similar term used in connexion with an instrument is its *sensitivity*, or change in reading for a given change in a measured quantity, e.g. number of scale divisions of a balance per kilogram. A spring balance having a large deflexion for each newton increase of force is said to be very sensitive. It will measure deflexion very precisely if supplied with a vernier scale, but will be accurate only if the scale is carefully marked, calibrated and set.

When an instrument, e.g. a dial gauge, relies on gears or other mechanism having friction or back-lash, readings should all be taken on an increasing scale or all on a decreasing scale. A reversal of the mechanism should be avoided if possible. For example, when measuring the load on a specimen in a testing machine by a movable poise the latter should be moved continually in one direction and never reversed, at least up to the maximum load. If by chance the poise overshoots, the investigator should wait until the pull on the specimen has caught up with the measured load.

(ii) The observer

Personal errors are due to the reaction or judgement of an observer.

They are sometimes constant, at least over a short period of time. For example, two observers each operating an accurate stop-clock will usually obtain a different time reading on receiving the same signal. The delay in stopping the clock is personal to the observer and can be taken into account. However, in starting and stopping the clock to obtain two consecutive readings the delay errors, if the same will cancel. It is usually advisable that a given set of readings be all taken by the same observer. Note, however, that the personal error may vary from day to day, or vary due to boredom and tiredness in a long experiment.

(iii) Experimental conditions

Accurate calibration of an instrument often depends on experimental conditions such as the temperature and barometric pressure. For this reason very accurate measurements and the checking of standard gauges are usually made in a room designed to remain at a constant temperature. When the instrument is used under conditions different from that in which it was calibrated a correction can often be made. For example, the change in length of a metal scale is proportional to the change in temperature.

Finally, an experiment is said to be accurately performed if it has small systematic errors.

(c) Random errors

If a measurement is repeated under similar conditions the values do not usually agree exactly. There is a scatter in the results about a mean value due to the accidental or random error.

A random error has the following properties: (a) a small error occurs more frequently than a large error; (b) a result is just as likely to be too large as too small.

Random errors may occur due to the following:

1. By an error of judgement; as when reading to 0.001 mm a micrometer scale divided at 0.01 mm intervals, without the aid of a vernier.
2. Unnoticed fluctuating conditions of temperature or pressure.
3. Small disturbances.
4. Lack of definition. For example the diameter of a rod of wood cannot be stated so precisely as that of a ground steel bar, even though the most accurate micrometer be used.

The effect of random errors on the result can be reduced by: (a) taking a mean value of a set of readings of the same measurement; (b) drawing a smooth curve through a set of points on a graph. Graphical methods are considered in paragraph 21.12.

An experiment which has small random errors is said to be performed precisely, but not necessarily accurately.

(d) Errors of calculation

For practical purposes the electronic calculator has largely eliminated errors of calculation in comparison with the obsolete slide rule and tables of logarithms but the following example is of interest.

Consider the calculation of z from the formula

$$z = x^4 - y^4$$

where

$x = 1.25000$ mm

$y = 1.06250$ mm

(*Note*: Since we have written the decimal figure to five places this implies x and y to be accurate to that degree.) Carrying out the calculation using a slide rule we obtain

$$z = 2.44 - 1.275$$

$$= \mathbf{1.165 \ mm^4}$$

A more accurate method is to apply our knowledge of algebra to write

$$z = 1.25^4 - 1.0625^4 = (1.25^2 - 1.0625^2)(1.25^2 + 1.0625^2)$$

(difference of two squares)
and factorizing the first bracket again

$$z = (1.25 - 1.0625)(1.25 + 1.0625)(1.25^2 + 1.0625^2)$$

$$= 0.1875 \times 2.3125 \times 2.691$$

$$= \mathbf{1.167 \ mm^4}, \text{ by slide rule}$$

Using six-figure logarithms we would obtain

$$z = \mathbf{1.16697 \ mm^4}$$

Using a ten-digit calculator would show an error due to the use of six-figure tables of about $0.0000069 \ mm^4$.

Arising from their speed and ease of use *mistakes* can easily be made with calculators. A rough check should always be made by rounding up the figures involved, the calculation should be repeated several times and the sequence of steps varied. Long calculations should be broken down as much as possible and particular care should be taken where mixed factors of numbers, squares, trigonometrical functions, etc. are involved.

21.4 Justifiable accuracy

In experimental work we must justify the accuracy of the results we give. Let Δx be a *small* error in the measurement of x in the example of the previous paragraph, then the corresponding error in z is given by

$$\Delta z \simeq \frac{dz}{dx} \times \Delta x$$

approximately, provided Δx is small compared with x. Since

$$z = x^4 - y^4$$

then $\dfrac{dz}{dx} = 4x^3$, if y is assumed constant

hence

$$\Delta z \simeq 4x^3 \Delta x$$

If the error in x be $\dfrac{1}{1,000} = 0.001$ mm, the error in z would be

$$\Delta z \simeq 4 \times 1.25^3 \times 0.001$$

$$= 0.0078$$

or about 0.01 mm^4

Hence for a measurement of x accurate to 0.001 mm we would be justified in giving an answer to two decimal places only, i.e.

$$z = 1.17 \text{ mm}^4$$

On the other hand, if the error in x is 0.015 mm, then

$$\Delta z = 4 \times 1.25^3 \times 0.015$$

$$= 0.12 \text{ mm}^4, \text{ approximately}$$

The error is now in the first decimal place and we would be justified in giving an answer to the first place only, i.e.

$$z = \mathbf{1.2 \ mm^4}$$

Evidently, for most practical purposes the use of a slide rule gives answers of a sufficient accuracy (to two significant figures). However, if the measured data is accurate to, say, four significant figures, the modern calculator gives the necessary accuracy but since a result is usually given to a larger number of decimal places it must always be rounded up to the significant figure justified by the least accurate of the original data.

21.5 Possible errors

It is necessary to make an estimate of the possible error involved in a particular measurement. For example, a stop-clock divided in 1 second intervals may involve a possible error of about 0.5 s; a stop-watch reading to 0.2 s may have an error of about 0.2 s. Similarly, a good micrometer having a scale divided into 0.001 mm intervals may have an error of 0.0002 mm; and a dial gauge, reading to 0.002 mm, may have an error of the same magnitude, possibly more, if not carefully calibrated. A useful exercise is to check a commercial dial gauge against an accurate set of gauge blocks.

21.6 Propagation of error, or derived error

An experimental result is often calculated from two or more measured quantities. An error in each measured value results in an error in the derived (calculated) value. For example, suppose an average speed v to be derived from measurements of distance s and time t by the formula:

$$v = \frac{s}{t}$$

If the time taken to travel 100 m is 20 s the derived speed is $\frac{100}{20} = 5$ m/s. If, however, the possible error in the time is ± 0.1 s, the derived speed would be $\frac{100}{20.1} = 4.975$ m/s, or $\frac{100}{19.9} = 5.025$ m/s, approximately.

Thus, if the time is 0.1 s too great the speed will be recorded as 0.025 m/s too low; and if the time is 0.1 s too short the speed will be recorded as 0.025 m/s too great.

21.7 Region of uncertainty

In the above example of a derived error, the (random) error in time t could be either positive (high), or negative (low), i.e.

$\Delta t = \pm 0.1$ s

Hence the error in velocity is more exactly expressed as:

$\Delta v = \pm 0.025$ m/s

(assuming error in s negligible). The result of the experiment is therefore stated as:

$v = 5 \pm 0.025$ m/s

Thus v lies between 5.025 and 4.975 m/s, i.e. v has some value in the *region of uncertainty* between these two values.

The magnitude of the term ± 0.025 indicates the reliability of the result.

21.8 Accepted value

If the value, $v = 5$ m/s is obtained by taking the average value of a number of tests it then represents the *best estimate* of the velocity and, subject therefore to the judgement and common sense of the experimenter, becomes the *accepted value* for that quantity.

21.9 Error derived from the sum of two quantities

Let a value z be derived from the sum of the measured quantities x and y. Thus

$z = x + y$

If Δx and Δy are errors in x and y the corresponding error Δz in z is given by:

$z + \Lambda z = (x + \Delta x) + (y + \Delta y)$

Subtracting these two equations gives:

$\Delta z = \Delta x + \Delta y$

Hence the (absolute) error in z is the sum of the errors in x and y.

Example   *A gauge block of thickness 10.00 mm has an error of ± 0.005 mm. It is combined with a similar block of thickness 20.00 mm, which has a possible error of 0.0075 mm. Find the possible error in the thickness.*

SOLUTION
The total thickness of the combined gauge block is

$(10.00 + 20.00) = 30.00$ mm

and the possible error in the thickness is therefore

$\pm (0.005 + 0.0075)$

$= \pm \mathbf{0.0125}$ **mm**

Note that the magnitudes of the errors have been added inside the bracket without regard to sign.

It is not so evident that if the values of two quantities are *subtracted* the error is still the *sum* of the errors (i.e. regions of uncertainty) of the two quantities. This is demonstrated in the following example.

Example   *The initial temperature of a thermometer is $12 \pm 0.2°$ C. The* *final temperature is 48 ± 0.4° C. What is the possible error in the rise of temperature?*

SOLUTION
The initial temperature lies between 11.8 and 12.2° C and the final temperature lies between 47.6 and 48.4° C, Fig. 21.1. Hence the rise in temperature lies between:

(*a*) greatest final value and least initial value;
(*b*) least final value and greatest initial value.

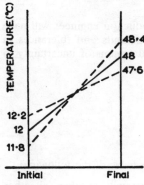

Fig. 21.1

Therefore rise in temperature lies between

$$48.4 - 11.8 = 36.6° C$$

and $47.6 - 12.2 = 35.4° C$

Thus

$$\text{mean rise} = \frac{36.6 + 35.4}{2}$$

$$= 36° C$$

$$\text{region of uncertainty} = \pm \frac{36.6 - 35.4}{2}$$

$$= \pm 0.6° C$$

Thus

rise in temperature = **36 ± 0.6° C**

The region of uncertainty could have been obtained immediately by *adding* the magnitude of the two separate errors; thus:

error of rise = ± (0.2 + 0.4)

= ± 0.6° C, as before

Note, however, that if the errors in the temperatures had been each of one sign only (+ *or* −) the error in the difference or rise in temperature would be found by *subtracting* the errors (due regard being paid to sign). Thus if

initial temperature = 12 + 0.2° C

and final temperature = 48 − 0.4° C

then

error in rise of temperature = − 0.4 − (+ 0.2)

$$= - \textbf{0.6}° \textbf{C}$$

Finally it may be remarked that the production engineer will have recognized a similarity between the ideas of limits and tolerances in practical gauging and the ideas of errors and regions of uncertainty in experimental work.

21.10 Graphical methods

The object of drawing a graph may be one or more of the following:

1. To show how one measured quantity varies with another (all other conditions remaining unaltered), e.g. the variation of the extension of a spring with load.
2. To determine a physical constant from the slope of the graph, e.g. stiffness of a spring.
3. To eliminate or reduce the effect of random errors on the result.
4. To derive a mathematical relationship between the measured quantities, and hence deduce a physical 'law'.

The types of graph which may be met with will be usually one of the following: (*a*) straight line graphs; (*b*) curves which may be reduced to a straight line graph by a suitable mathematical method; (*c*) empirical curves, whose shape is initially unknown but is determined by using the experimental results alone.

21.11 The straight line graph

The construction and use of a simple straight line graph will be illustrated in the following example.

Table 21.1 gives the extension (x mm) of a spring for various values of the load (W N) It is required to find a value for the stiffness of the spring.

Table 21.1

Load (N)	0	2	4	6	8	10	12	13	14	15	16
Extension x (mm)	0	0.55	0.95	1.50	1.95	2.575	3.15	3.75	4.5	5.75	6.6

In mathematics it is customary to plot the *independent variable* horizontally, along the base, and the *dependent* quantity is plotted on

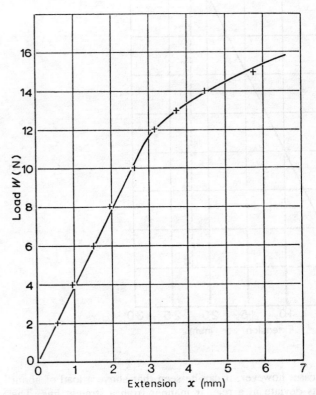

Fig. 21.2

the vertical ordinate. In this example the extension x obviously depends on the load W, which may have any independent value. Hence we should normally plot W along the base and x along the vertical axis. However, it is conventional for engineers to plot a load–extension diagram as in Fig. 21.2, i.e. with the extension plotted along the base. As will be seen this allows the spring stiffness to be obtained directly from the slope of the graph.

Through the points plotted in Fig. 21.2 is drawn the *best straight line* such that as many points lie on one side of the line as lie on the

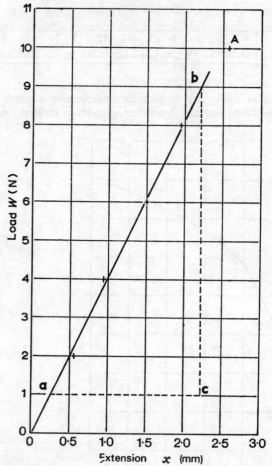

Fig. 21.3

other. In this case, however, it can be seen that above a load of about 12 N the points deviate in a regular manner from a straight line. The spring has evidently been overloaded. If this curved portion is of interest a smooth curve is drawn through the experimental points so that any *scatter* occurs evenly about the curve. The stiffness S of a spring is defined for an elastic spring as the ratio

$$S = \frac{\text{load}}{\text{extension}} = \frac{W}{x}$$

This refers only to the portion of the load-extension graph which obeys Hooke's law, i.e the straight-line portion.

Evidently each experimental point is in error by some amount (i.e. does not lie on the straight line), hence the stiffness calculated from the values for each point would be in error. The best result for the stiffness is therefore obtained from the gradient of the straight line, and it should be noted that the line need not pass through the origin.

To obtain the most accurate value of the gradient the results are re-plotted in Fig. 21.3 to a larger scale. The point A has been disregarded as it is doubtful whether it lies on the straight or curved portion of the graph. The right-angled triangle **abc** is completed and the stiffness calculated from the slope, thus:

$$S = \frac{W}{x} = \frac{\mathbf{bc}}{\mathbf{ac}}$$

$$= \frac{8(\text{N})}{1.975 \ (\text{mm})}$$

$$= 4.05 \ \text{N/mm} = \mathbf{4.05 \ kN/m}$$

Note that for accuracy points **a** and **b** should be *as far apart as possible and on the straight line.*.

Since the plotted points lie close to, and fairly on either side of, the straight line, the error in the slope is probably small.

It may be remarked that if there had been appreciable scatter in the results it would have been insufficiently accurate to use only four or five points to determine the straight line.

21.12 Equation to a straight line

If for a set of values of a certain quantity x there corresponds a single set of values of another quantity y (such that when y is plotted against x a smooth curve is obtained), then y is said to a *function* of x. That is, there is a functional relationship between y and x. If the graph of y against x is a straight line there is said to be a *linear* relationship between y and x.

It may be shown that the equation of a straight line is of the form

$$y = mx + c$$

where m is gradient **bc/ac** of the line (Fig. 21.4),and c is the value of y at which the straight line cuts the vertical axis.

In the special case when $c = 0$, and the line passes through the origin, we have

$$y = mx$$

or $\dfrac{y}{x} = m$, a constant

We now say that y *is proportional to x*, i.e. doubling the value of x will result in doubling the value of y. Note that this is only true when the straight line passes through the origin.

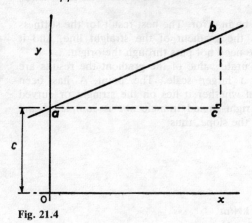

Fig. 21.4

21.13 Equations which may be reduced to a straight line

If theory suggests a possible relationship between x and y such that

$$y = ax^n$$

where a and n are *unknown* constants, then this equation may be put into the form of a straight line as follows: Taking logs of both sides of the equation

$$\log y = n \log x + \log a$$

or putting $Y = \log y$, $X = \log x$, and $C = \log a$, then

$$Y = nX + C$$

This is the equation of a straight line. Hence we plot Y against X to obtain the straight line shown, Fig. 21.5. As before, the slope of the line is n and the intercept with the Y axis is C. Hence a and n may be calculated.

Again suppose

$$y = ae^{nx}$$

where $\quad$ e $= 2.7128\ldots$

then taking logs to base 10

$$\log_{10} y = nx \log_{10} e + \log_{10} a$$

therefore $\qquad Y = 0.4343nx + C$

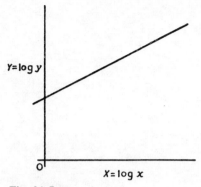

Fig. 21.5

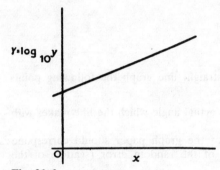

Fig. 21.6

where $Y = \log_{10}y$, and $C = \log_{10}a$. Hence a straight line is obtained by plotting $\log_{10}y$ against x, Fig. 21.6. The gradient is now $0.4343\,n$, and the intercept is $\log_{10}a$.

Note that the above method does not apply to curves of the form

$$y = mx^n + c$$

In this case the value of the index n must usually be known or guessed beforehand. For example, if $n = \frac{1}{2}$, then

$$y = mx^{\frac{1}{2}} + c$$

$$= m\sqrt{x} + c$$

$$= mX + C$$

where $X = \sqrt{x}$. This is the equation of a straight line. Hence, if our choice $n = \frac{1}{2}$ is correct, a graph of y against $\sqrt{x}$ will be a straight line, Fig. 21.7.

Note that unless the equation connecting y and x is reduced to a straight line form by some algebraic device it is usually difficult to obtain the value of the constants m, n, a, etc.

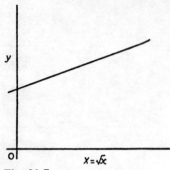

Fig. 21.7

21.14 Choice of axes

When choosing the axes of a straight line graph the following points should be borne in mind:

1. Choose the scales so that the actual angle which the line makes with the axis is approximately 45°.
2. The smallest scale division of the graph paper should correspond roughly with the magnitude of the random error (scatter) of the plotted values.
3. The origin should generally be shown.

There is an important exception to this last point, however. For example, Table 21.2 gives corresponding values of two quantities x and y. These values are plotted in Fig. 21.8. As shown, the points are too close together and the slope of a straight line drawn through them cannot be determined accurately.

Table 21.2

x	6	6.5	7	7.5	8	8.5	9
y	4.10	4.25	4.38	4.54	4.66	4.80	4.96

The values are re-plotted correctly in Fig. 21.9. As a warning the words 'false zero' are added at the intersection of the axes. This is often necessary since, when comparing two graphs from different sources, the shape and position of each curve is the first thing that strikes the eye and a misleading impression can be obtained.

To obtain the equation of the straight line from the graph of Fig. 21.9 the method is as follows: Choose the points **a**, **b** on the line and read off the corresponding values of x and y.

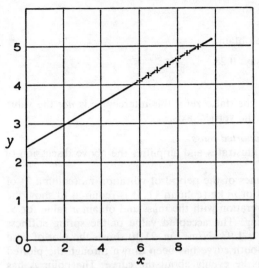

Fig. 21.8 Incorrect graph.

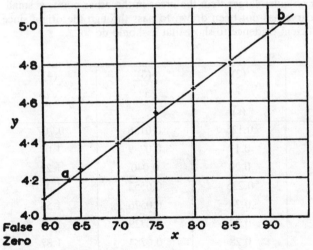

Fig. 21.9 Correct graph.

Thus

At **a** $x = 6.35, y = 4.2$

At **b** $x = 9.17, y = 5.0$

Since both pairs of values satisfy the equation of a straight line, $y = mx + c$, we must have the pair of equations:

$$4.2 = 6.35\,m + c$$

$$5.0 = 9.17\,m + c$$

Solving for m and c, we obtain

slope, $m = 0.284$, say. 0.28

intercept, $c = 2.40$

Note that owing to the 'false zero' this intercept c is *not* the value of y where the line cuts the vertical axis.

Vibration of a spring-supported mass
The following example illustrates and amplifies the above discussion of graphical methods.

Table 21.3 gives values of the period of vibration τ s (column 2) of a spring-supported mass of m kg (column 1). It is required to show the variation of period of vibration with the mass and obtain a value for S, the stiffness of the spring. The accepted value of the spring stiffness from other experiments is 1.93 kN/m. Fig. 21.10 shows the plot of time τ against mass m. A smooth curve has been drawn through the plotted points so that the points lie evenly about the curve. The point A has been disregarded when drawing the curve since the likelihood of a random error so much larger than the error in the other points is small. The plotted curve has not been drawn to pass through the origin since there is insufficient evidence to show that it should do so.

Table 21.3

(1) m (kg)	(2) τ (s)	(3) τ^2	(4) $\sqrt{m}$
$\frac{1}{2}$	0.125	0.0156	0.707
1	0.17	0.0289	1.00
$1\frac{1}{2}$	0.20	0.040	1.22
2	0.235	0.0552	1.41
$2\frac{1}{2}$	0.24	0.0576	1.58
3	0.27	0.0729	1.73
$3\frac{1}{2}$	0.28	0.0784	1.87
4	0.30	0.09	2.00
$4\frac{1}{2}$	0.32	0.1024	2.12
5	0.33	0.109	2.24
$5\frac{1}{2}$	0.35	0.1225	2.35
6	0.37	0.137	2.45

The variation of τ with m is shown by the graph of Fig. 21.10, but as drawn, the graph does not allow us to obtain a mathematical relation

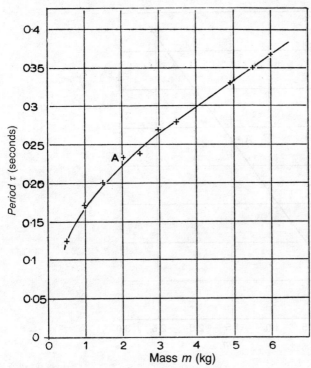

Fig. 21.10

between the quantities. The theoretical relation between τ and m, however, is

$$\tau = 2\pi \sqrt{\left(\frac{m}{S}\right)}$$

Thus

$$\tau = \frac{2\pi}{\sqrt{(S)}} \times \sqrt{m}$$

$$= k\sqrt{m}$$

where

$$k = \frac{2\pi}{\sqrt{S}}, \text{ a constant}$$

Hence if τ is plotted against $\sqrt{m}$ a straight line should be obtained, and the constant k is given by the slope of the line.

Values of $\sqrt{m}$ are shown in column (4) of Table 21.3., and the graph of $\tau - \sqrt{m}$ is shown in Fig. 21.11. The best straight line has been drawn through the plotted points. The latter lie fairly evenly about the line, hence it may be deduced that there is a linear relation

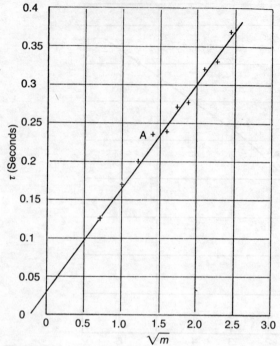

Fig. 21.11

between τ and $\sqrt{m}$, in agreement with theory.

The line cuts the $\sqrt{m}$ axis to the left of the zero point. The theory requires that the line should go through the origin. The discrepancy between theory and experiment is due to the (constant) error in not allowing for the mass of the spring. The 'error' is of course in the theory. The value of m where the line cuts the horizontal axis represents the mass to be added to each value of m to allow for the mass of the spring.

A value for the stiffness of the spring may be found as follows:

From

$$k = \frac{2\pi}{\sqrt{S}}$$

$$S = \frac{4\pi^2}{k^2}$$

From the graph, since k is given by the slope of the line:

$$k = \frac{0.37}{2.72} = 0.136 \text{ s/kg}^{\frac{1}{2}}$$

therefore

$$k^2 = 0.0185 \text{ s}^2/\text{kg}$$

hence

$$S = \frac{4\pi^2}{0.0185}$$

$$= \textbf{2,130 N/m}$$

This estimate of S is not very good, so we try an alternative method. From

$$\tau = 2\pi \sqrt{\left(\frac{m}{S}\right)}$$

Squaring both sides

$$\tau^2 = \frac{4\pi^2}{S} \times m$$

$$= cm$$

where

$$c = \frac{4\pi^2}{S}, \text{ a constant}$$

Hence if we plot τ^2 against m a straight line should result. The values of τ^2 are given in column (3) of Table 21.3 and the resulting graph of τ^2 against m is shown in Fig. 21.12.

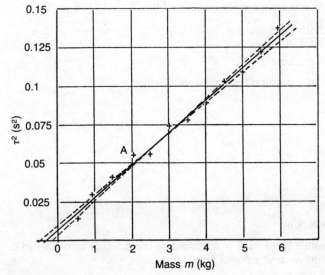

Fig. 21.12

The best straight line through the plotted points is shown as a full line on the figure. It is seen that the scatter about the line is greater than in the plot of Fig. 21.11. This method of plotting shows up the random errors. Point A has been disregarded when drawing the line since the likelihood of an error of the magnitude shown is probably small. By comparison with Fig. 21.11 it is seen that the points are more evenly spaced, i.e. at $\frac{1}{2}$-kg intervals. This method of plotting is therefore preferable.

In addition to the best straight line, drawn in full, two other broken lines have been drawn. These represent a fair estimate of the greatest and least slopes of a line through the plotted points. The possible error in the slope due to the method of plotting may be found as follows: From the graph

$$\text{maximum slope} = \frac{0.136}{6.2} = 0.0219$$

$$\text{minimum slope} = \frac{0.129}{6.5} = 0.01985$$

$$\text{slope of best straight line} = \frac{0.1325}{6.32} = \mathbf{0.021} = \mathbf{c}$$

$$\text{error of maximum slope} = 0.0219 - 0.021 = 0.0009$$

$$\text{error of minimum slope} = 0.01985 - 0.021 = -0.00115$$

$$\text{mean error} = \pm \frac{(0.0009 + 0.00115)}{2}$$

$$= \pm 0.00103$$

Thus possible percentage error in slope of best straight line is

$$\pm \frac{0.00103}{0.021} \times 100$$

$$= \pm 4.9 \text{ per cent}$$

We are now in a position to recalculate S, since

$$\frac{4\pi^2}{S} = c$$

then

$$S = \frac{4\pi^2}{c}$$

$$= \frac{4\pi^2}{0.021}$$

$$= \mathbf{1,878 \ N/m}$$

The possible error in S is the same as the possible error in c, i.e. ± 4.9 per cent, or ± 91.8 N/m. Thus

$$S = 1,878 \pm 91.8 \text{ N/m}$$

The accepted value of S was given as 1.93 kN/m. The error in our mean value is therefore

$$1{,}930 - 1{,}878 = 52 \text{ N/m}$$

which is well within the error involved in drawing the best straight line. Note, however, that the value 1,878 N/m is much better than that obtained from the straight line of Fig. 21.11, i.e. 2,130 N/m.

Index